Flow Control Techniques and Applications

Providing comprehensive coverage, this is the first book to systematically introduce different flow control techniques. With a dedicated chapter for each technique, all of the most important, typical, and up-to-date methods are discussed, including the vortex generator, biological techniques, the jet and synthetic jet, the plasma actuator, and closed-loop control. Understand their key characteristics and control mechanisms, and learn about their applications in different fields such as aviation and aerospace, mechanical engineering, and building construction. The necessary background on flow control is provided, including the history of the discipline, and the definition, classification, and development of each technique, making this essential reading for graduate students, researchers, and engineers working in the field.

Jinjun Wang is the Director of the Institute of Fluid Mechanics at Beijing University of Aeronautics and Astronautics (BUAA), and the Director of the Fluid Mechanics Key Laboratory at the Ministry of Education of China. He was the Head of the Department of Aircraft Design and Applied Mechanics at BUAA between 1997 and 2003.

Lihao Feng is an associate professor at Beijing University of Aeronautics and Astronautics (BUAA), and an editorial board member of the *Journal of Flow Control, Measurement & Visualization*.

Flow Control Techniques and Applications

JINJUN WANG

Beijing University of Aeronautics and Astronautics

LIHAO FENG

Beijing University of Aeronautics and Astronautics

CAMBRIDGE
UNIVERSITY PRESS

CAMBRIDGE
UNIVERSITY PRESS

University Printing House, Cambridge CB2 8BS, United Kingdom

One Liberty Plaza, 20th Floor, New York, NY 10006, USA

477 Williamstown Road, Port Melbourne, VIC 3207, Australia

314–321, 3rd Floor, Plot 3, Splendor Forum, Jasola District Centre, New Delhi – 110025, India

79 Anson Road, #06–04/06, Singapore 079906

Cambridge University Press is part of the University of Cambridge.

It furthers the University's mission by disseminating knowledge in the pursuit of education, learning, and research at the highest international levels of excellence.

www.cambridge.org
Information on this title: www.cambridge.org/9781107161566
DOI: 10.1017/9781316676448

First published 2019

Printed and bound in Great Britain by Clays Ltd, Elcograf S.p.A.

A catalogue record for this publication is available from the British Library.

Library of Congress Cataloging-in-Publication Data
Names: Wang, Jinjun, author. | Feng, Lihao, author.
Title: Flow control techniques and applications / Jinjun Wang, Beijing University of Aeronautics and Astronautics, Lihao Feng, Beijing University of Aeronautics and Astronautics.
Description: Cambridge, United Kingdom ; New York, NY, USA : Cambridge University Press , [2018] | Series: Cambridge aerospace series | Includes bibliographical references.
Identifiers: LCCN 2018027182 | ISBN 9781107161566
Subjects: LCSH: Air flow. | Aeronautics.
Classification: LCC TL574.F5 W36 2018 | DDC 629.132/32–dc23
LC record available at https://lccn.loc.gov/2018027182

ISBN 978-1-107-16156-6 Hardback

Contents

Preface

Flow control is one of the frontier subjects in fluid mechanics and aerodynamics. It attempts to introduce perturbations into the flow field to alter the original flow development path into an ideal state, and thus to achieve the desired goals, such as lift enhancement drag reduction, vibration suppression, noise reduction, fuel and heat transfer enhancement, etc. There are many different kinds of flow control techniques, which could be classified into passive control and active control based on energy input. Passive control needs no energy and it is easy to implement. It could be further distinguished as the fundamental techniques that are developed based on our understanding of flow physics, and the biomimetic techniques which are derived from nature. Active control needs extra energy, and can be adjusted during the control process and thus be implemented in real-time and unsteady flow control. Thus, control effectiveness and efficiency for the active techniques is usually more significant than that for passive ones.

Flow control has developed in association with the development of the aviation industry. In particular, it has made great contributions to the advancement of aviation. Some control techniques have been used in aircraft up to now, such as high-lift airfoils, vortex generators, wing tips, etc. Flow control could also be widely used in space, turbomachinery, combustion, building, transportation, etc. Implementation of flow control is usually related to complex flow phenomena. Thus, flow control is of great significance for not only the development of the subject of fluid mechanics but also has great potential applications in engineering.

The primary focus of this book is to introduce the most important and typical flow control techniques. Chapter 1 offers a general introduction to flow control, including the background, classification, and features of various passive and active techniques. Then, each chapter will mainly discuss one typical control method, including its fundamental characteristics, applications in various fields and control mechanisms. The book is organized in this way to enable readers to better understand and master flow control.

Chapter 2 introduces the Gurney flap, which is a small device attached to the pressure surface of the airfoil to effectively increase the lift coefficient. Chapter 3 introduces the vortex generator, which is a simple device to induce streamwise vortex for separation control, thus to reduce the pressure-difference drag. Chapter 4 introduces roughness, which could be used to delay flow transition and influence the coherent structures in the turbulent boundary layer, thus to reduce friction drag. Chapter 5 introduces polymer, which is also an effective control approach for turbulent drag reduction.

Chapter 6 introduces biological control techniques, where four of them are presented. Hairy coating helps to reduce flow separation, and the leading-edge tubercles on the wing improve the aerodynamic performance at high angles of attack. The riblets on a flat plate contribute to a reduction in the friction drag, while the grooves on a bluff body can reduce lift fluctuation.

Chapter 7 introduces the jet, which can enhance momentum mixing between inner and outer boundary layer, thus to delay or eliminate flow separation. In addition to its conventional applications, some novel control approaches are also introduced, such as circulation control, jet vortex generator, and jet Gurney flap. Chapter 8 introduces the synthetic jet, which combines the blowing and suction to induce vortical structures periodically for flow control. The synthetic jet is found to be more efficient than the jet, and thus shows great potential for application in various engineering fields.

Chapter 9 introduces the plasma actuator, which is a technique based on electronics. The features and applications of various plasma actuators are presented. In particular, some novel control conceptions combining the characteristics of the DBD plasma actuator and conventional techniques are presented, including the plasma synthetic jet, the plasma Gurney flap, plasma circulation control, and the plasma vortex generator. Chapter 10 introduces Lorentz force, which is a technique based on electromagnetism. The uses of streamwise and spanwise Lorentz forces for the control of boundary layer and flow around airfoils and bluff bodies are presented. Chapter 11 introduces closed-loop control, which is a method of controlling active techniques. The main ideas of closed-loop control and some control examples based on the reduced-order model and measured variables are presented.

This book could be developed for courses aimed at postgraduate students in subjects of fluid mechanics, aerodynamics, aviation propulsion, fluid machinery, aircraft design, and wind engineering. It is also useful for researchers and engineers involved in the related fields.

Although this book has only two authors, it represents the efforts of our research group of Turbulence and Flow Control Laboratory at the Beijing University of Aeronautics and Astronautics. Zhenyao Li did the typesetting work and picture editing, and undertook the arduous task of seeking figure permissions. Liyang Liu also assisted with the picture editing. Chong Pan, Qi Gao, Guosheng He, Yang Xu, Di Wu, Yuan Qu, and Jiang-Sheng Wang put forward helpful suggestions. Lei Wang, Tuo Chen, Tingshen Liu, Guangyao Cui, and Guopeng Cui helped to prepare materials. We appreciate everyone's efforts.

The contents presented in this book are mainly based on our own research and existing literature. We have attempted to present a comprehensive introduction to the features, mechanisms, and applications of the flow control. However, we are aware that some techniques may not be covered as fully as others and some works may not be included due to our limited knowledge.

1 Introduction

1.1 Background

Flow control attempts to introduce perturbations into the flow field to alter the original flow development path to an ideal state. Thus, one can achieve desired goals, such as lift enhancement, drag reduction, vibration suppression, noise reduction, fuel and heat transfer enhancement, etc. In addition, the implementation of flow control is usually related to complex flow phenomena. Thus, flow control is of great significance for not only the development of the subject of fluid mechanics but also has great potential applications in engineering.

The most important application of flow control is in aviation, which is aimed to improve the performance of aircrafts. The origin of modern flow control could be traced back to 1904, when the boundary layer theory was developed by Prandtl. He was also considered to be the first one to use steady suction control to delay flow separation of diffusers and a circular cylinder (Joslin and Miller 2009). Since then flow control has developed in association with the development of the aviation industry. In particular, it has made great contributions to the advancement of aviation.

Flow control can improve the aerodynamic performance, safety and environmental performance of the aircraft. There are substantial data to prove this. A 1% reduction in the cruise drag of a large passenger aircraft could lead to about 1600 kilograms reduction of airplane empty weight or an increase of about 10 more passengers (Chen et al. 2009). In particular, a 1% drag reduction is equivalent to saving 57 000 liters aviation fuel for B737 and 380 000 liters for B747 per year (Ma and Cui 2007). Researches in Boeing Company (Garner et al. 1991) indicate that for a commercial jet transport, a 0.1 increase in lift coefficient is equivalent to saving airplane empty weight of about 630 kilograms at the constant angle of attack, a 1.5% increase in maximum lift coefficient is equivalent to about 3000 kilograms increase in payload at a fixed approach speed, and a 1% increase in the lift-to-drag ratio during takeoff is equivalent to about 1270 kilograms increase in payload.

Thus, flow control has attracted much attention by many research institutions. Many significant research projects have been conducted recently to advance the understanding and applications of flow control.

In 2001, the Group of Personalities formed by Philippe Busquin, who was the European Commissioner for Research, published a landmark report on "European

Aeronautics: A Vision for 2020," in order to make aviation better serve society's needs and for European Aeronautics to become a global leader in the field. In this report, several goals for performance of aircraft in 2020 were proposed, including a 50% reduction of noise, a 50% cut in CO_2 emissions per passenger kilometre, and an 80% cut in nitrogen oxide emissions. Flow control is definitely needed to achieve these goals. Thus, the European Union conducted a series of research projects associated with flow control in succession, including "AEROMEMS (Advanced Aerodynamic Flow Control using MEMS)," "AEROMEMS II," "AVERT (Aerodynamic Validation of Emission Reducing Technologies)," "PLASMAERO (Useful Plasmas for Aerodynamic Control)." Many institutes and universities were involved in the projects to advance the applications and validations of flow control techniques.

In 2006, the National Aeronautics and Space Administration (NASA) requested the National Research Council (NRC) to undertake a decadal survey of civil aeronautical research and technology priorities that would help NASA fulfill its responsibility to preserve USA leadership in aeronautical technology. This report called "Decadal Survey of Civil Aeronautics: Foundation for the Future" presented a set of strategic objectives for the next decade of research and technology. It listed the top 11 challenges for aerodynamics and aeroacoustics, and the second and the fourth are closely related to flow control, namely "Aerodynamic performance improvement through transition, boundary layer, and separation control" and "Aerodynamic designs and flow control schemes to reduce aircraft and rotor noise." Under the heading of aeronautical technology, NASA established a "Vehicle Systems Program" to conduct fundamental research on advanced technologies for future flight vehicles (Washburn et al. 2002). It was indicated that several components of this program, such as "Breakthrough Vehicle Technologies Program (BVT)," "Ultra Efficient Engine Technology Program (UEET)," and "21st Century Aircraft Technology Program (TCAT)," needed the support of flow control. Following these projects, the NASA Langley Research Center, Air Force Research Laboratory, Boeing Company conducted a series of investigations that supported improved high-lift capability, drag reduction, noise suppression, propulsion/airframe integration, and maneuvering performance.

After 100 years of development of the modern aircraft, it has become more and more difficult to reduce drag and noise by optimizing aircraft design only. Thus, flow control provides an approach to further improve the aircraft performance. Though it meets some difficulties in both techniques and technologies for engineering applications, a few flow control techniques have been used in aircrafts up to now, such as high-lift airfoils, vortex generators, and wing tips, etc. It is suggested that flow control is of importance because of its great potential to significantly advance the development of aviation.

1.2   Classification

There are many different kinds of flow control techniques, which could be classified into different groups according to certain criteria. For the purpose of flow control, techniques could be used to increase lift, reduce drag, suppress vortex-induced-vibration, reduce

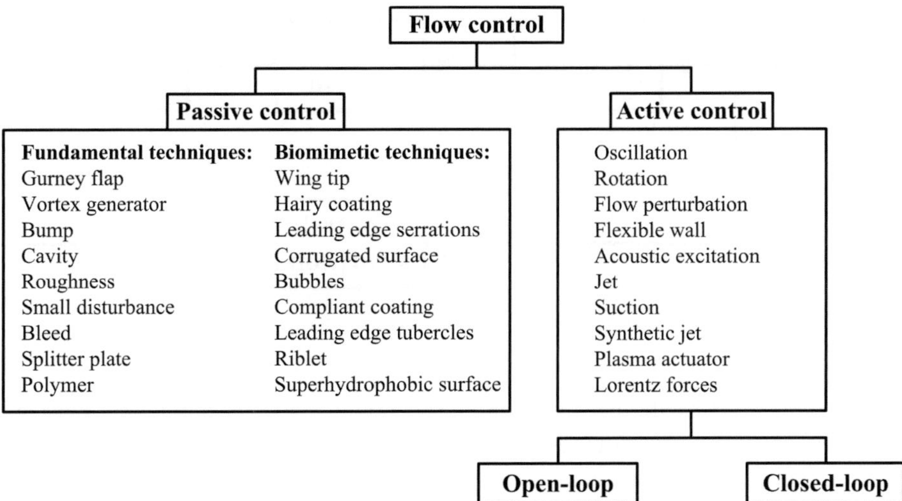

Figure 1.1 Categorization of flow control techniques.

noise, enhance mixing, etc. Some of the goals are related with each other. We can increase the lift coefficient of airfoils by either reducing flow separation or increasing the circulation. The former is only effective when there is a flow separation, for example at post stall angles of attack, meanwhile a reduction in drag could also be achieved. On the other hand, circulation control can increase the lift coefficient from low to high angles of attack. Pressure difference drag could be reduced by delaying flow separation, while the reduction of friction drag is related with boundary layer control. When the intensity of wake vortex shedding is weakened, it might be beneficial for reducing vortex-induced-vibration and noise. In comparison, stronger vortical structures could enhance mixing and heat transfer.

We can also classify flow control techniques according to the application fields, such as aviation, space, turbomachinery, combustion, building, transportation, etc. The appropriate flow control techniques might be different for various fields. However, it is essential that we should first understand the fundamental characteristics of the techniques and then propose the optimal control strategy.

Besides the categories mentioned above, flow control techniques are most frequently classified into passive and active based on energy input, as shown in Figure 1.1. Passive flow control techniques need no energy, are easy to implement, but cannot be changed during the control process. Active techniques need extra energy, and can be adjusted during the control process and thus be implemented in real-time and unsteady flow control. Thus, effectiveness and efficiency for the active techniques is usually more significant than that for passive ones.

There exist different kinds of passive flow control techniques. In general, we can further distinguish them into different groups based on their origin. One group is developed based on our understanding of flow physics, which is called the fundamental technique. For example, we already know that vortical structures may be

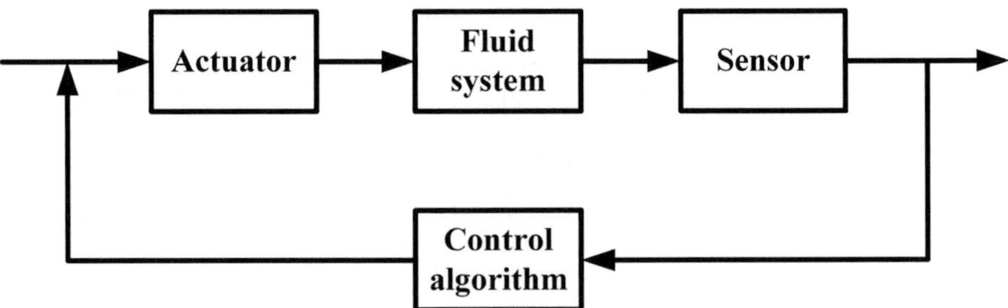

Figure 1.2 Schematic of the closed-loop control algorithm.

induced by the instability of the separated shear layer for flow over an object with adverse pressure gradient, thus a small inclined plate placed in the crossflow could induce the streamwise vortex. The idea of control by vortex generators has therefore been proposed.

The other group may be derived from nature, from animals to plants. Through natural selection living organisms develop their own shapes and organs with specific features, that are expected to be optimal and adaptive for organisms in certain environments. Thus, we can learn from them to copy or mimic these features to improve the performance of man-made systems. Accordingly, different kinds of biomimetic flow control methods have been proposed.

The active flow control techniques are not only related to fluid mechanics but also to other disciplines. The essential factor of active flow control is to introduce perturbations into the flow fields, which could be achieved by methods based on mechanics, electronics, or electromagnetism, etc. For example, the traditional free jet is usually produced by a mechanical air bump. The synthetic jet could be produced by either a mechanical device or the combination of mechanics and electronics, namely MEMS (Micro-electro-mechanical system). The plasma actuator is related to electronics, and the Lorentz force to electromagnetism. In addition, acoustics, thermodynamics, and optics are also useful in developing effective active flow control techniques.

When active flow control is performed, it could be applied either in an open-loop or closed-loop approach. The open control implements actuation based on the scheduled program and the control parameters are not influenced by the control results. For the closed-loop control, however, the control parameters depend on information from the control system which in turn depends on the control. The schematic of the closed-loop control system is shown in Figure 1.2. A control system includes fluid system, actuator, sensor, and control algorithm. The fluid system stands for the flow field to be controlled. The actuator refers to the active flow control techniques, which could introduce excitation to the fluid system. The sensor measures some variables from the controlled flow, which are used as a reference to adjust the actuator parameters. The control algorithm is used to guide the operation of the actuator. Such a control approach is also referred as feedback control. It is different from feedforward control, where the sensor signal is based on the oncoming flow but not the controlled flow.

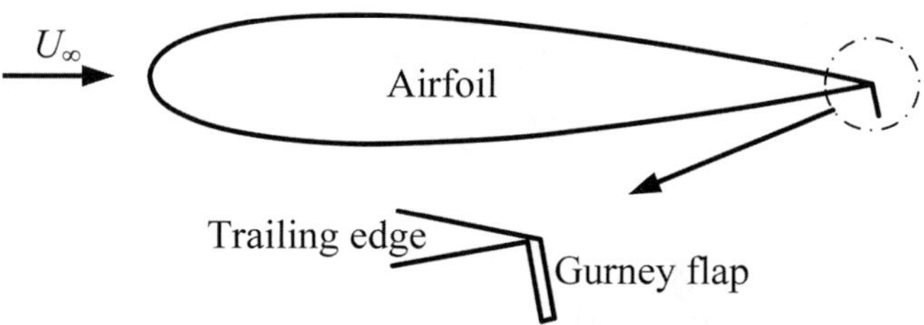

Figure 1.3 Schematic of a Gurney flap mounted on the pressure surface of an airfoil.

We will first present most of the passive and active flow control techniques briefly in this introductory chapter, and then some typical and important techniques will be introduced in detail in the main chapters.

1.3 Passive Flow Control

1.3.1 Gurney Flap

The Gurney flap is a simple device, such as a flat plate, which can be easily attached to the pressure surface of an airfoil, as shown in Figure 1.3. The size of the Gurney flap is only of the order of about 1% chord length of the airfoil, but can significantly increase the lift coefficient. There have been numerous works to show its effects on airfoils, wings and aircrafts, which has been reviewed by Wang et al. (2008). Under certain conditions, both the lift coefficient and lift-to-drag ratio could be increased, though there is a drag penalty. Thus, the Gurney flap shows great potential application to shorten the takeoff/landing distance of aircraft.

1.3.2 Vortex Generator

A vortex generator is usually a small device placed onto the wall, which can induce a streamwise vortex. There are many devices that can be used as a vortex generator, such as flat plate, wishbone, doublet, airfoil, wedge, ramp, etc. The devices are usually arranged in one row along the spanwise direction to induce a set of vortices, either in a co-rotating manner where each device is inclined in the same direction (Figure 1.4(a)), or a counter-rotating manner where a pair of devices induces vortices with different rotation directions (Figure 1.4(b)). The induced streamwise vortices have the great ability to enhance momentum mixing, which is beneficial for separation flow control. Thus, vortex generators have been applied in various fields for flow control, ranging from low-speed to high-speed fields, which has been reviewed by Lin (2002) and Lu et al. (2012).

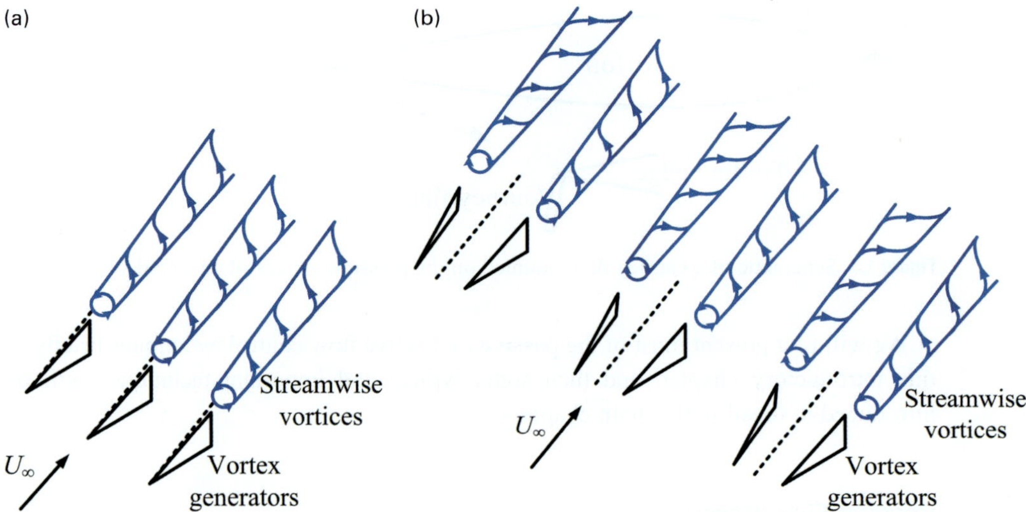

Figure 1.4 Schematic of vortex generators. (a) Co-rotating arrangement; (b) counter-rotating arrangement.

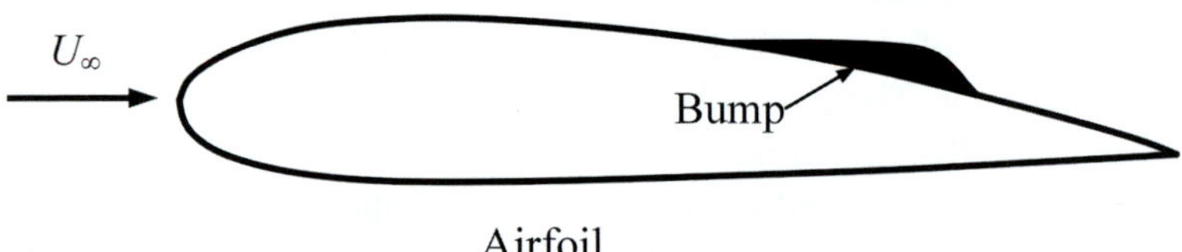

Figure 1.5 Schematic of an airfoil with a bump.

1.3.3   Bump

A bump is usually used to control the shock wave of a transonic airfoil and thus to reduce its drag. It is placed in the downstream of the normal shock wave over the airfoil suction surface, as shown in Figure 1.5. The bump could induce λ-shock structures that can reduce the impact of shock waves by replacing a single normal shock wave with several shock legs. The total pressure loss through a series of oblique shock waves is usually smaller than that across a normal shock wave, thus the wave drag can be greatly reduced. Milholen and Owens (2005) and Li et al. (2011) found that drag reduction of around 12%–15% could be achieved at Mach numbers of 0.73 to 0.78. Besides, the bump could also be used to delay buffet onset and expand buffet boundary (such as Tian et al. 2011).

1.3.4   Cavity

A cavity with trapped vortex is usually used for the control of airfoils. A cavity of suitable shape is positioned along the spanwise direction over the suction surface of

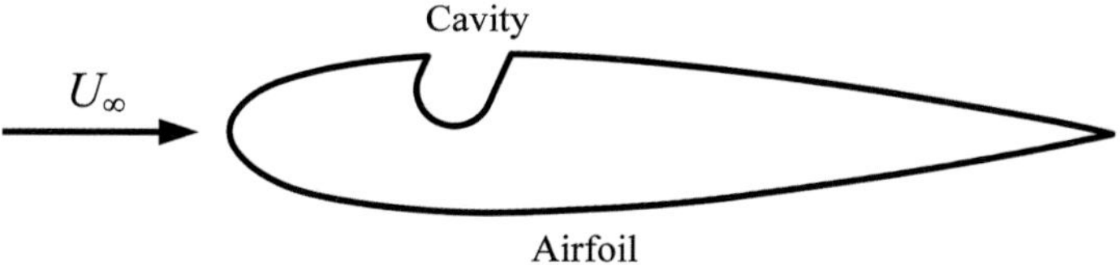

Figure 1.6 Schematic of an airfoil with a cavity.

the airfoil, as shown in Figure 1.6. A large-scale vortex will be induced in the cavity, creating a recirculation region closed by the dividing streamline (Lasagna et al. 2011). The flow over the airfoil suction surface will be modified. In particular, the flow over the airfoil with a cavity separates before the forward edge of the cavity at high angles of attack. The separated flow displays a strong interaction with the cavity, causing the flow to shed smaller-scale vortical structures than the airfoil without a cavity. Thus, the near wake becomes narrower and the lift-to-drag ratio is increased with cavity control, in comparison with the clean airfoil (Olsman and Colonius 2011). In order to stabilize the trapped vortex, additional blowing or suction control could be used, such as that conducted by Iollo and Zannetti (2001) and Olsman et al. (2011).

1.3.5 Roughness

There have been numerous works to show the effects of roughness on laminar flow, flow transition, and turbulent flow, such as those reviewed by Jiménez (2004). It is well known that roughness may promote flow transition, and one example is shown in Figure 1.7(a) for the golf ball, which has global dimples to make it a roughness surface. The dimples over the golf surface could accelerate flow transition to turbulence, which enhances momentum mixing to enable fluids to resist adverse pressure gradient. Thus, the flow may separate at around 120° from the front stagnation point, in comparison with the separation angle of about 80° for the laminar flow over a smooth ball. Thus, the golf ball has a smaller pressure-difference drag and could move further than a smooth one.

However, pioneer work conducted by Fransson et al. (2006) indicated that a row of cylindrical roughness elements placed on a flat plate (Figure 1.7(b)) with specific height and spacing could effectively delay flow transition. In particular, an original turbulent flow could be changed to a laminar flow. Subsequent studies have also indicated the effect of the roughness elements on the bypass transition and the turbulent flow.

1.3.6 Small Disturbance

A small disturbance may trigger the instability of the local flow to further influence the global field. Some examples of disturbance are the trip wire (Figure 1.8(a)) and the small

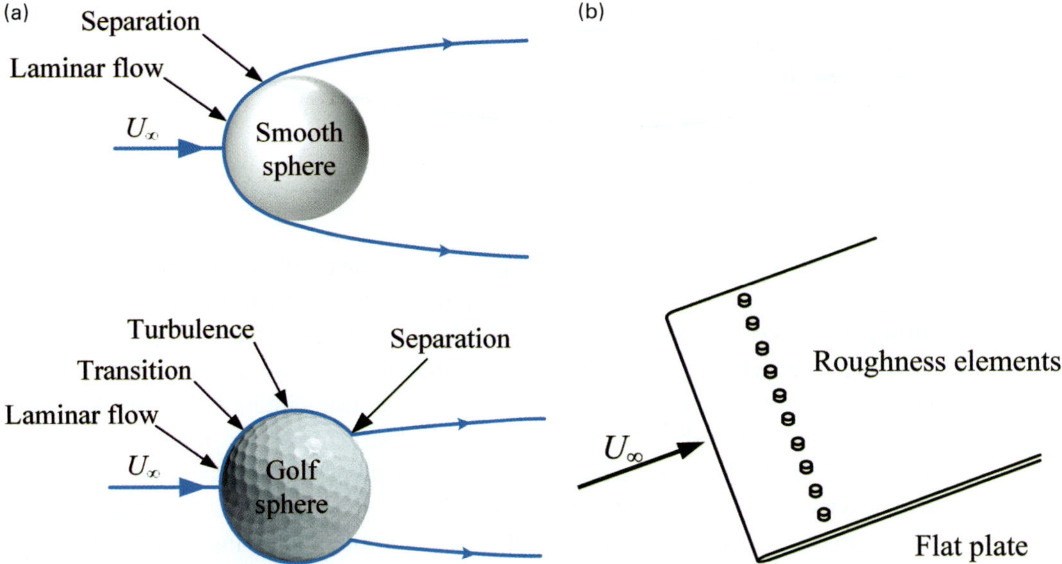

Figure 1.7 (a) Effect of dimples on a sphere. (b) Schematic of roughness elements placed on a flat plate.

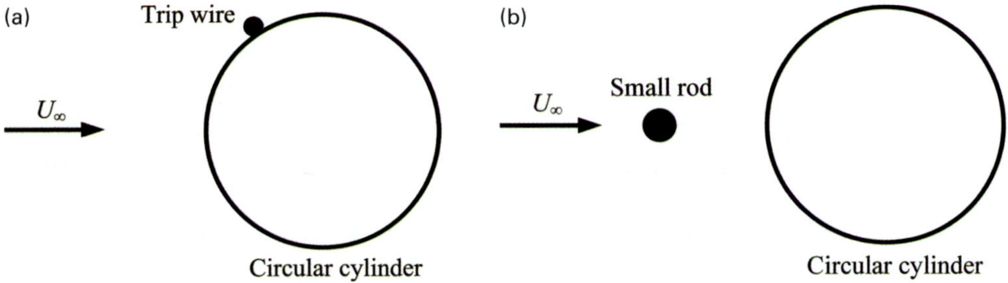

Figure 1.8 Schematic of a circular cylinder with a trip wire (a) and a small rod in the upstream position (b).

rod (Figure 1.8(b)). A trip wire is usually used to promote flow transition to turbulence for different objects, including a cylinder (such as Ekmekci and Rockwell 2010) and a sphere (such as Son et al. 2011). However, it is mostly used for the boundary layer experiment, when a turbulent flow is needed.

A rod is usually a much smaller cylinder in comparison with the scale of the controlled bluff bodies, including a square cylinder (such as Zhang et al. 2005), a circular cylinder (such as Wang et al. 2006; Zhang et al. 2006a), a disc (such as Zhang et al. 2006b; Wang et al. 2013b), etc. The small rod could be placed upstream or downstream of the bluff body with a staggered angle. The wake induced by the small rod may interact with the flow over the bluff body, which could result in a global variation of the bluff body's wake and a reduction in drag. For example, Zhang et al. (2005) and Wang et al. (2006) found that there were six different wake

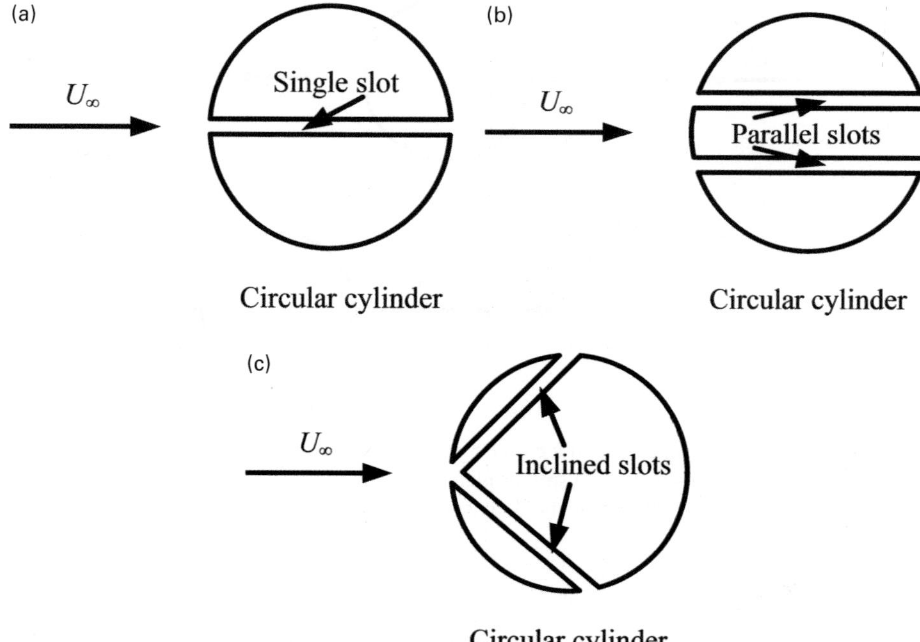

Figure 1.9 Schematic of a circular cylinder with bleed control. (a) One slot across the front and rear stagnations points; (b) two parallel slots; and (c) two inclined slots.

modes induced by a small rod upstream of the bluff body, namely cavity flow, wake impinging, wake merging, wake splitting, weak boundary layer interaction mode and negligible interaction. In particular, a maximum drag reduction by about 98% was found for the cavity flow mode.

1.3.7 Bleed

Bleed control is achieved by machining narrow slots across the controlled body. Fluids would flow across the slot and form a localized jet from the exit. One application is to control the circular or square cylinder, and there are several different arrangements of the slot. One slot could be across the front and rear stagnation points (such as Fu and Rockwell 2005; Baek and Karniadakis 2009), as shown in Figure 1.9(a). Also, two slots can be used, which could be parallel to the streamwise direction (such as Aydın et al. 2010), as shown in Figure 1.9(b), or with some inclined angle (such as Shi and Feng 2015). For all cases, the jets issuing from the slots will interact with the boundary layer or shear layer, leading to an ideal control effect. Baek and Karniadakis (2009) and Shi and Feng (2015) indicated that the bleed control could convert the original asymmetric vortex shedding mode to the symmetric one, which is beneficial for the suppression of vortex-induced vibration.

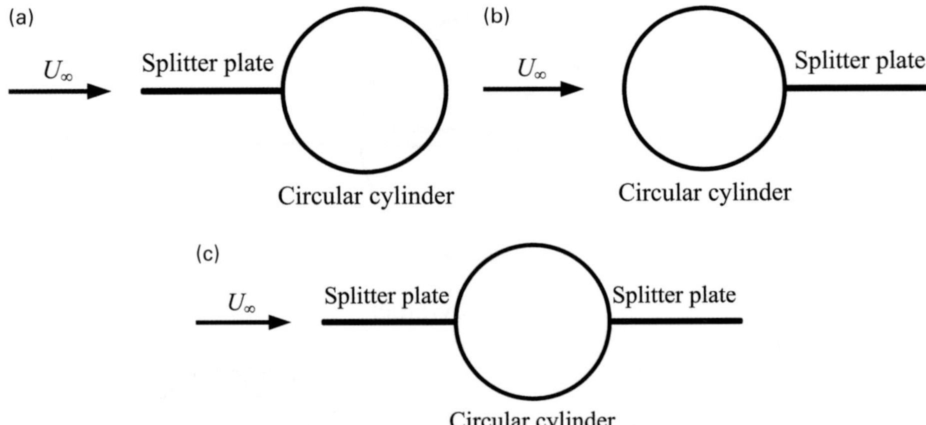

Figure 1.10 Schematic of a circular cylinder with splitter plates. (a) A splitter plate placed at upstream position; (b) a splitter plate placed at downstream position; and (c) two splitter plates placed at both upstream and downstream positions.

1.3.8   Splitter Plate

A splitter plate is a flat plate placed upstream (Figure 1.10(a)), downstream (Figure 1.10(b)), or both upstream and downstream (Figure 1.10(c)) of the object along the streamwise direction. Note that it is not necessary to connect the splitter plate to the controlled body surface; there could be a distance between them. Celik et al. (2008) indicated that the splitter plate upstream of the cylinder changed the flow dynamics of the downstream cylinder in the formation region. The splitter plate downstream of the cylinder may restrict the mutual interaction between the upper and lower shear layer, resulting in an elongated recirculation region, a larger vortex formation length, a weaker wake vortex strength, and a reduced drag coefficient, which have been found by Hwang et al. (2003), Akilli et al. (2005), Serson et al. (2014), among others. Hwang and Yang (2007) indicated that the upstream splitter plate reduced the stagnation pressure, while the downstream one increased the base pressure by suppressing vortex shedding. Thus, the combined effect with both upstream and downstream plates caused a significant drag reduction on the cylinder. Note that the splitter plate could be either rigid or flexible.

1.3.9   Polymer

Polymer is a transitional passive control technique for turbulent drag reduction. When the polymer additives are dropped in the fluids, the turbulent flow in the near-wall region might force the polymer to roll up chains, which are stretched in the mean flow direction, as shown in Figure 1.11. In this case, polymer chains exhibit characteristic length scales associated with the turbulent structures, which is related to drag reduction. It was found that minute concentrations of polymers could reduce the drag in turbulent flows by up to 80% (Procaccia et al. 2008). Thus, the study of drag reduction by polymer is important

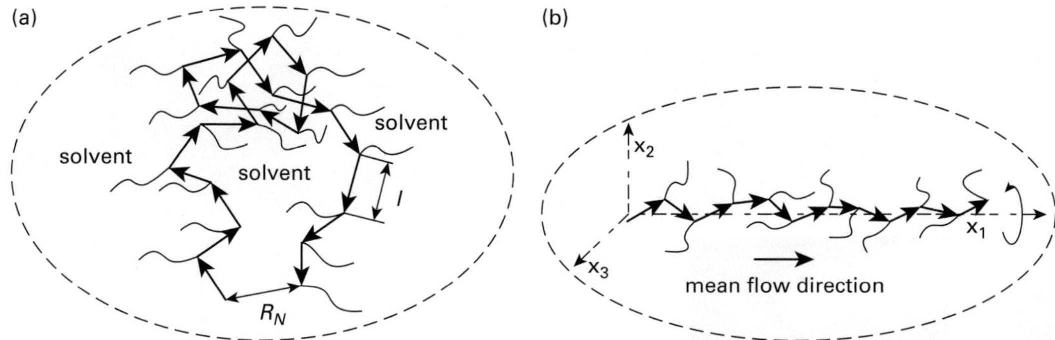

Figure 1.11 Schematic of polymer in solution at equilibrium (a) and its response to stretching by turbulent motions (b) (Jovanović et al. 2006). Reproduced with permission of The American Society of Mechanical Engineers.

for potential applications in engineering, such as oil transportation, heating and cooling systems, etc.

1.3.10   Biomimetic Techniques

Some specific features on organisms are shown in Figure 1.12. They might be related to high aerodynamic performance, and some biomimetic flow control techniques are proposed accordingly.

One can observe the deflection of the wing tip of a bird during its cruise stage (Figure 1.12(a)) and the raising of a bird's feathers during its landing (Figure 1.12(b)), which are considered to be the reason why birds can improve their aerodynamic performance. It has been found that the wing tip could reduce the lift-induced drag of the aircraft, which has been widely used in modern airplanes as a well-developed flow control technique. On the other hand, researchers have proved the effects of hairy coating on bluff bodies and airfoils, confirming the function of raising feathers during bird landing.

The owl has a special structure of combed serrations at the leading edge of its feathers (Figure 1.12(c)), which are not observed in most other birds. The combed serrations are found to be of importance to reduce noise by functioning as a vortex generator to attach the flow on the wing suction surface. Thus, the owl has a great ability to fly silently. The leading-edge serration could be used as an effective control approach for noise reduction, as has been indicated by Ito (2009) and Narayanan et al. (2015), and reviewed by Choi et al. (2012).

Insect wings are usually not smooth but corrugated, as shown in Figure 1.12(d). The corrugation allows the insect wing to have the advantages of low mass, high stiffness, and low membrane stress (Meng and Sun 2013). Some researchers, such as Hu and Tamai (2008) and Barnes and Visbal (2013), have pointed out that a corrugation airfoil could have a larger lift coefficient at post stall angles of attack. Thus, the corrugation configuration could be used as an approach to improve the aerodynamic performance of micro air vehicles (MAVs).

Figure 1.12 Some specific features on organisms that might be related to high aerodynamic performance. (a) Wing tip of an albatross. (b) Self-adaptive hairy flaps of a falcon (Brücker and Weidner 2014), reproduced with permission, copyright © 2014 Elsevier Ltd. All rights reserved. (c) Combed serrations on the leading edge of the owl feather (Bachmann and Wagner 2011), reproduced with permission from John Wiley and Sons. (d) Corrugation of a dragonfly forewing (Kesel 2000), reproduced with permission, copyright © The Company of Biologists Limited 2000. (e) Air bubbles releasing from a penguin. (f) Compliant skin of a dolphin (Allen and Bridges 2003), reproduced with permission, © John Wiley and Sons, 2003. (g) Tubercles on the leading edge of a humpback whale. (h) Riblet on sharkskin (Liu and Li 2012), reproduced with permission from Elsevier. (i) Riblets on a saguaro (Talley et al. 2001), reproduced with permission. (j) Water droplet sitting on the lotus leaf (upper) and micrographs of lotus leaf surface to show the microstructures (Bhushan and Jung 2011), reproduced with permission, copyright © 2011 Elsevier Ltd. All rights reserved.

Figure 1.12 (*cont*).

A penguin may release air bubbles to reduce drag and thus promotes swimming speed (Figure 1.12(e)). Many researches, such as Xu et al. (2002), Elbing et al. (2008), and Jacob et al. (2010), have indicated that the air bubbles injected to the boundary layer could lead to a reduction in the friction drag. Thus, the air bubbles show great potential application for drag reduction in underwater vehicles, such as submarines.

It is found that the skin of a dolphin is compliant (Figure 1.12(f)), which is suggested to be effective in reducing the drag of the dolphin when swimming. Accordingly, a flow control technique, namely compliant coating, has been proposed. There have been many researches to show that compliant coating is capable of substantially delaying laminar-to -turbulence transition and reducing turbulent activities in the turbulent boundary layer. For reviews of this field please refer to Riley et al. (1988) and Gad-el-Hak (2002).

The leading edge of the flippers of humpback whales is also not smooth and has bumpy tubercles (Figure 1.12(g)). Experimental and numerical studies have suggested that the leading-edge tubercles are beneficial for aerodynamic performance at high angles of attack. A recent review on this field has been made by Aftab et al. (2016).

From a close-view, we can see that the sharkskin is comprised of small riblet structures (Figure 1.12(h)), which might be related to the drag reduction of shark swimming. Thus, riblets have been used as a useful drag reduction technique in the past, such as in the so-called sharkskin swimsuit.

Similarly, riblets also occur on one kind of saguaro with typical diameter of the order of 0.5 m and height of 10 m (Figure 1.12(i)). It is suggested that riblets are related to

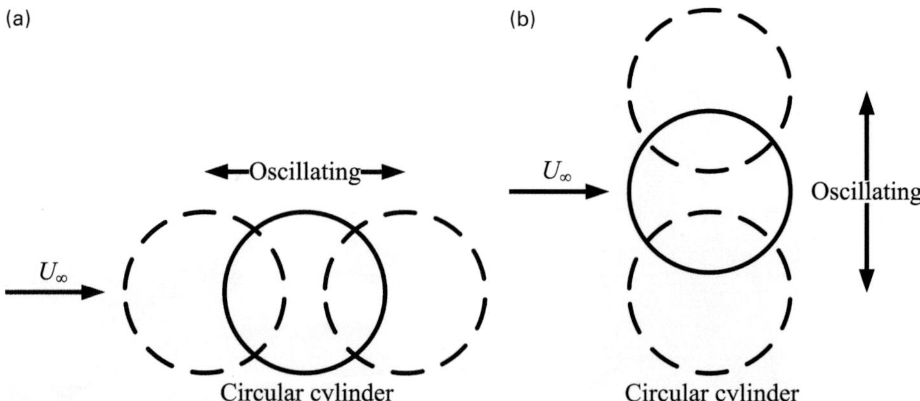

Figure 1.13 Schematic of a circular cylinder oscillating in the streamwise (a) and vertical (b) directions.

suppression of vortex-induced-vibration, which is beneficial for the saguaro in living up to 150 years of age with high wind velocity in their natural habitat.

It is well known that a water drop has a high static contact angle on the lotus leaf, as shown in Figure 1.12(j). The micrograph shows that the lotus leaf surface is very rough with microstructures, which are formed by papillose epidermal cells covered with epicuticular wax tubules. Such features enable the lotus leaf to self-clean and have low adhesion. Thus, the control idea of superhydrophobic surface is developed, where the artificial surface can be made by microstructures. It has been shown that the superhydrophobic surface could lead to a significant reduction in friction drag for both laminar (such as Busse et al. 2013) and turbulent (such as Daniello et al. 2009) flows. More details to show the effects can be found in the review papers by Rothstein (2010) and Bhushan and Jung (2011).

1.4   Active Flow Control

1.4.1   Oscillation and Flow Perturbation

Oscillation and perturbation in flow are two control approaches that may result in similar effects, that are usually imposed on bluff bodies and airfoils. The oscillation is achieved by forcing the controlled body to move in the streamwise (Figure 1.13(a)) or vertical (Figure 1.13(b)) direction. Similarly, the perturbations imposed in the flow could also be in the streamwise (Figure 1.14(a)) or vertical (Figure 1.14(b)) direction. Some primary conclusions have been made for the flow around an oscillating cylinder or the perturbed flow around a stationary cylinder, as has been summarized by Feng and Wang (2010, 2014).

First, vortex-synchronization could be induced by both methods, where the vortex shedding frequency is closely related to the excitation frequency. It could be categorized into two groups: vortex synchronization at the excitation frequency and at the subharmonic excitation frequency. Vertical oscillation and vertical flow perturbation may result

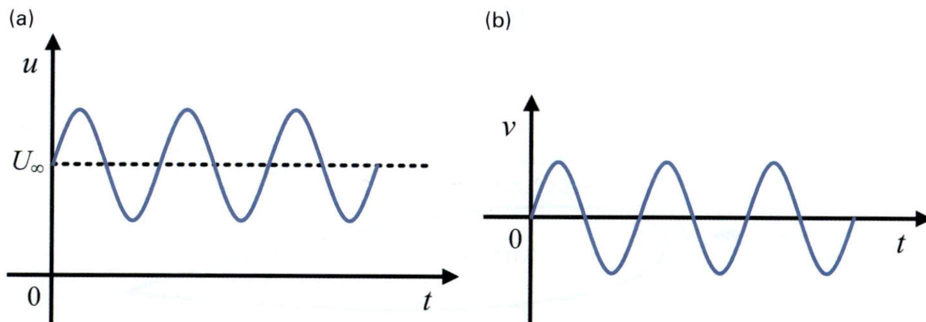

Figure 1.14 Schematic of flow perturbation in the streamwise (a) and vertical (b) directions.

in vortex synchronization at the excitation frequency, while streamwise oscillation and streamwise flow perturbation usually result in vortex synchronization at the subharmonic excitation frequency.

Second, the vortex-synchronization regime is related to both the excitation frequency and amplitude. The fundamental trend is that a higher excitation amplitude can induce a much larger vortex-synchronization regime. The vortex-synchronization regime induced by vertical oscillation and vertical flow perturbation is usually around the natural frequency, while that for the streamwise cases is usually around twice of the natural frequency.

Thirdly, the near wake dynamics could be modified by oscillation and flow perturbation. In particular, the symmetric vortex shedding mode could be induced by symmetric excitation, such as streamwise oscillation and streamwise flow perturbation.

Similar to the oscillation approach, rotation is another control method for the bluff body, as has been studied by Navrose et al. (2015). The oscillation method can also be used with a flat plate to control the boundary layer, where the drag could be reduced by up to 45%, as indicated by Choi (2002). Similarly, flexible wall oscillation is another approach for boundary layer control, as conducted by Shen et al. (2003).

1.4.2 Acoustic Excitation

Acoustic excitation is a control method based on a sound wave, which could be achieved by a loudspeaker, as shown in Figure 1.15. The basis for such control is the boundary layer receptivity. The acoustic pressure wave interacts with the boundary layer/shear layer and triggers the instability which results in global variations of the flow field. There have been studies to show that the acoustic excitation could increase both the lift and lift-to-drag ratio of the airfoil in the close and post stall regions (such as Yarusevych et al. 2007; Yang and Spedding 2013). In addition, the effects of acoustic excitation on the flow around circular cylinders (such as Mohany and Ziada 2009) and on the development of vortical structures (such as Leblanc 2001) and jets (such as Huang et al. 2013) have also been found. The sound press level, excitation frequency, as well as the Reynolds number, determine the effects. However, the experimental conditions, such as the test-section geometry and acoustic resonance, are also of great importance in achieving the ideal effects.

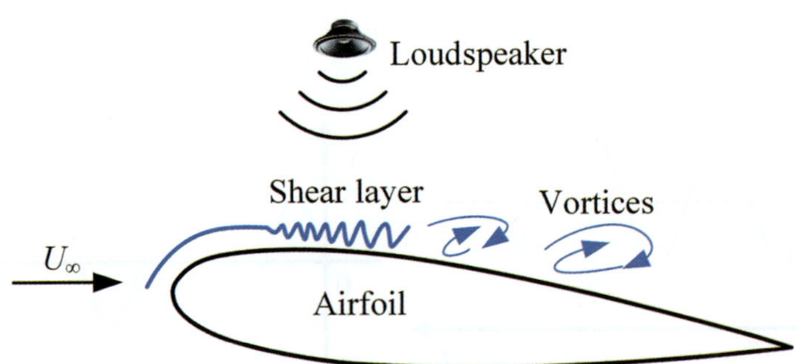

Figure 1.15 Schematic of flow around an airfoil under acoustic excitation.

(a)

(b)

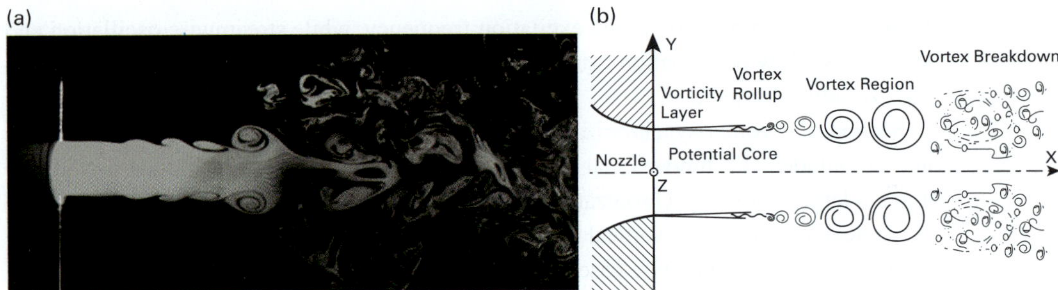

Figure 1.16 (a) Flow pattern of a jet at $Re_j = 1620$. (b) Schematic of a jet issuing from a nozzle (Todde et al. 2009), reproduced with permission, copyright © Springer 2009.

1.4.3   Jet

The jet, also known as free jet, steady jet or continuous jet, is one of the earliest techniques used for boundary layer flow control. When fluids are issuing from an orifice, flow separates from the edges to form a separated shear layer, which gradually rolls up into vortical structures, as shown in Figure 1.16. The vortices are undergoing flow transition as convecting downstream, resulting in the turbulent flow. The jet can enhance momentum mixing between inner and outer boundary layers, which is beneficial for separation delay. Similarly, suction control is another approach by drawing the low momentum fluids away from the near-wall region. Both blowing and suction can be used for flow control in various fields.

1.4.4   Synthetic Jet

The synthetic jet has been widely investigated since the 1990s. A typical synthetic jet actuator includes a cavity, an orifice and an oscillating device, such as a piezoceramic diaphragm or piston, as shown in Figure 1.17. An excitation signal drives the diaphragm or piston to move forward to and backward from the orifice, and thus the fluids are ejected from the orifice and sucked into the cavity, periodically. During the blowing cycle, the separated shear layer formed from the orifice edge grows and develops into

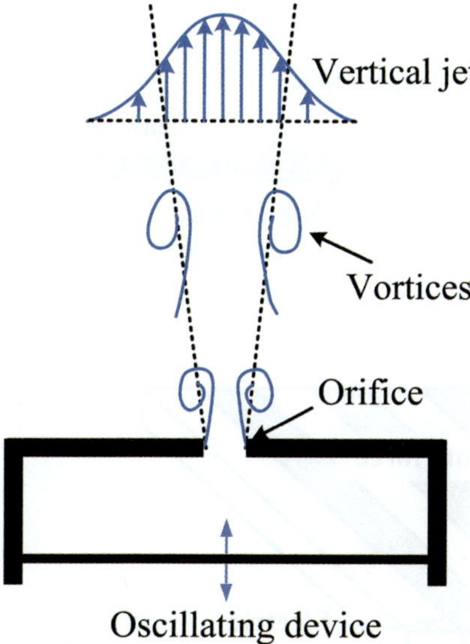

Figure 1.17 Schematic of a synthetic jet actuator.

vortical structures that are convecting downstream gradually due to the self-induced velocity. During the suction cycle, the vortical structures formed during the previous blowing cycles have moved further downstream and they are hardly influenced by the suction process. Thus, a series of vortical structures are formed periodically during the blowing and suction processes, which could enhance momentum mixing efficiency. The synthetic jet is also called the zero-net-mass-flux jet, as the net flow flux during one period is zero though the net momentum flux is not zero. The synthetic jet has become one of the most important active flow control techniques, and it has shown significant effects in various fields, as have been reviewed by Glezer and Amitay (2002), Luo and Xia (2005), and Zhang et al. (2008).

1.4.5 Plasma Actuator

Wide investigation of the plasma actuator for flow control started in the 1990s. The dielectric barrier discharge (DBD) plasma actuator is representative, though several other types have been developed, such as corona discharge actuator and plasma spark-jet actuator. A typical DBD plasma actuator consists of an exposed electrode and an embedded electrode, separated by a dielectric sheet, as shown in Figure 1.18. When the electrodes are supplied with high voltage and frequency, the air over the embedded electrode is ionized and thus a wall jet forms. Accompanying the wall jet, the flow separates from the wall and rolls up into a vortex. These are the two main features of the DBD plasma actuator: inducing a wall jet and a starting vortex. The advantages of plasma control include electronic design with no moving parts, extremely fast response and real-time control ability, low power consumption, low mass, and simple to use, making the plasma actuator one of the most popoular techniques for

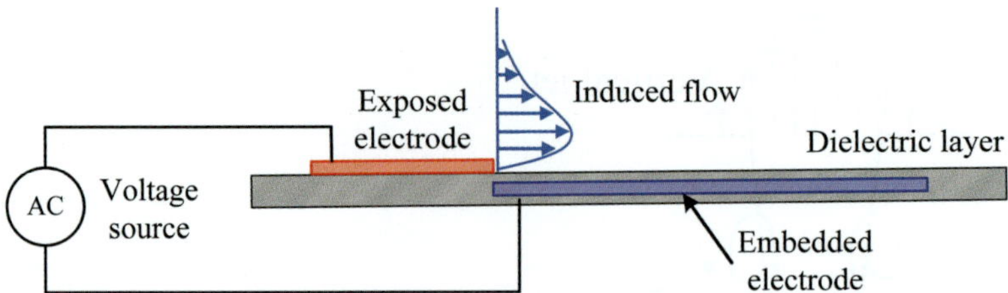

Figure 1.18 Schematic of a DBD plasma actuator.

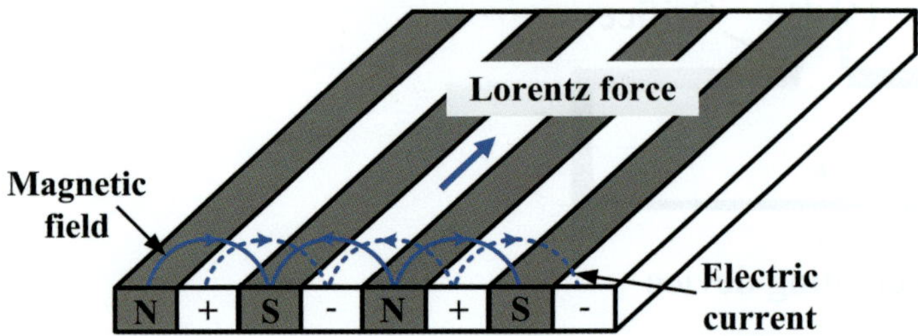

Figure 1.19 Schematic of an arrangement for the creation of the Lorentz force (Adapted from Berger et al. 2000). Reproduced with permission from AIP Publishing.

flow control applications. The main developments can also be found in the review papers by Moreau (2007), Corke et al. (2010), Wang et al. (2013a), Wu and Li (2015).

1.4.6   Lorentz Force

The subject of flow control with Lorentz force (or electromagnetic body force) originates from electromagnetism. When a charged particle is moving in the presence of an electric field and a magnetic field, it will experience a Lorentz force. The Lorentz force required for flow control can be created by placing electrodes and magnets side by side, parallel to one another, as shown in Figure 1.19. The direction of the Lorentz force can be changed by varying the direction of either electric field or magnetic field, and its magnitude can be adjusted by varying the strength of either electric or magnetic field. Thus, the Lorentz force could be produced in a sinusoidal waveform and in either spanwise or streamwise direction relative to the free stream. The spanwise Lorentz force could reduce the friction drag of the turbulent boundary layer, though the streamwise Lorentz force may increase it. Moreover, the Lorentz force could reduce flow separation and thus increase the lift coefficient while reducing the drag coefficient of airfoils.

1.5   Concluding Remarks

Flow control attempts to introduce perturbations into the flow field to alter the original flow development path into an ideal state, and thus to achieve the desired goals, such as lift enhancement, drag reduction, vibration suppression, noise reduction, fuel and heat transfer enhancement, etc. It is of great significance for not only the development of the subject of fluid mechanics but also has great potential applications in engineering. Thus, flow control has attracted much attention. A book about flow control is also very useful for postgraduate students, researchers and engineers.

This book introduces most of the important and typical flow control techniques, including the most recent developments. It firstly introduces most of passive and active flow control techniques briefly in this introduction chapter. Then, the main content of the book follows different kinds of flow control techniques. Each chapter will mainly discuss one typical control method, including its fundamental characteristics, applications in various fields, and control mechanisms. In this way, it is expected that the readers will be able to better understand and master flow control.

References

Aftab, S. M. A., Razak, N. A., Rafie, A. S. M., and Ahmad, K. A. Mimicking the humpback whale: an aerodynamic perspective. *Progress in Aerospace Sciences*, 2016, 84: 48–69

Akilli, H., Sahin, B., and Tumen, N. F. Suppression of vortex shedding of circular cylinder in shallow water by a splitter plate. *Flow Measurement and Instrumentation*, 2005, 16(4): 211–219

Allen, L. and Bridges, T. J. Flow past a swept wing with a compliant surface: stabilizing the attachment-line boundary layer. *Studies in Applied Mathematics*, 2003, 110(4): 333–349

Report of the Group of Personalities formed by Busquin P. European Aeronautics: A Vision for 2020. Published by the European Commission, 2001

Aydin, B. T., Cetiner, O., and Unal, M. F. Effect of self-issuing jets along the span on the near-wake of a square cylinder. *Experiments in Fluids*, 2010, 48(6): 1081–1094

Bachmann, T. and Wagner, H. The three-dimensional shape of serrations at barn owl wings: towards a typical natural serration as a role model for biomimetic applications. *Journal of Anatomy*, 2011, 219(2): 192–202

Baek, H. and Karniadakis, G. E. Suppressing vortex-induced vibrations via passive means. *Journal of Fluids and Structures*, 2009, 25(5): 848–866

Barnes, C. J. and Visbal, M. R. Numerical exploration of the origin of aerodynamic enhancements in low-Reynolds number corrugated airfoils. *Physics of Fluids*, 2013, 25(11): 115106

Berger, T. W., Kim, J., Lee, C., and Lim, J. Turbulent boundary layer control utilizing the Lorentz force. *Physics of Fluids*, 2000, 12(3): 631–649

Bhushan, B. and Jung, Y. C. Natural and biomimetic artificial surfaces for super hydrophobicity, self-cleaning, low adhesion, and drag reduction. *Progress in Materials Science*, 2011, 56(1): 1–108

Brücker, C. and Weidner, C. Influence of self-adaptive hairy flaps on the stall delay of an airfoil in ramp-up motion. *Journal of Fluids and Structures*, 2014, 47: 31–40

Busse, A., Sandham, N. D., McHale, G., and Newton, M. I. Change in drag, apparent slip and optimum air layer thickness for laminar flow over an idealised superhydrophobic surface. *Journal of Fluid Mechanics*, 2013, 727: 488–508

Celik, B., Akdag, U., Gunes, S., and Beskok, A. Flow past an oscillating circular cylinder in a channel with an upstream splitter plate. *Physics of Fluids*, 2008, 20(10): 103603

Chen, Y. C., Liu, H., Zhang, B. Q., and Zhu, Z. Q. A review on drag reduction study and application for large aircraft. *The Application and Development of CFD in Large Civil Aircraft*, Shanghai Jiao Tong University Press, 2009, 32–47 (in Chinese)

Choi, K. S. Near-wall structure of turbulent boundary layer with spanwise-wall oscillation. *Physics of Fluids*, 2002, 14(7): 2530–2542

Choi, H., Park, H., Sagong, W., and Lee, S. Biomimetic flow control based on morphological features of living creaturesa. *Physics of Fluids*, 2012, 24(12): 121302

Corke, T. C., Enloe, C. L., and Wilkinson, S. P. Dielectric barrier discharge plasma actuators for flow control. *Annual Review of Fluid Mechanics*, 2010, 42: 505–529

Daniello, R. J., Waterhouse, N. E., and Rothstein, J. P. Drag reduction in turbulent flows over superhydrophobic surfaces. *Physics of Fluids*, 2009, 21(8): 085103

Ekmekci, A. and Rockwell, D. Effects of a geometrical surface disturbance on flow past a circular cylinder: a large-scale spanwise wire. *Journal of Fluid Mechanics*, 2010, 665: 120–157

Elbing, B. R., Winkel, E. S., Lay, K. A., Ceccio, S. L., Dowling, D. R., and Perlin, M. Bubble-induced skin-friction drag reduction and the abrupt transition to air-layer drag reduction. *Journal of Fluid Mechanics*, 2008, 612: 201–236

Feng, L. H. and Wang, J. J. Circular cylinder vortex-synchronization control with a synthetic jet positioned at the rear stagnation point. *Journal of Fluid Mechanics*, 2010, 662: 232–259

Feng, L. H. and Wang, J. J. Modification of a circular cylinder wake with synthetic jet: Vortex shedding modes and mechanism. *European Journal of Mechanics-B/Fluids*, 2014, 43: 14–32

Fransson, J. H. M., Talamelli, A., Brandt, L., and Cossu, C. Delaying transition to turbulence by a passive mechanism. *Physical Review Letters*, 2006, 96(6): 064501

Fu, H. and Rockwell, D. Shallow flow past a cylinder: control of the near wake. *Journal of Fluid Mechanics*, 2005, 539: 1–24

Gad-el-Hak, M. Compliant coatings for drag reduction. *Progress in Aerospace Sciences*, 2002, 38 (1): 77–99

Garner, P. L., Meredith, P. T., and Stoner, R. C. Areas for future CFD development as illustrated by transport aircraft applications. AIAA Paper 1991–1527

Glezer, A. and Amitay, M. Synthetic jets. *Annual Review of Fluid Mechanics*, 2002, 34: 503–529

Hu, H. and Tamai, M. Bioinspired corrugated airfoil at low Reynolds numbers. *Journal of Aircraft*, 2008, 45(6): 2068–2077

Huang, R. F., Jufar, S. R., and Hsu, C. M. Flow and mixing characteristics of swirling double-concentric jets subject to acoustic excitation. *Experiments in Fluids*, 2013, 54(1): 1421

Hwang, J. Y., Yang, K. S., and Sun, S. H. Reduction of flow-induced forces on circular cylinder using a detached splitter plate. *Physics of Fluids*, 2003, 15(8): 2433–2436

Hwang, J. Y. and Yang, K. S. Drag reduction on a circular cylinder using dual detached splitter plates. *Journal of Wind Engineering and Industrial Aerodynamics*, 2007, 95(7): 551–564

Iollo, A. and Zannetti, L. Trapped vortex optimal control by suction and blowing at the wall. *European Journal of Mechanics-B/Fluids*, 2001, 20(1): 7–24

Ito, S. Aerodynamic influence of leading-edge serrations on an airfoil in a low Reynolds number – a study of an owl wing with leading edge serrations. *Journal of Biomechanical Science and Engineering*, 2009, 4(1): 117–123

Jacob, B., Olivieri, A., Miozzi, M., Campana, E. F., and Piva, R. Drag reduction by microbubbles in a turbulent boundary layer. *Physics of Fluids*, 2010, 22(11): 115104.

Jimenez, J. Turbulent flows over rough walls. *Annual Review of Fluid Mechanics*, 2004, 36: 173–196

Joslin, R. D. and Miller, D. N. Fundamentals and applications of modern flow control. American Institute of Aeronautics and Astronautics, 2009

Jovanović, J., Pashtrapanska, M., Frohnapfel, B., Durst, F. K. J., and Koskinen, K. On the mechanism responsible for turbulent drag reduction by dilute addition of high polymers: theory, experiments, simulations, and predictions. *Journal of Fluids Engineering*, 2006, 128(1): 118–130

Kesel, A. B. Aerodynamic characteristics of dragonfly wing sections compared with technical aerofoils. *Journal of Experimental Biology*, 2000, 203(20): 3125–3135

Lasagna, D., Donelli, R., De Gregorio, F., and Luso, G. Effects of a trapped vortex cell on a thick wing airfoil. *Experiments in Fluids*, 2011, 51(5): 1369–1384

Leblanc, S. Acoustic excitation of vortex instabilities. *Physics of Fluids*, 2001, 13(11): 3496–3499

Li, P.F., Zhang, B. Q., Chen, Y. C., and Chen, Z. L. Wave drag reduction of airfoil with shock control bump. *Acta Aeronautica et Astronautica Sinica*, 2011, 32(6): 971–977 (in Chinese)

Lin, J. C. Review of research on low-profile vortex generators to control boundary-layer separation. *Progress in Aerospace Sciences*, 2002, 38(4): 389–420

Liu, Y. and Li, G. A new method for producing "Lotus Effect" on a biomimetic shark skin. *Journal of Colloid and Interface Science*, 2012, 388(1): 235–242

Lu, F. K., Li, Q., and Liu, C. Microvortex generators in high-speed flow. *Progress in Aerospace Sciences*, 2012, 53: 30–45

Luo, Z. B. and Xia, Z. X. Advances in synthetic jet technology and applications in flow control. *Advances in Mechanics*, 2005, 35(2): 221–234

Ma, H. D. and Cui, E. J. Drag prediction and reduction for civil transportation. *Mechanics in Engineering*. 2007, 29(2): 1–8

Meng. X. G. and Sun, M. Aerodynamic effects of wing corrugation at gliding flight at low Reynolds numbers. *Physics of Fluids*, 2013, 25(7): 071905

Milholen, W. E. and Owens, L. R. On the application of contour bumps for transonic drag reduction. AIAA Paper 2005–462

Mohany, A. and Ziada, S. Effect of acoustic resonance on the dynamic lift forces acting on two tandem cylinders in cross-flow. *Journal of Fluids and Structures*, 2009, 25(3): 461–478

Moreau, E. Airflow control by non-thermal plasma actuators. *Journal of Physics D: Applied Physics*, 2007, 40(3): 605–636

Steering Committee for the Decadal Survey of Civil Aeronautics, National Research Council. *Decadal Survey of Civil Aeronautics: Foundation for the Future*. Published by the National Academies Press, 2006

Narayanan, S., Chaitanya, P., Haeri, S., Joseph, P., Kim, J. W., and Polacsek, C. Airfoil noise reductions through leading edge serrations. *Physics of Fluids*, 2015, 27(2): 025109

Navrose, Meena, J. and Mittal, S. Three-dimensional flow past a rotating cylinder. *Journal of Fluid Mechanics*, 2015, 766: 28–53

Olsman, W. F. and Colonius, T. Numerical simulation of flow over an airfoil with a cavity. *AIAA Journal*, 2011, 49(1): 143–149

Olsman, W. F., Willems, J. F., Hirschberg, A., Colonius, T., and Trieling, R. R. Flow around a NACA0018 airfoil with a cavity and its dynamical response to acoustic forcing. *Experiments in Fluids*, 2011, 51(2): 493–509

Procaccia, I., L'vov, V. S., and Benzi, R. Colloquium: Theory of drag reduction by polymers in wall-bounded turbulence. *Reviews of Modern Physics*, 2008, 80(1): 225–247

Riley, J. J., Gad-El-Hak, M., and Metcalfe, R. W. Compliant coatings. *Annual Review of Fluid Mechanics*, 1988, 20: 393–420

Rothstein, J. P. Slip on superhydrophobic surfaces. *Annual Review of Fluid Mechanics*, 2010, 42: 89–109

Serson, D., Meneghini, J. R., Carmo, B. S., Volpe, E. V., and Gioria, R. S. Wake transition in the flow around a circular cylinder with a splitter plate. *Journal of Fluid Mechanics*, 2014, 755: 582–602

Shen, L., Zhang, X., Yue, D. K., and Triantafyllou, M. S. Turbulent flow over a flexible wall undergoing a streamwise travelling wave motion. *Journal of Fluid Mechanics*, 2003, 484: 197–221

Shi, X. D. and Feng, L. H. Control of flow around a circular cylinder by bleed near the separation points. *Experiments in Fluids*, 2015, 56(12): 214

Son, K., Choi, J., Jeon, W.P., and Choi, H. Mechanism of drag reduction by a surface trip wire on a sphere. *Journal of Fluid Mechanics*, 2011, 672: 411–427

Talley, S., Iaccarino, G., Mungal, G., and Mansour, N. An experimental and computational investigation of flow past cacti. Annual Research Briefs. Center for Turbulence Research, NASA Ames/Stanford University, 2001: 51–63

Tian, Y., Liu, P., and Peng, J. Using shock control bump to improve transonic buffet boundary of airfoil. *Acta Aeronautica et Astronautica Sinica*, 2011, 32(8): 1421–1428 (in Chinese)

Todde, V., Spazzini, P. G., and Sandberg, M. Experimental analysis of low-Reynolds number free jets. *Experiments in Fluids*, 2009, 47(2): 279–294

Wang, J. J., Li, Y. C., and Choi, K. S. Gurney flap – Lift enhancement, mechanisms and applications. *Progress in Aerospace Sciences*, 2008, 44(1): 22–47

Wang, J. J., Zhang, P. F., Lu, S. F., and Wu, K. Drag reduction of a circular cylinder using an upstream rod. *Flow, Turbulence and Combustion*, 2006, 76(1): 83–101

Wang, J. J., Choi, K. S., Feng, L. H., Jukes, T.N., and Whalley, R. D. Recent developments in DBD plasma flow control. *Progress in Aerospace Sciences*, 2013a, 62: 52–78

Wang, J. J., Pan, C., Choi, K. S., Gao, L., and Lian, Q. X. Formation, growth and instability of vortex pairs in an axisymmetric stagnation flow. *Journal of Fluid Mechanics*, 2013b, 725: 681–708

Washburn, A. E., Gorton, S. A., and Anders, S. G. Snapshot of active flow control research at NASA Langley. AIAA Paper 2002–3155

Wu, Y. and Li, Y. Progress and outlook of plasma flow control. *ACTA Aeronautica et Astronautica Sinica*, 2015, 36: 381–405 (in Chinese)

Xu, J., Maxey, M. R., and Karniadakis, G. E. Numerical simulation of turbulent drag reduction using micro-bubbles. *Journal of Fluid Mechanics*, 2002, 468: 271–281

Yang, S. L. and Spedding, G. R. Separation control by external acoustic excitation at low Reynolds numbers. *AIAA Journal*, 2013, 51(6): 1506–1515

Yarusevych, S., Sullivan, P. E., and Kawall, J. G. Effect of acoustic excitation amplitude on airfoil boundary layer and wake development. *AIAA Journal*, 2007, 45(4): 760–771

Zhang, P. F., Wang, J. J., and Feng, L. H. Review of zero-net-mass-flux jet and its application in separation flow control. *Science in China Series E: Technological Sciences*, 2008, 51(9): 1315–1344

Zhang, P. F., Wang, J. J., and Huang, L. X. Numerical simulation of flow around cylinder with an upstream rod in tandem at low Reynolds numbers. *Applied Ocean Research*, 2006a, 28(3): 183–192

Zhang, P. F., Wang, J. J., Lu, S. F., and Mi, J. Aerodynamic characteristics of a square cylinder with a rod in a staggered arrangement. *Experiments in Fluids*, 2005, 38(4): 494–502

Zhang, P., Gao, L., and Wang, J. J. Drag reduction of a disk with an upstream rod. *Wind and Structures*, 2006b, 9(3): 245–254

2 Gurney Flap

2.1 Background

Flaps have been widely used on aircraft to improve their aerodynamic performance and to provide flight control. They are huge control surfaces hinged on the trailing edge of the main wing to achieve effective control. In comparison, the Gurney flap is a much smaller device, with its size only about 1% chord length of the airfoil, but can significantly increase the lift coefficient of the airfoils, wings, and aircraft. It is easy to attach the Gurney flap onto the pressure surface near the trailing edge of wings.

Originally, the Gurney flap was installed at the trailing edge of the racing car wing to improve the downforce. Liebeck (1978) conducted a wind-tunnel experiment with Gurney flaps and introduced this concept to the aerodynamic community. Over the last four decades since then the advantages of the Gurney flap have attracted much attention. A review of its applications can be found elsewhere (Wang et al. 2008).

In this chapter, we will introduce the effects of Gurney flaps on the airfoils, wings, and aircraft. The influence of different parameters will be compared and the control mechanism will be revealed. Finally, some suggestions for engineering applications will be given, as it is expected that Gurney flaps could actually be used in the near future.

2.2 Control of Airfoil

As a simple flow control device, the dominant parameters that may influence the effect of the Gurney flap include the Gurney flap height, the mounting angle, the distance to the trailing edge, and the shape of the Gurney flap, etc. These have been illustrated in Figure 2.1. In this section we will introduce the individual effects of these parameters on an airfoil.

2.2.1 Effects of the Gurney Flap Height

Typical effects on a symmetric airfoil with different Gurney flap heights at a Reynolds number of $Re = 2.1 \times 10^6$ are shown in Figure 2.2. The Gurney flap can increase the lift

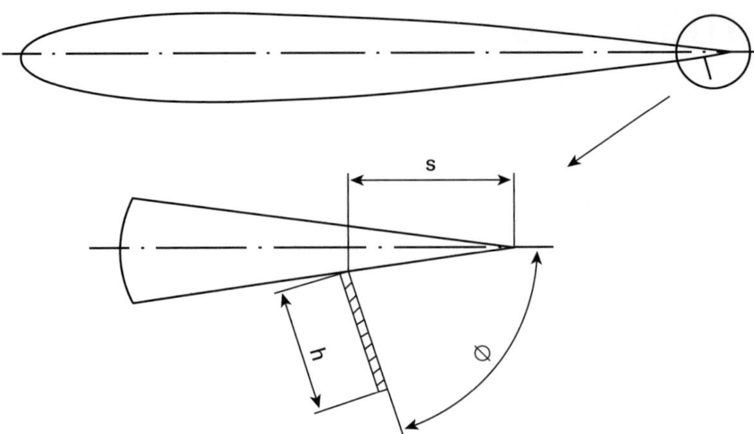

Figure 2.1 Schematic of a Gurney flap mounted on the pressure surface of an airfoil (Li et al. 2003a). Reproduced with permission.

coefficient by shifting the lift curve upwards before stall angle, and the increment increases with the Gurney flap height (Figure 2.2(a)). As a result, the maximum lift coefficient is increased by about 10%, 11%, 18%, 21%, and 27% for the Gurney flap heights 0.5%c, 1%c, 1.5%c, 2%c, and 3%c, respectively. It is also shown that the stall angle is decreased while the zero lift angle of attack becomes more negative with larger Gurney flap heights. Even after stall, the Gurney flap can still increase the lift coefficient, though the influence of the Gurney flap height is not evident.

As well as the increase in the lift coefficient, the Gurney flap also increases the drag coefficient of the airfoil (Figure 2.2(b)). It is shown that the drag is also shifted upwards with Gurney flap control, and the increment increases with the Gurney flap height.

Due to the effects on both lift and drag coefficients, the lift-to-drag ratio is also changed (Figure 2.2(c)). At low-to-moderate lift coefficients, the Gurney flap control decreases the lift-to-drag ratio, and the decrement increases with the flap height. In particular, the maximum lift-to-drag ratio is reduced, while the maximum value among the control cases is obtained by the $h/c = 0.5\%$ case. It is suggested that there is an optimal height to improve the aerodynamic performance of the airfoil. At higher lift coefficients, however, the lift-to-drag ratio can be significantly increased. As a result, the effect of the Gurney flap on the maximum lift-to-drag ratio is small, but the lift coefficient for a given lift-to-drag ratio is significantly increased.

The Gurney flap also increases the nose-down pitching moment (Figure 2.2(d)). The increment appears to be proportional to the lift coefficient increment, which increases with the Gurney flap height. Such characteristics, however, may increase the difficulty of flap application for aircraft control.

The Gurney flap is also found to be effective for a supercritical airfoil at high Mach number $Ma = 0.7$, as shown in Figure 2.3. The lift curves are shifted upwards with the Gurney flap control (Figure 2.3(a)). The stall angle and zero lift angle are both slightly

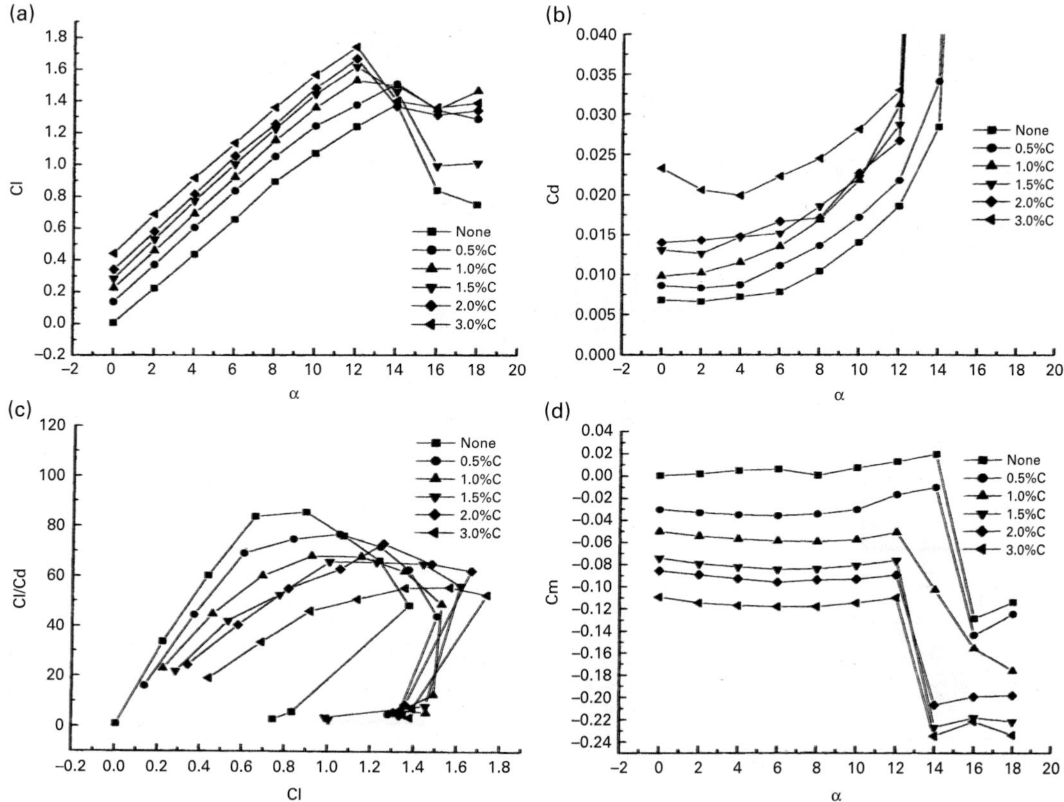

Figure 2.2 Aerodynamic forces on a symmetric airfoil for different Gurney flap heights. (a) Lift coefficient; (b) drag coefficient; (c) lift-to-drag ratio; and (d) pitching moment coefficient (Li et al. 2002a). Reproduced with permission, copyright © Springer Nature 2002.

decreased. Compared with the clean airfoil, the 0.5%c, 1.0%c, 1.5%c, and 2.0%c height Gurney flaps increase the maximum lift coefficient by about 10%, 17%, 23%, and 29%, respectively. In particular, significant lift enhancement is also obtained after stall, which is slightly different with the effects on a symmetric airfoil at low Mach number. The Gurney flap has little influence on the drag coefficient at small angles of attack $\alpha \leq 2°$, while it increases the drag coefficient slightly at $\alpha > 2°$ (Figure 2.3(b)). As a result, the Gurney flap can increase the lift-to-drag ratio for a supercritical airfoil, as shown in Figure 2.3(c), which is obviously different from that for a low speed airfoil. The increment seems to increase with the increase of Gurney flap height, and the maximum lift-to-drag ratio is increased by about 14%, 23%, 21%, and 41% with 0.5% c, 1.0%c, 1.5%c, and 2.0%c height Gurney flaps, respectively. Moreover, the Gurney flap increases the nose-down pitching moment, which increases with increasing the Gurney flap height (Figure 2.3(d)).

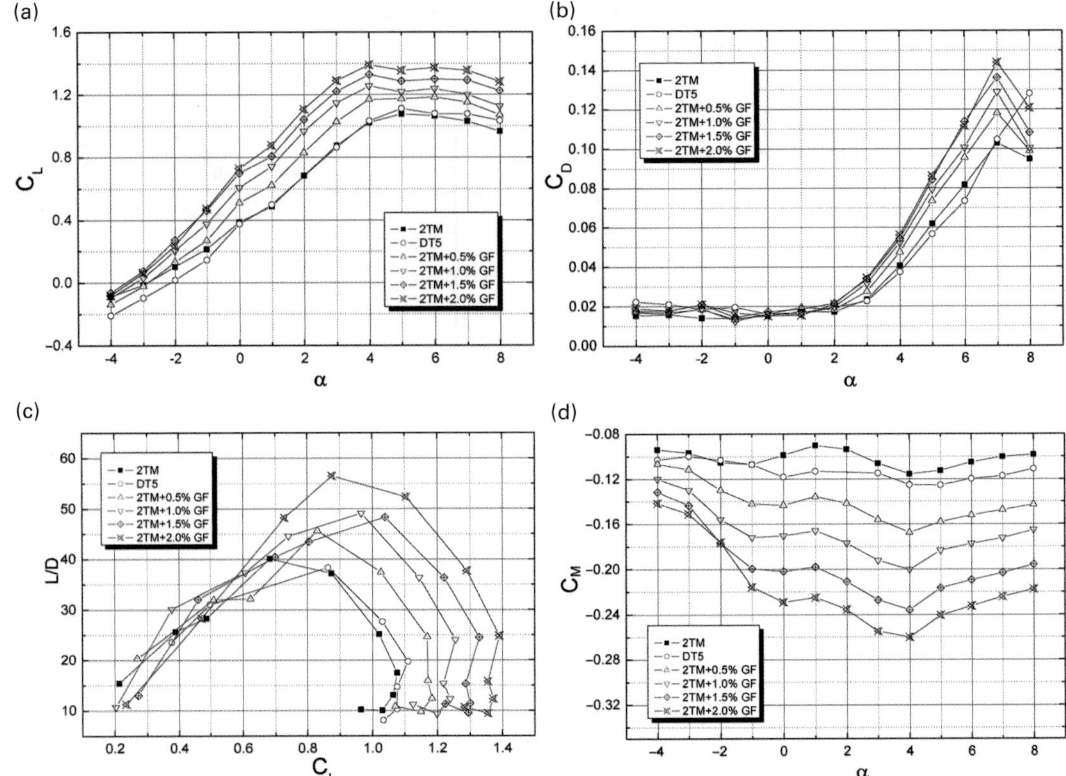

Figure 2.3 Aerodynamic forces on a supercritical airfoil for different Gurney flap heights. (a) Lift coefficient; (b) drag coefficient; (c) lift-to-drag ratio; and (d) pitching moment coefficient (Li et al. 2007). Reproduced with permission, copyright © 2006 Elsevier Masson SAS. All rights reserved.

For practical application, the Gurney flap is most useful during the takeoff and landing stages, since it can increase the lift coefficient and thus the takeoff/landing distance. However, during the cruise stage, there is actually no acquirement for additional lift enhancement and the Gurney flap even induces a drag penalty. Yu et al. (2011) discussed this problem and proposed a solution, and they obtained the aerodynamic forces for the supercritical airfoil at $U_\infty = 85$ m/s, $\alpha = 10°$, which is usually the takeoff angle of general civilian aircraft. A 0.5%c height Gurney flap could increase the lift by about 18.2% and lift-to-drag ratio by about 3.4%. So it could improve the takeoff and landing performance of civil aircraft.

On the other hand, a comparison of the lift and lift-to-drag ratio of the RAE-2822 airfoil without and with the 0.25%c height Gurney flap at $\alpha = 1°$ and $2°$, $Ma = 0.729$ are listed in Table 2.1. It is supposed that the cruising angle is $\alpha = 2°$. It is found that a 0.25% c height Gurney flap on the airfoil at $\alpha = 1°$ has a larger lift coefficient than the clean airfoil at $\alpha = 2°$, and even with much higher lift-to-drag ratio. Thus, it is suggested that the cruising angle of the aircraft can be slightly decreased when the Gurney flap is used,

Table 2.1 Lift coefficient and lift-to-drag ratio of airfoil without and with 0.25% *c* Gurney flap control at *Ma* = 0.729 (Yu et al. 2011).

	Natural case ($\alpha = 2°$)	Control case ($\alpha = 1°$)	Difference
C_L	0.648	0.649	+0.15%
C_L/C_D	50.4	55.8	+10.7%

(a)　　　　　　　　　　　　　　　　　　(b)

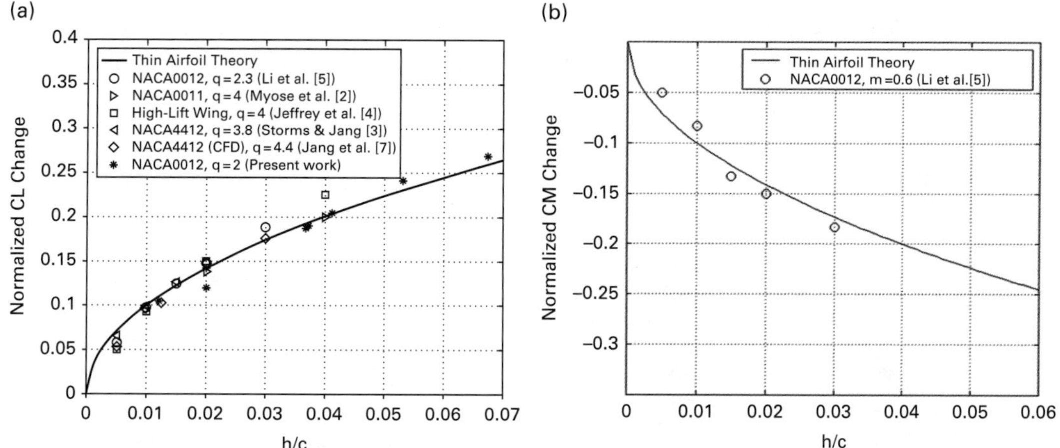

Figure 2.4 (a) Lift coefficient enhancement as a function of normalized Gurney flap height; and (b) pitching moment coefficient change as a function of normalized Gurney flap height (Liu and Montefort 2007). Reproduced with permission, copyright © The American Institute of Aeronautics and Astronautics.

in order to maintain the cruising lift but not increase the drag. Such results are of great value for the engineering applications of the Gurney flap.

We can see the basic trend that both the lift coefficient and the nose-down pitching moment are increased with the increase of Gurney flap height. Numerical studies have shown that there might be a function to simulate this variation. Based on the thin-airfoil theory, numerical, and experimental data, Liu and Montefort (2007), Yu et al. (2011), and Feng et al. (2012) indicated that the lift coefficient increment between with and without Gurney flap correlated well with the square root of the Gurney flap height, yielding $\Delta C_L = q\sqrt{h/c}$, where the factor q is an arbitrary value depending on the airfoil model and test condition. Similarly, there is also a function of $\Delta C_M = m\sqrt{h/c}$ to show the relationship between the pitching moment and the Gurney flap height, where m is also an arbitrary value. The theoretical curves are shown in Figure 2.4, which are also validated by data from the literature. Furthermore, it can be deduced that there is an obvious linear relation between the lift coefficient increment and the pitching moment increment.

2.2.2   Effects of the Gurney Flap Location

Li et al. (2003a) tested the influence of the mounting location of the Gurney flap on the airfoil performance. Four typical locations were studied for a $1.5\%c$ height Gurney flap vertically mounted onto the pressure surface, with the distance between the mounting location and the airfoil trailing edge of $s = 0$, $2\%c$, $4\%c$, and $6\%c$. For all control cases, the lift coefficient is increased while the stall angle is decreased in comparison with the clean airfoil (Figure 2.5(a)). However, the lift enhancement decreases as the Gurney flap moves away from the airfoil trailing edge. The maximum lift coefficient is increased by about 17.4%, 17.2%, 16.9%, and 13.6% for the Gurney flaps with the mounting location

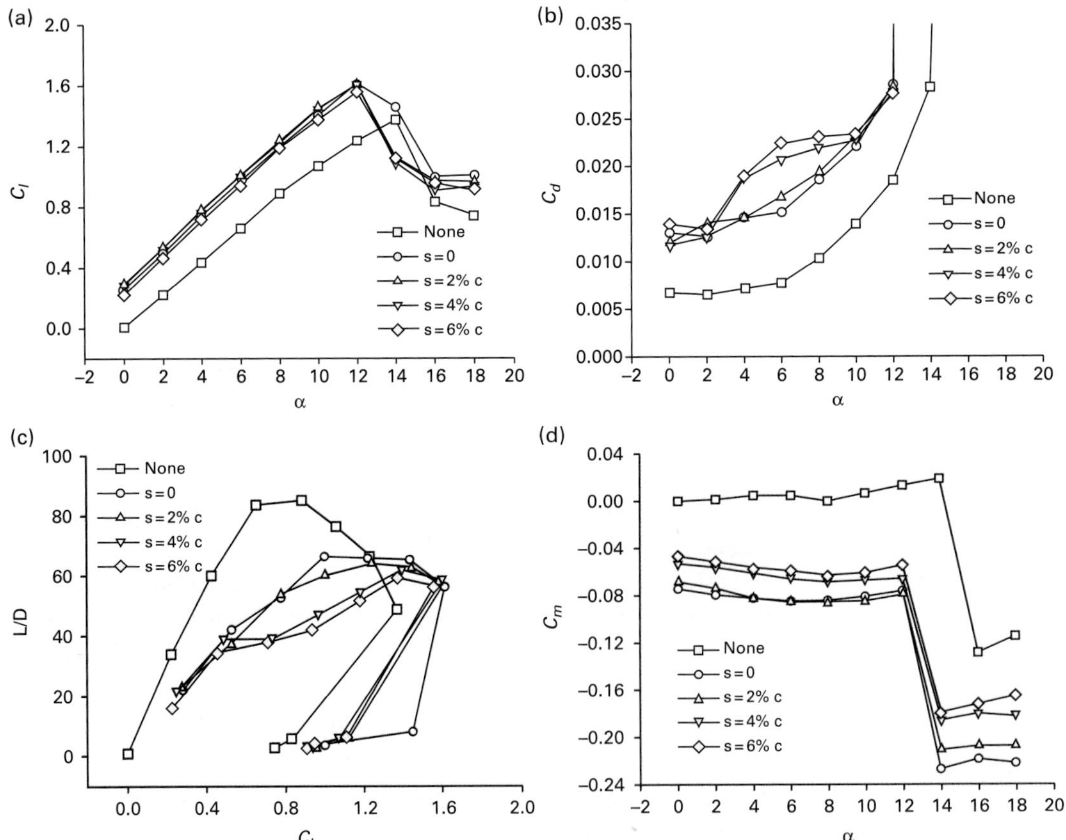

Figure 2.5 Aerodynamic forces on a symmetric airfoil for different mounting locations of the Gurney flap. (a) Lift coefficient; (b) drag coefficient; (c) lift-to-drag ratio; and (d) pitching moment coefficient (Li et al. 2003a). Reproduced with permission.

0, 2%c, 4%c, and 6%c from the trailing edge, respectively. On the other hand, the Gurney flap also increases the drag coefficient, while the drag increment increases with the increase of mounting distance from the trailing edge (Figure 2.5(b)). Although the lift-to-drag ratio is reduced with Gurney flap, a larger lift-to-drag ratio is expected for the $s = 0$ case (Figure 2.5(c)). The nose-down pitching moment decreases with the forward positioning of the Gurney flap, but is still more negative than that of the clean airfoil (Figure 2.5(d)).

2.2.3 Effects of the Gurney Flap Mounting Angle

The influence of the mounting angle of the Gurney flap relative to the airfoil pressure surface on the aerodynamic performance is shown in Figure 2.6. The three typical installed angles of the Gurney flap all increase the lift coefficient significantly before

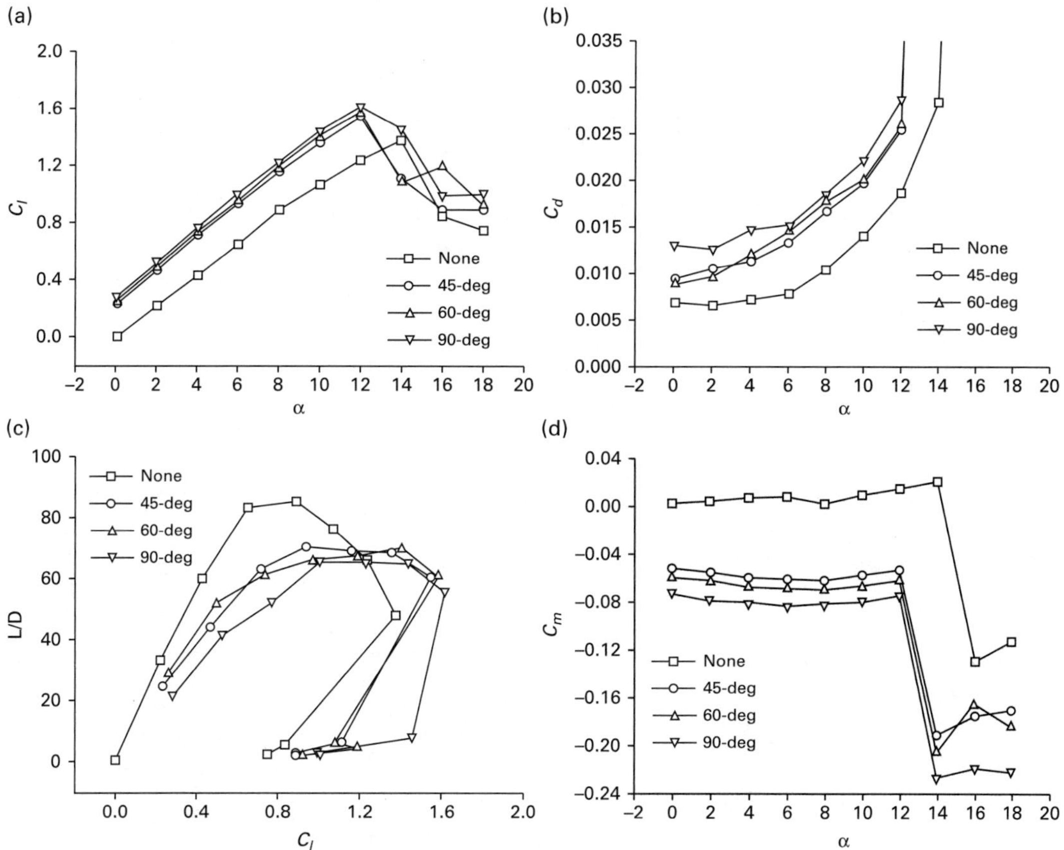

Figure 2.6 Aerodynamic forces on a symmetric airfoil for different mounting angles of the Gurney flap. (a) Lift coefficient; (b) drag coefficient; (c) lift-to-drag ratio; and (d) pitching moment coefficient (Li et al. 2003a). Reproduced with permission.

stall angle. However, the best effect appears in the case with mounting angle 90°, followed by the cases with 60° and 45°. The maximum lift coefficient is increased by about 12%, 15%, and 17% for the cases of 45°, 60°, and 90°, respectively. However, the stall angle is reduced to 12° from the original 14°. Similar effects happen for the drag coefficient. As a result, the 90°control case may lead to the smallest lift-to-drag ratio. The nose-down pitching moment increases with the increase of mounting angle.

2.2.4 Effects of the Gurney Flap Configuration

Most of the Gurney flaps used are flat plates, but Li et al. (2003b) tested the effect of a sawtoothed plate, as shown in Figure 2.7(d). Figure 2.7 shows the variation of lift and drag coefficients for 1.5%c height Gurney flaps. The lift enhancement for the sawtoothed cases is not as evident as the flat plate case, but they can obtain

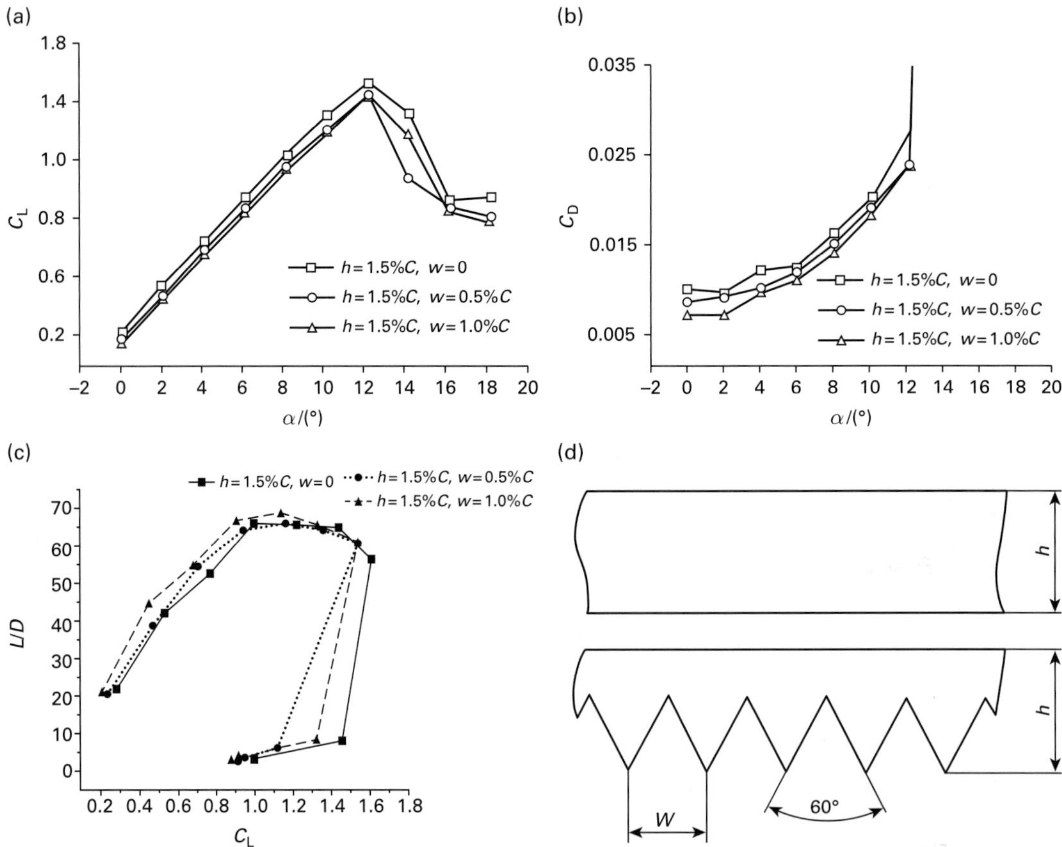

Figure 2.7 Aerodynamic forces on a symmetric airfoil for different configurations of the Gurney flap. (a) Lift coefficient; (b) drag coefficient; (c) lift-to-drag ratio; and (d) different configurations of the Gurney flap (Li et al. 2003b). Reproduced with permission.

a smaller drag penalty. As a result, all sawtoothed cases can increase the maximum lift-to-drag ratio in comparison with the flat plate case. Van Dam et al. (1999) and Lee and Ko (2009) also obtained similar results for sawtoothed and perforated Gurney flaps, respectively. The variations are attributed to a reduction in the effective windward area of the Gurney flap.

2.2.5 Lift Enhancement Mechanism

The flow around the Gurney flap is similar to that over a bluff body. Actually, as the fluids flow over the airfoil upper surface near the trailing edge and the Gurney flap lower edge, flow separates and thus a shear layer develops from both sides. Spanwise vorticity concentrates near the tails of the shear layers. The vorticity separates from the main shear layers when it is large enough, forming vortical structures shedding downstream individually. Such a process can be clearly seen in Figure 2.8.

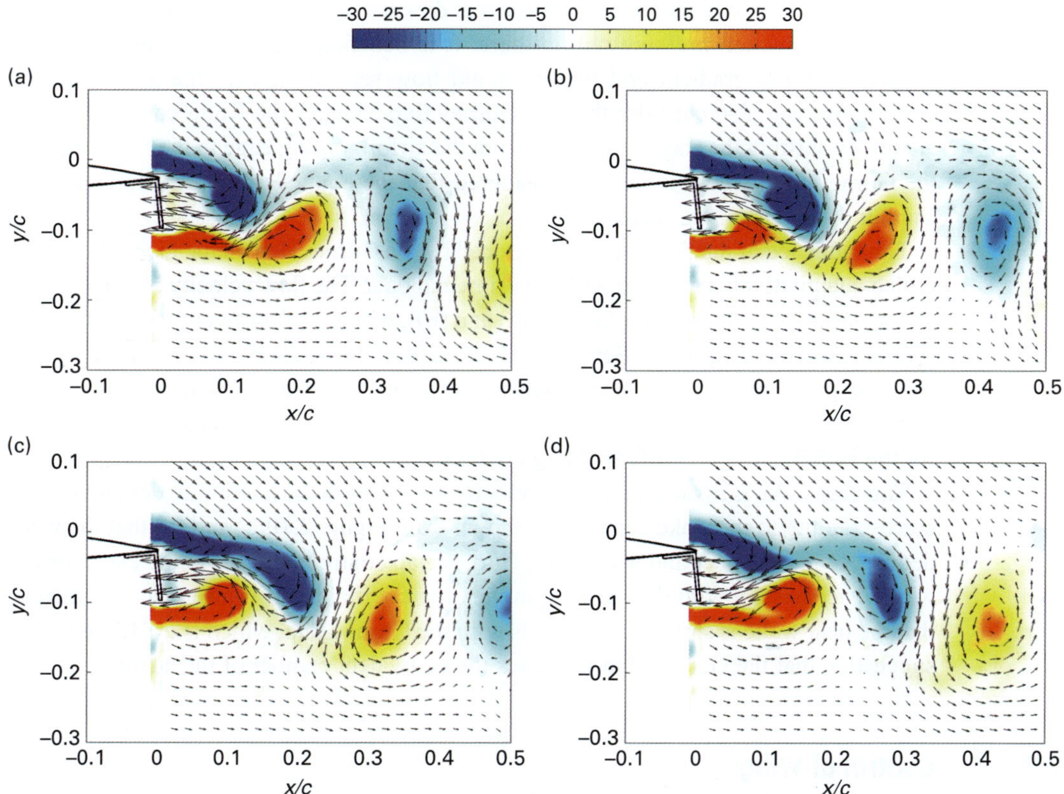

Figure 2.8 Evolution of the phase-averaged spanwise vorticity field $\omega_z c/U_\infty$ superposed with velocity vectors (with $0.8U_\infty$ subtracted from streamwise velocity) at four phases 0 (a), 0.5π (b), π (c), and 1.5π (d) for the airfoil installed $h/c = 7.0\%$ Gurney flap at $\alpha = 2°$, $Re = 20\,000$ (Feng et al. 2012). Reproduced with permission, copyright © Springer-Verlag 2012.

In comparison with the clean airfoil, the Gurney flap decreases the dominant shedding frequency, while the reduction of the shedding frequency increases with an increase of the flap height. This is due to the increased distance between the two separating shear layers, requiring more time for the opposite shear layer to cross the wake, which has been suggested by Lee and Ko (2009) and Feng et al. (2012). Accordingly, the distance l between upper and lower shear layers can be used as a characteristic length to calculate the dimensionless frequency fl/U_∞. It was found by Feng et al. (2012) that most of the dimensionless frequencies were collapsed onto a range from 0.35 to 0.38, despite the differences in the Gurney flap height.

The variation in the vortex dynamics results in a significant change in the mean flow topology. Typical flow topology around the airfoil trailing edge without and with a Gurney flap is shown in Figure 2.9. The flow over the airfoil is attached at small angles of attack, while there might be a separation region over the airfoil suction surface at higher angles of attack. Flow topology changes with Gurney flap control. One recirculation region forms over the upstream of the Gurney flap induced by the adverse pressure gradient, which significantly increases the pressure of the lower surface. In addition, two recirculation regions form at downstream of the Gurney flap, inducing a significant suction pressure region there. It is beneficial for fluids to overcome the adverse pressure gradient and thus to avoid flow separation over the airfoil suction surface. We can also consider that the Gurney flap changes the Kutta condition of the airfoil, by shifting the stagnation point from the airfoil trailing edge to the lower edge of the Gurney flap, where it can be considered that the equivalent chord length and the camber of the airfoil are increased.

The influence of the flow topology induced by the Gurney flap on the global flow field is also quantitatively validated with experimental data. The flow velocity over the airfoil upper surface is increased (Figure 2.10(a) and (b)), while the flow over the lower surface is reduced. According to the Bernoulli equation, both the suction over the upper surface and the pressure over the lower surface will be increased, which is validated in Figure 2.11. Thus, the pressure difference over both surfaces is increased, accounting for the increment in both lift and drag coefficients.

The near wake is associated with a velocity deficit, with the maximum deficit velocity corresponding to the wake center. It is shown from Figure 2.10(c) and (d) that the wake deficit becomes deeper and wider with Gurney flap control than that in the clean airfoil, which is consistent with the increase in the drag coefficient. It is also clearly indicated that the near wake is shifted downwards with the Gurney flap, which is evidence for the suggestion that the Gurney flap increases the equivalent camber of the airfoil.

2.3   Control of Wing

2.3.1   Finite Wing

For flow over a finite wing, a tip vortex forms near the wing tip, which induces lift-induced drag to deteriorate the aerodynamics performance. Lee (2011) placed Gurney flaps with different heights on a finite wing with aspect ratio of 0.87. The experiment was conducted at $Re = 1.05 \times 10^5$. The streamwise vorticity contours for different cases at $x/c = 2.5$ and $\alpha = 6°$ are shown in Figure 2.12. The Gurney flap may increase the local maximum vorticity for the 1%c height case, but decrease the maximum vorticity for 2%c and 3.2%c cases, which can also be observed from Figure 2.13(a). However, the Gurney flap can increase the total circulation of the tip vortex for all control cases (Figure 2.13(b)). Such variation actually provides us with a method to manipulate the strength of tip vortex as required.

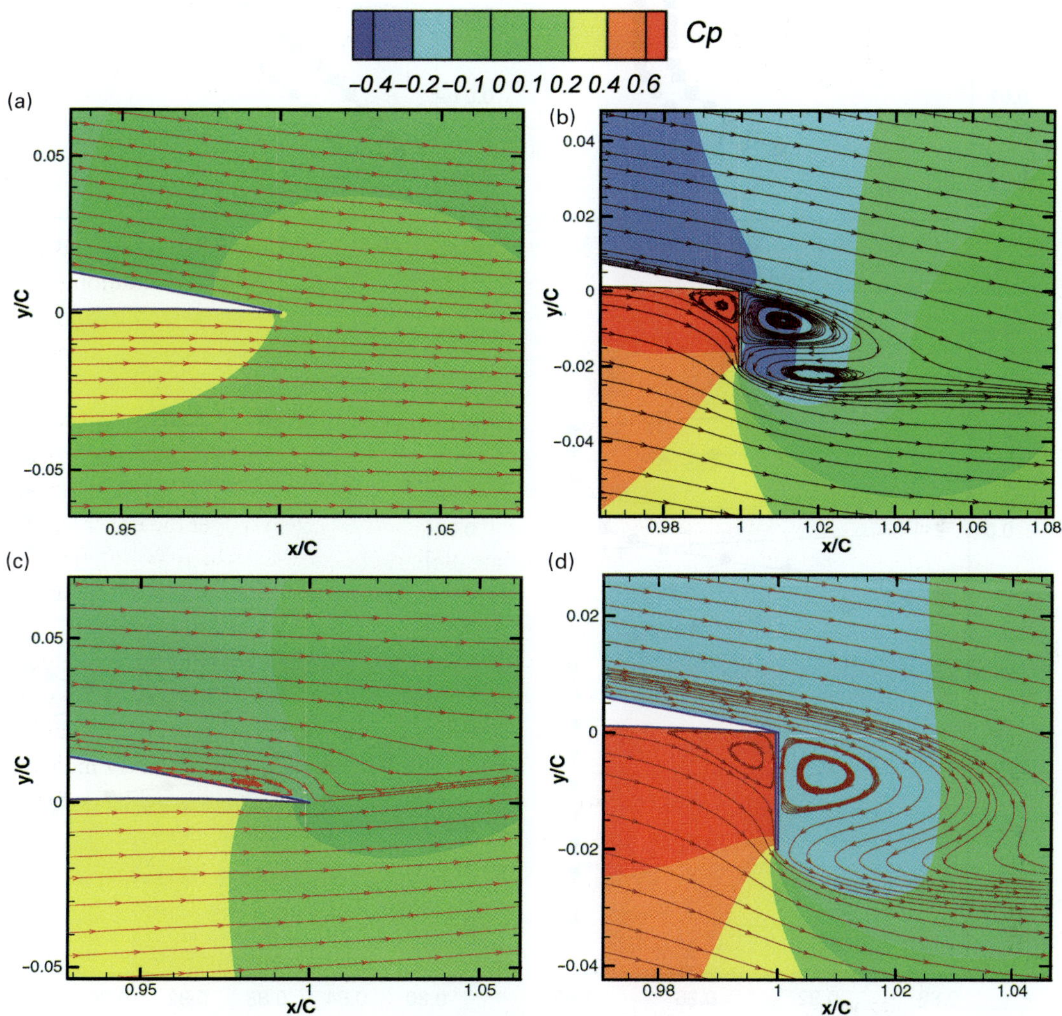

Figure 2.9 Pressure contours superposed with streamlines of RAE-2822 airfoil without (a, c) and with (b, d) 2%c height Gurney flap control at $\alpha = 6°$ (a, b) and 14° (c, d) (Yu et al. 2011). Reproduced with permission, copyright © 2011 by the authors

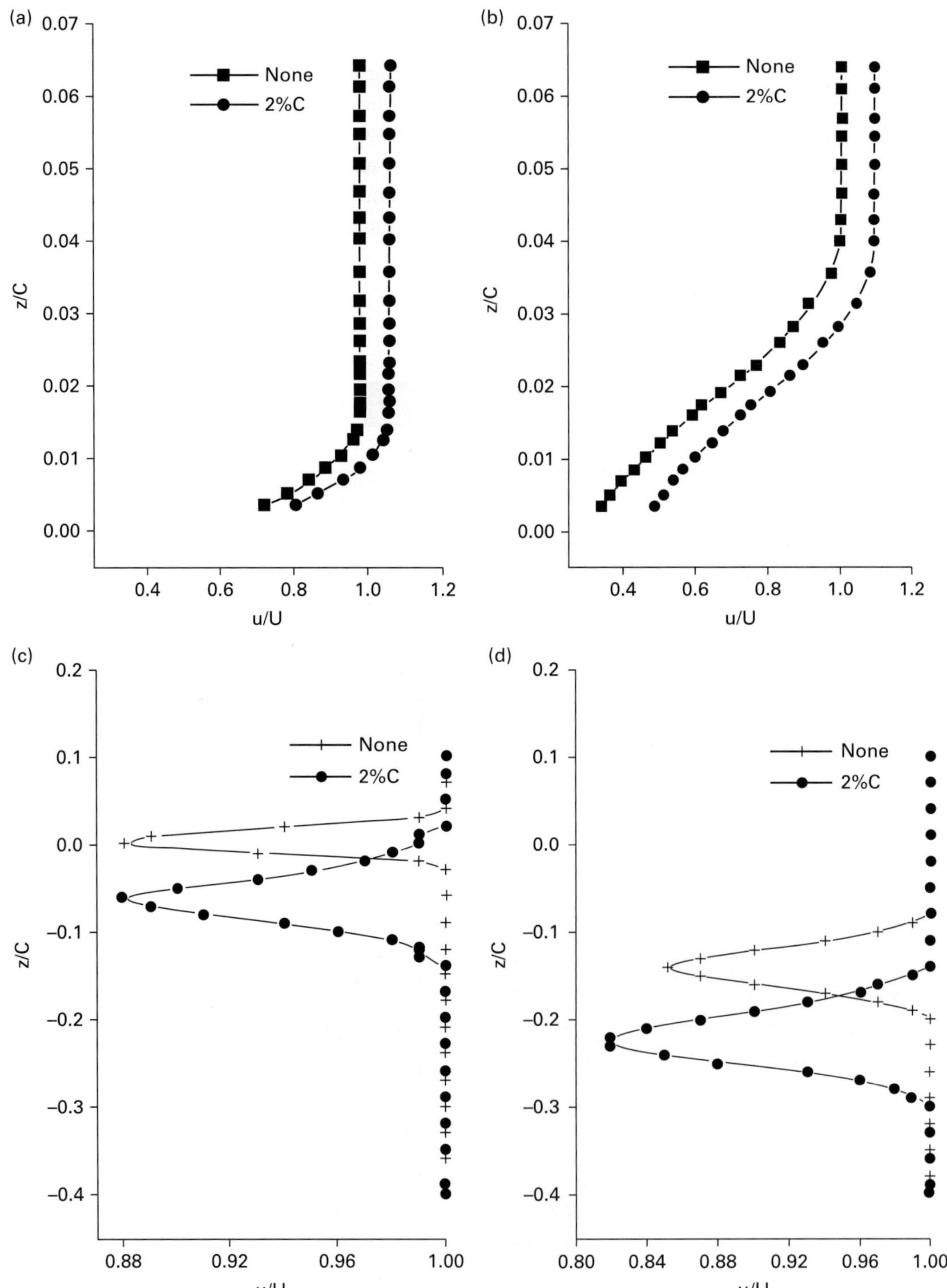

Figure 2.10 Velocity profiles of the airfoil at $x/c = 0.9$ (a, b) and 1.7 (c, d) from the leading edge without and with 2%c height Gurney flap control at $\alpha = 0°$ (a, c) and 10° (b, d) (Li et al. 2002a). Reproduced with permission, copyright © Springer Nature 2002.

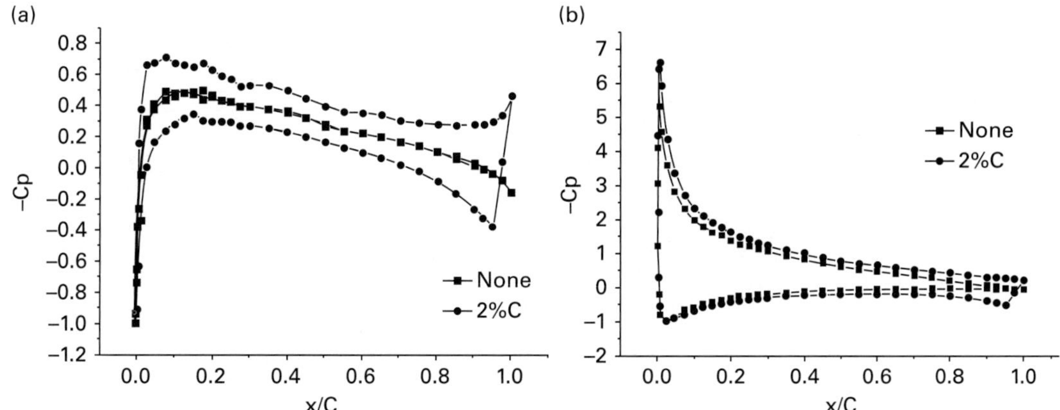

Figure 2.11 Pressure distribution over the airfoil without and with 2%c height Gurney flap control. (a) $\alpha = 0°$; and (b) $\alpha = 10°$ (Li et al. 2002a). Reproduced with permission, copyright © Springer Nature 2002.

2.3.2   Delta Wing

The delta wing is the basic configuration of a modern fighter aircraft. Thus, there is also a requirement to improve its aerodynamic performance. One useful technique is the Gurney flap. Li et al. (2002b) used the Gurney flap to control a cropped non-slender delta wing with a leading edge of 40°, as shown in Figure 2.14. The Gurney flaps with four different heights, $h/c = 1\%$, 2%, 3%, and 5%, were used, denoted as G1, G2, G3, and G5, respectively.

Similar to the effects on a low speed airfoil, the Gurney flap can also increase the lift coefficient, drag coefficient, and the nose-down pitching moment and the increment increases with the flap height, as shown in Figure 2.15. The maximum lift coefficient is increased by about 16%, 27%, 27%, and 33% for the $1\%c$, $2\%c$, $3\%c$, and $5\%c$ Gurney flaps, respectively. It is found that only the $1\%c$ Gurney flap can increase the maximum lift-to-drag ratio in comparison with the clean wing in their investigation. Also, at high lift coefficient, all Gurney flaps can increase the lift-to-drag ratio. A higher lift coefficient can be obtained at the same lift-to-drag ratio as the clean wing. Similar effects were also found by Wang and Li (2007) for a double delta wing at a low freestream speed.

2.3.3   Aircraft

Aiming at real applications, Gurney flaps have also been tested on an aircraft model. One example is presented by Albertani (2008) to control a micro aerial vehicle (MAV), as shown in Figure 2.16. The mean aerodynamic chord c was 0.1073 m, the wing span b was 0.377 m, and the measured wing area was 0.0376 m². The Gurney flap had a height of $2.8\%c$, and two spanwise lengths of $50\%b$ and $75\%b$ were tested. One wind tunnel test was made at a Reynolds number of $Re = 95200$.

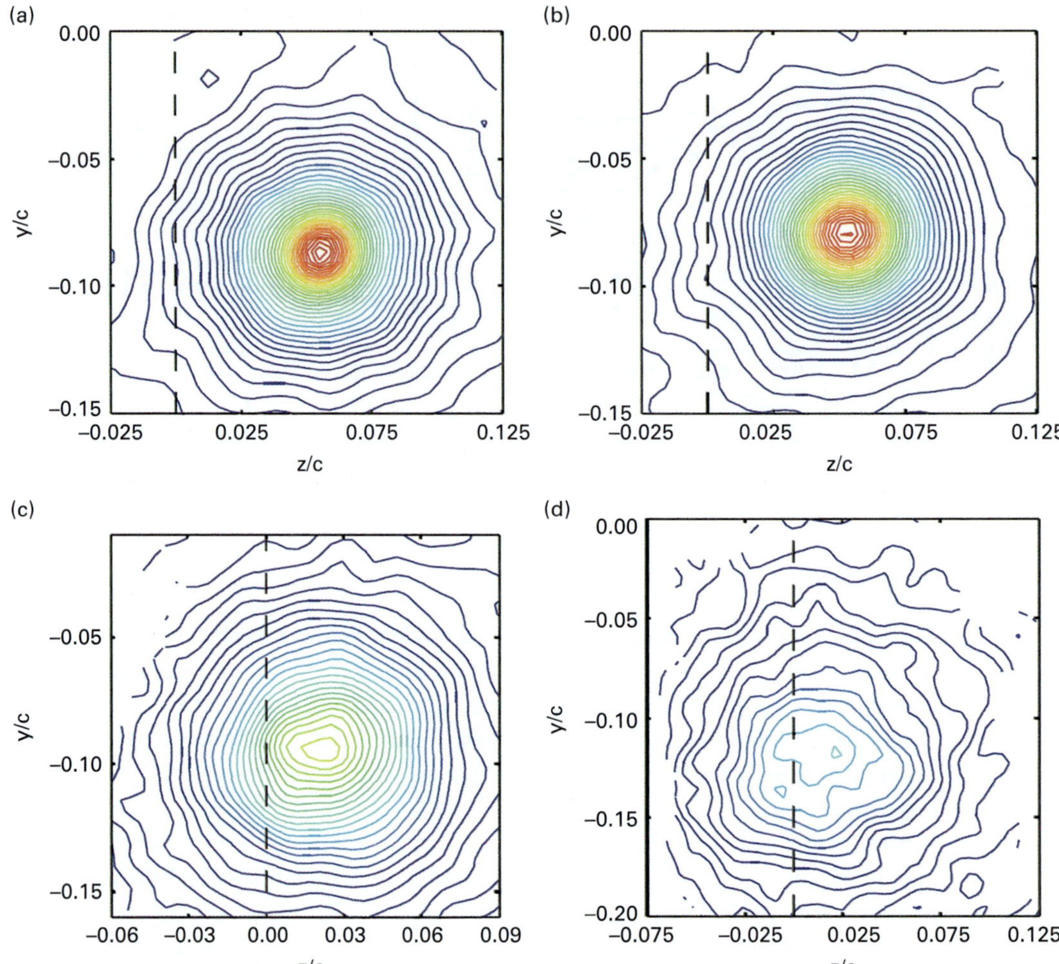

Figure 2.12 Normalized iso-vorticity contours at $x/c = 2.5$ and $\alpha = 6°$ for a finite wing without and with Gurney flap control. (a) Natural case; (b) $h = 1\%c$; (c) $h = 3.2\%c$; and (d) $h = 4.8\%c$ (Lee 2011). Reproduced with permission, copyright © Springer-Verlag 2010.

Figure 2.17(a) shows the significant increment of the lift coefficient of the MAV by Gurney flap control. The maximum lift coefficient is increased by about 18% and 27% for the 50%b and 75%b Gurney flaps, respectively. We can also see the benefit of the Gurney flap control from the C_L–C_D curve. It is indicated that smaller drag than the no control case is induced by control for the same lift coefficient (Figure 2.17(b)). The nose-down pitching moment is increasing with the Gurney flap span (Figure 2.17(c)), namely the flap area, which is similar with the effects on airfoils and wings.

Wang et al. (2006) used Gurney flaps to control a simplified forward-swept aircraft model, which consisted of a fuselage, a forward-swept wing, and a canard wing, as

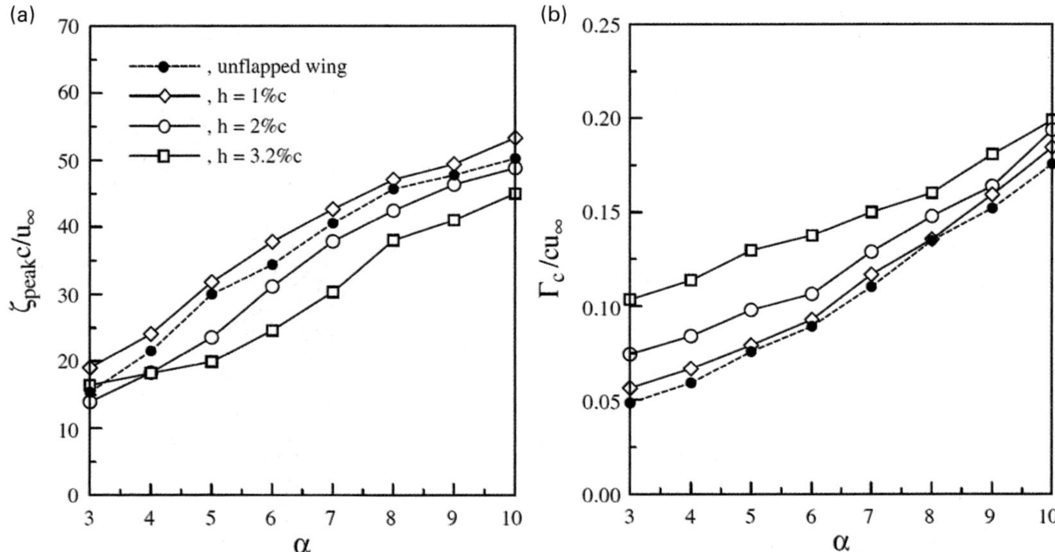

Figure 2.13 Variation of peak vorticity (a) and vortex circulation (b) with the angle of attack at x/c = 2.5 for a finite wing without and with Gurney flap control (Lee 2011). Reproduced with permission, copyright © Springer-Verlag 2010, from Lee 2011.

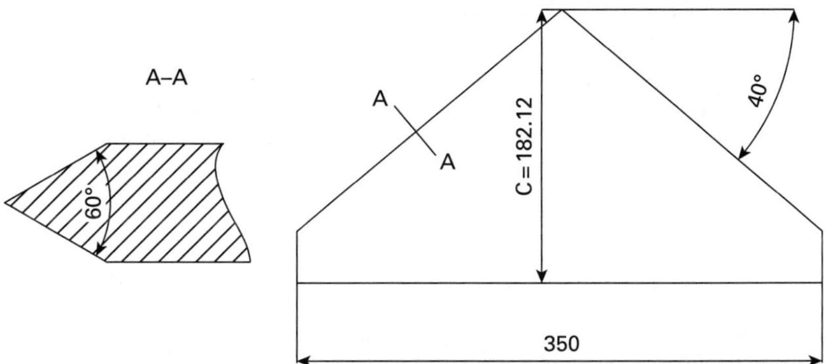

Figure 2.14 Schematic of a delta wing used for Gurney flap control (Li et al. 2002b). Reproduced with permission, copyright © Springer Nature 2002.

shown in Figure 2.18. The parameters of the model are also shown. The Gurney flaps were made of rectangular flat plates with 1 mm thick, 200 mm long, and height ranging from 1.2 to 4.8 mm, corresponding to 1%–4% of the mean chord length of the main wing. They were perpendicularly glued on the pressure surface of the forward-swept wing near the trailing edge. The free-stream velocity was fixed at 20 m/s, corresponding to $Re = 1.44 \times 10^5$ based on the mean chord length.

The aerodynamic forces on the aircraft without and with Gurney flaps are shown in Figure 2.19. The lift coefficient, drag coefficient, and nose-down pitching moment all

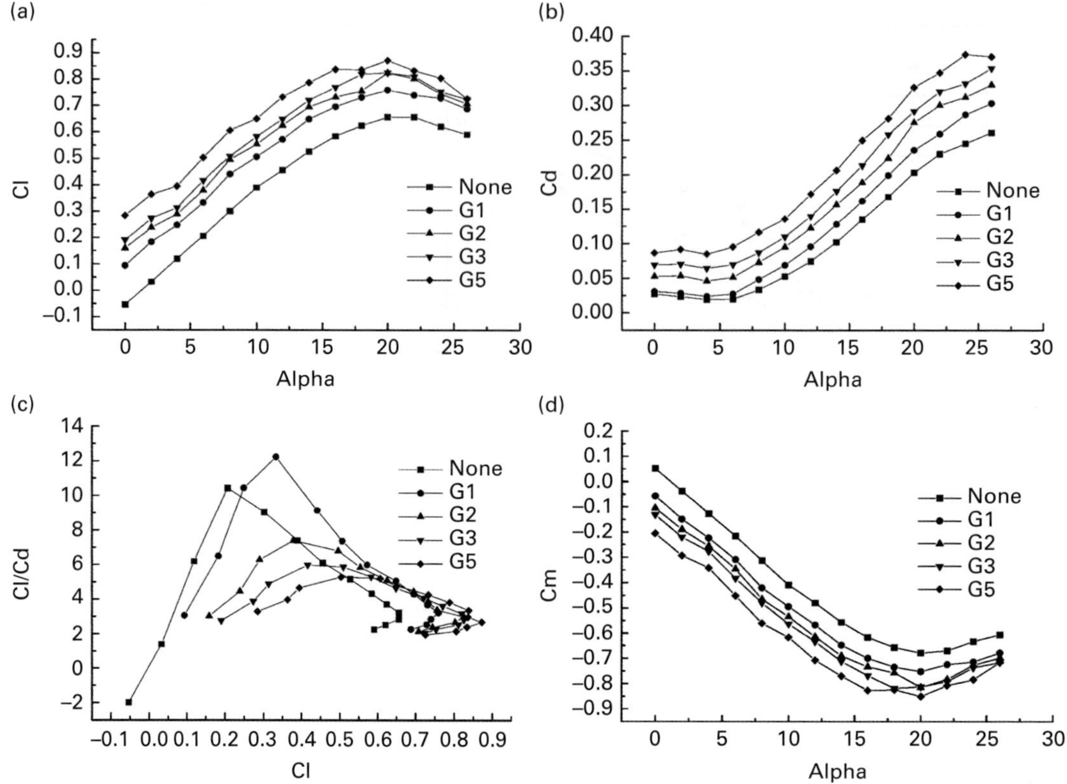

Figure 2.15 Aerodynamic forces on a delta wing for different Gurney flap heights. (a) Lift coefficient; (b) drag coefficient; (c) lift-to-drag ratio; and (d) pitching moment coefficient (Li et al. 2002b). Reproduced with permission, copyright © Springer Nature 2002.

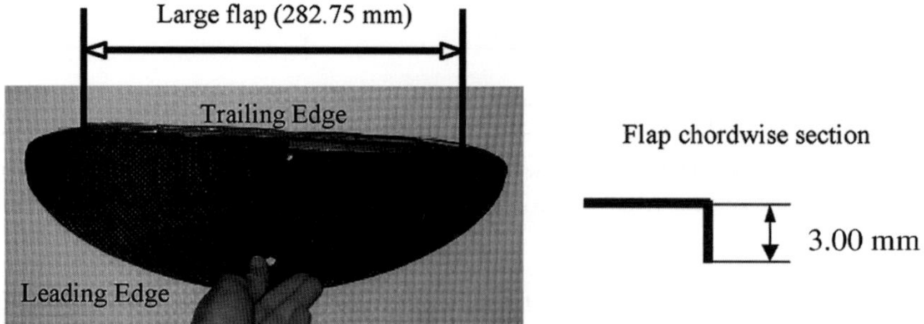

Figure 2.16 Schematic of a MAV used for Gurney flap control (Albertani 2008). Reproduced with permission, copyright © The American Institute of Aeronautics and Astronautics.

increase with the increase of the Gurney flap height. The maximum lift coefficient is increased for all control cases, though the stall angle is reduced to 36° from the original 38°. It is also indicated that the lift-to-drag ratio is increased by the Gurney flaps at small

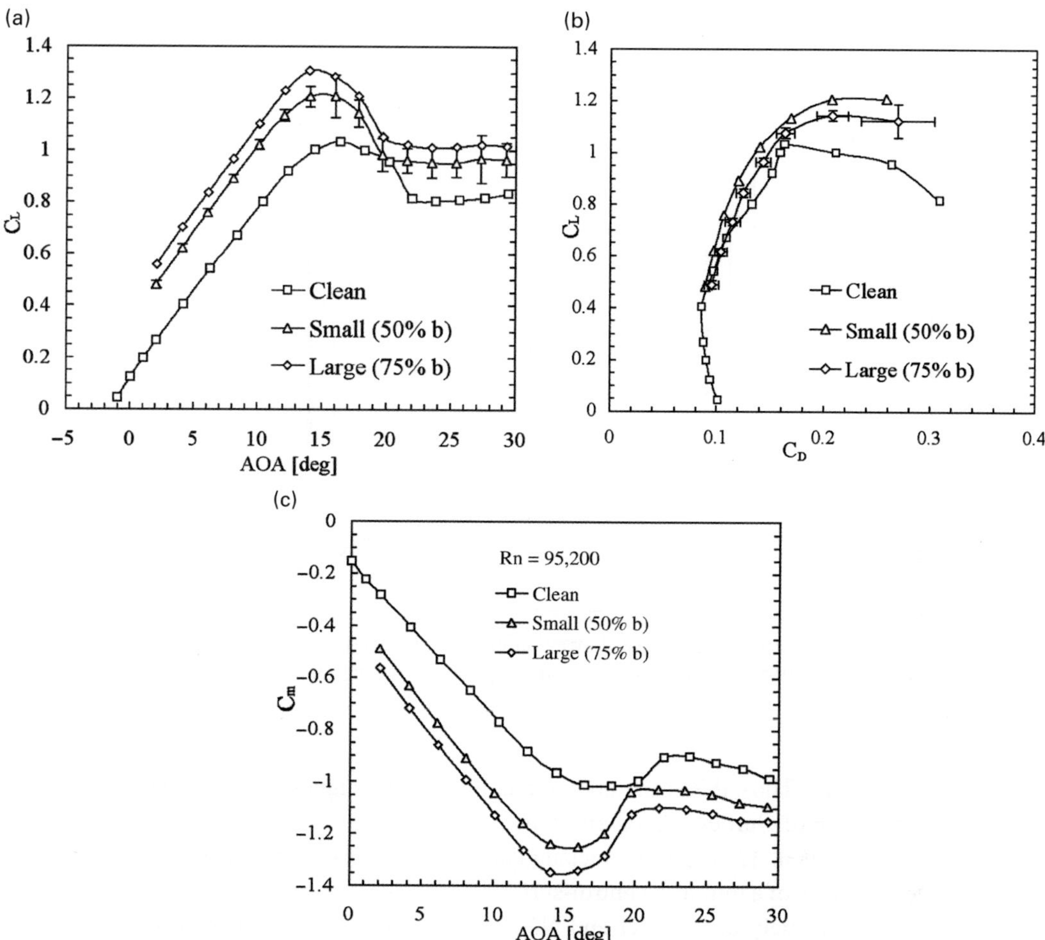

Figure 2.17 Aerodynamic forces of the MAV for different Gurney flap spans. (a) Lift coefficient versus angle of attack; (b) lift coefficient versus drag coefficient; and (c) pitching moment coefficient versus angle of attack (Albertani 2008). Reproduced with permission, copyright © The American Institute of Aeronautics and Astronautics.

angles of attack, because the associated increment in the drag coefficient is relatively small. There is a significant finding that the lift-to-drag ratio increases with flap height for all cases tested. It is indicated that the maximum lift-to-drag ratio is increased by about 8%, 16%, 13%, and 24% for the 1%c, 2%c, 3%c, and 4%c Gurney flaps, respectively. That is quite different from the applications in airfoils and wings, as a number of researches indicated that the Gurney flap could not increase the lift-to-drag ratio for airfoils and wings unless at a certain small height. Wang et al. (2006) indicated that the drag of the main wing was a small portion of the total drag of the aircraft, and thus the drag increase of the main wing due to the application of Gurney flap did not contribute very much to the total drag increase. On the other hand, the main wing contributed much to the total lift of the aircraft, which was greatly increased by the

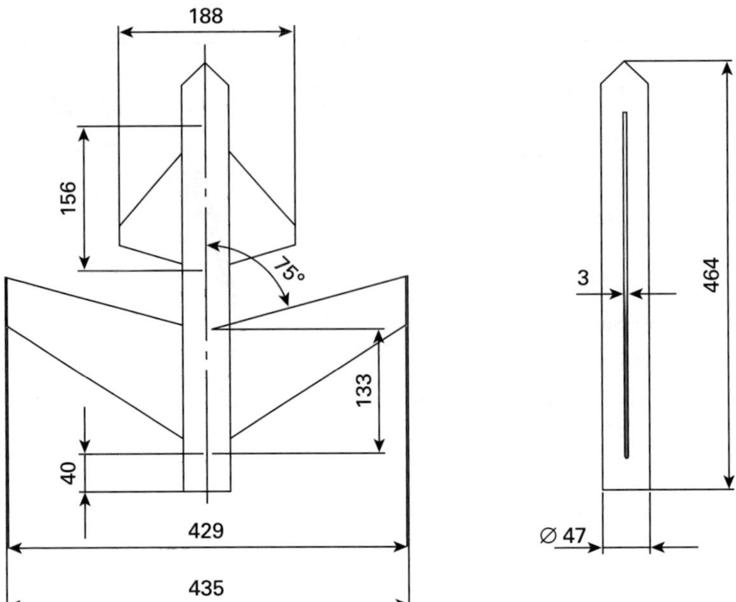

Figure 2.18 Schematic of a forward-swept aircraft model used for Gurney flap control (Wang et al. 2006). Reproduced with permission, copyright © The American Institute of Aeronautics and Astronautics.

Gurney flap. Thus, this is an important finding, which suggests a promising future for the application of Gurney flap on aircrafts.

Wang et al. (2010) used rectangular Gurney flaps to improve the aerodynamic performance of a multiple-control-surface tailless aircraft model. There were ten control surfaces, namely elevons N1 and N2, flaperons N3 and N4, canard wing N5, as seen from the plan view in Figure 2.20. Gurney flaps were made of rectangular flat plate with 1 mm thickness and heights ranging among 1 mm, 3 mm, and 5 mm, corresponding to 0.6%, 1.8%, and 3.0% of the mean chord length of the main wing, respectively. Their length was the same as the trailing-edge span of individual control surfaces. The Gurney flaps were glued perpendicularly onto the bottom surface at the trailing edge of different control surfaces. The experimental speed was kept at 20 m/s, corresponding to the Reynolds number of 2.3×10^5 based on the mean chord length.

It was indicated that the elevons and flaperons with Gurney flaps could improve the lift coefficient, while the nose-down pitching moment coefficient also increased inevitably. The increase of the pitching moment usually led to terrible uncontrollability of the aircraft. On the contrary, the canard wing with Gurney flaps could decrease the nose-down pitching moment with little lift enhancement or even lift reduction. Therefore, if the Gurney flaps were installed on combinations of the canard wing and either the elevons or flaperons, it might lead to an increase in the lift coefficient without any additional increment of the pitching moment or with less increment. Accordingly,

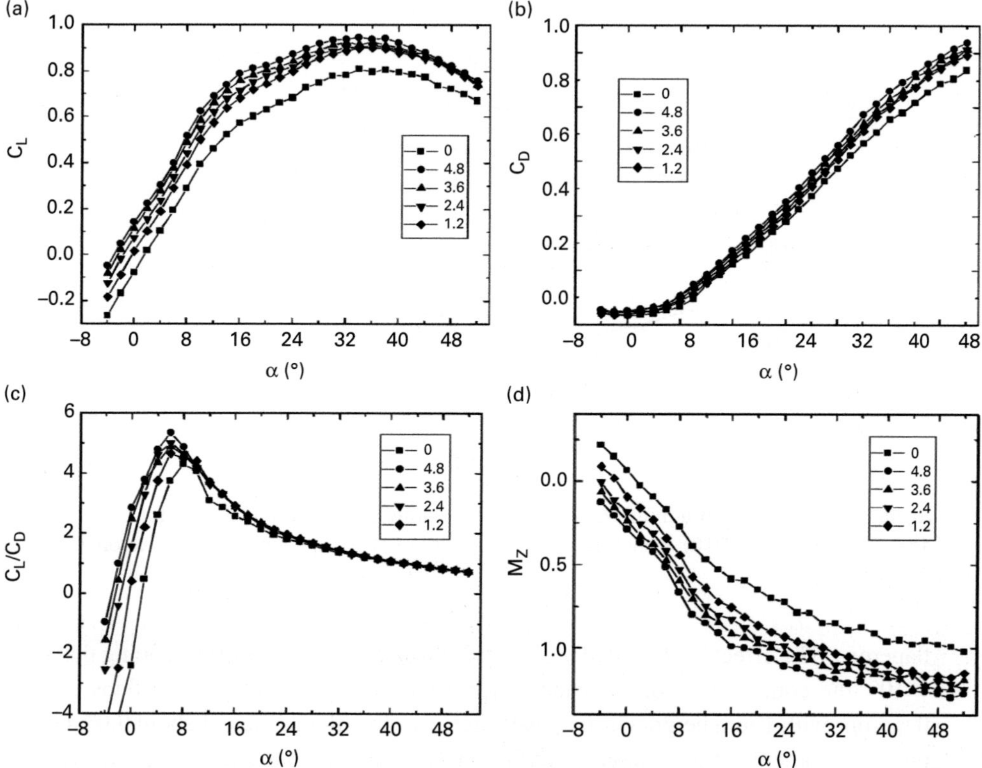

Figure 2.19 Aerodynamic forces on a forward-swept aircraft for different Gurney flap heights. (a) Lift coefficient; (b) drag coefficient; (c) lift-to-drag ratio; and (d) pitching moment coefficient (Wang et al. 2006). Reproduced with permission, copyright © The American Institute of Aeronautics and Astronautics.

Wang et al. (2010) proposed an effective control strategy for the tailless aircraft with Gurney flaps.

Different combinations of the canard wing and either N1 or N2 installed with Gurney flaps were tested. The height of the Gurney flaps on the canard wing was 5 mm, while it changed among 1 mm, 3 mm, and 5 mm on elevons N1 and N2. Figure 2.21(a) and (b) presents the lift and pitching moment coefficients increment compared with the clean aircraft model. When it is 1 mm on the elevons, the lift coefficient increases about 0.02 at the linear stage for $\alpha < 10°$, while the pitching moment coefficient enhancement is nearly zero. However, other cases will still increase the nose-down pitching moment coefficient when improving the lift coefficient. Moreover, compared with the cases for the elevons with Gurney flaps and the canard wing without, the combination of the canard wing and either N1 or N2 installed 5 mm Gurney flaps makes the lift coefficient increase about 0.01, while the nose-down pitching moment coefficient decreases about 0.01 for $\alpha < 10°$.

Then the combinations of the canard wing and either the elevons or flaperons with Gurney flaps were tested. The height of the Gurney flaps on the canard wing was still kept at 5 mm, while it changed among 1 mm, 3 mm, and 5 mm on both the elevons and

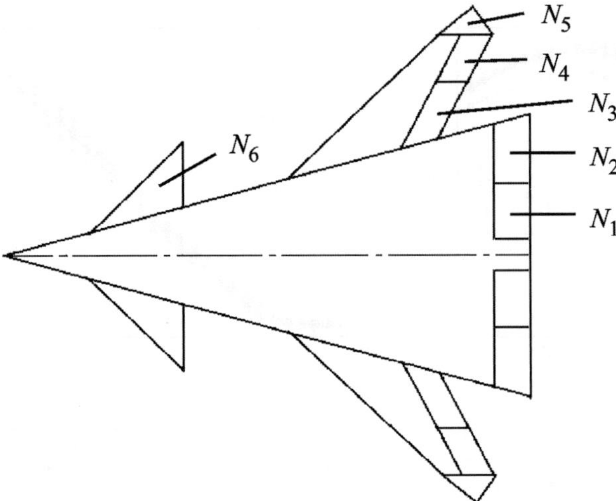

Figure 2.20 Schematic of a multiple-control-surface tailless aircraft model used for Gurney flap control (Wang et al. 2010). Reproduced with permission of Beijing University of Aeronautics and Astronautics.

flaperons. The effect is presented in Figure 2.21(c) and (d). It is worth noticing that for α < 10°, the combination of the canard wing with 5 mm height Gurney flaps and the flaperons with 1 mm height Gurney flaps increases the lift coefficient about 0.03, but the pitching moment hardly changes. In addition, compared with the cases when the Gurney flaps are only installed onto the elevons and flaperons, the combinations of the canard wing and either the elevons or flaperons make the lift coefficient increase slightly, while the nose-down pitching moment coefficient decreases at the linear stage for $\alpha < 20$; after $\alpha > 20°$, the lift improvement decreases a little, while the nose-down pitching moment coefficient is reduced evidently.

We have only presented investigations for the applications of Gurney flaps on a few aircraft. Nowadays, the effects of Gurney flaps have also been validated on other configurations, such as microrotorcraft (Nelson and Koratkar 2005), annular wing aircraft (Traub 2009), and target drone (Zhao et al. 2010).

2.4   Dynamic Flow Control

The Gurney flap has proved its ability to improve the aerodynamic performance of stationary airfoils and wings. Some effects have also been made to test the dynamic control ability of the Gurney flap. There are usually two main methods. One is the control of a dynamic body by using a stationary Gurney flap, and the other is to control a stationary object by a moveable Gurney flap. For both cases, the flow is highly unsteady. Thus, we classify them as Gurney flap dynamic flow control.

Lee and his co-workers from McGill University conducted a series of studies into the control of flow around an oscillating airfoil by using Gurney flaps (Lee and Gerontakos

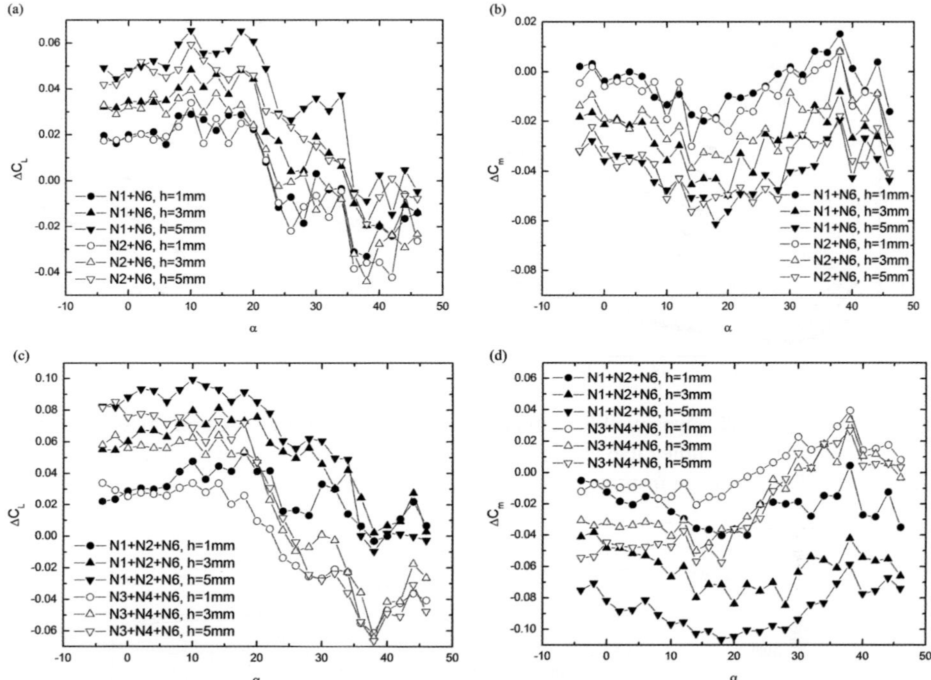

Figure 2.21 Lift coefficient (a, c) and pitching moment (b, d) coefficient increments for the combinations of the canard wing and either N1 or N2 with Gurney flaps (a, b) and the combinations of the canard wing and either the elevons or flaperons with Gurney flaps (c, d) (Wang et al. 2010). Reproduced with permission of Beijing University of Aeronautics and Astronautics.

2006; Lee and Lee 2007). A NACA 0015 airfoil was forced oscillating around the axis at 25% chord, following a function of $\alpha = 14° + 8°\sin2\pi ft$, where the reduced frequency was $\pi fc/U_\infty = 0.09$. Two Gurney flap heights, $h = 2.5\%c$ and $5\%c$, were both in regular and inverted arrangements, where the results are shown in Figure 2.22. The C_L-α loop is shifted vertically upward with the regular Gurney flap, which is attributed to positive camber effects. Thus, the maximum lift coefficient is also increased, and the increment increases with the Gurney flap height. When the Gurney flap is installed in an inverted arrangement, the lift is shifted downwards during the pitching-up motion. However, there is nearly no difference between the control and clean airfoils around the stall angle and during the pitching-down motion. The total vortex circulation generally follows the variations of the lift coefficient during the pitching motion.

On the other hand, Tang and Dowell (2007) controlled a NACA 0012 airfoil by an oscillating Gurney flap (Figure 2.23). The airfoil had a chord length of $c = 0.2554$ m, and a span of $b = 0.52$ m. The free-stream velocity was $U_\infty = 20$ m/s, corresponding to the airfoil Reynolds number of $Re = 3.48 \times 10^5$. The oscillating Gurney flap was made of an aluminum sheet, with height of $h_s = 6.35$ mm (0.25 in. or 2.5%c), length of 0.508 m, and thickness of 0.51 mm, which was located at 90%c from the airfoil leading edge.

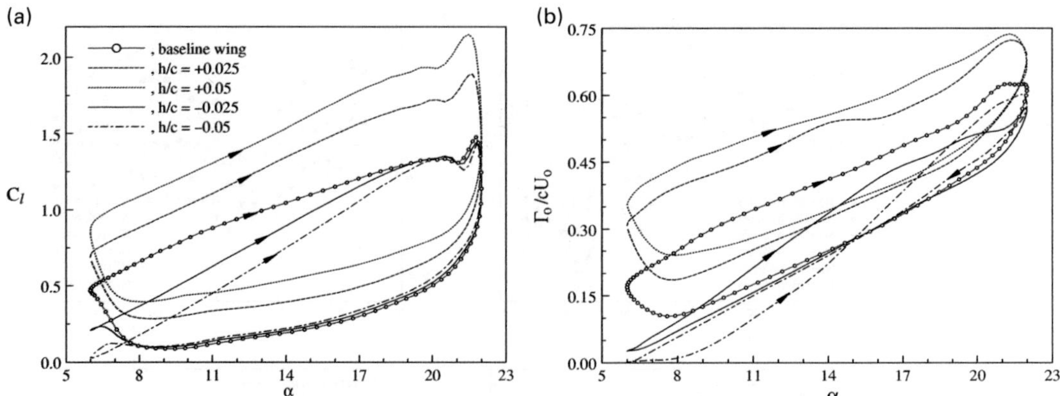

Figure 2.22 Dynamic lift coefficient (a) and total vortex circulation (b) loops of the airfoil without and with Gurney flap control (Lee and Lee 2007). Reproduced with permission, copyright © The American Institute of Aeronautics and Astronautics.

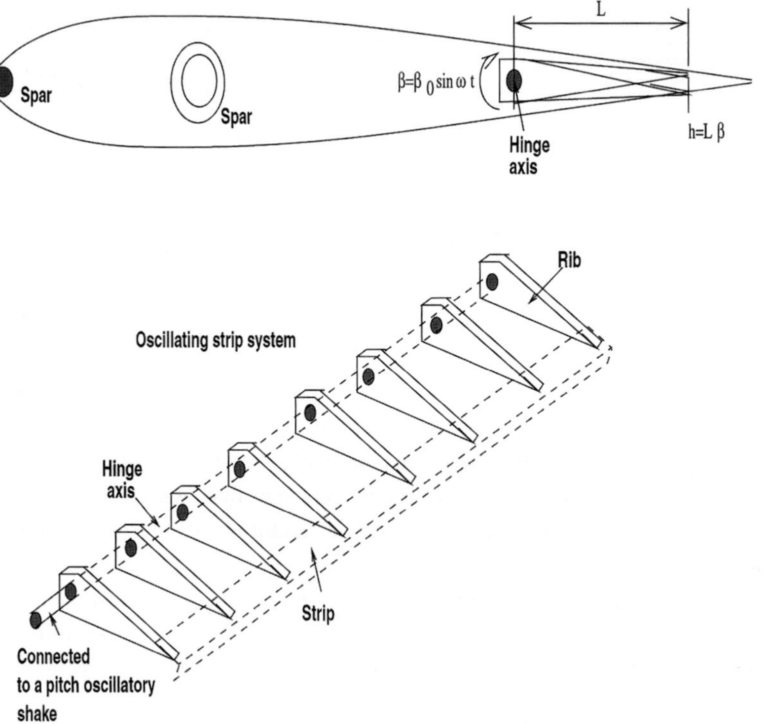

Figure 2.23 Schematic of the oscillating Gurney flap on an airfoil (Tang and Dowell 2007). Reproduced with permission, copyright © The American Institute of Aeronautics and Astronautics.

The Gurney flap was fixed to the end of the eight special rib elements, which were bolted to the hinge axis located at 69%c. When the hinge axis was torsionally oscillated, the Gurney flap was also forced to oscillate. The instantaneous height of the oscillating

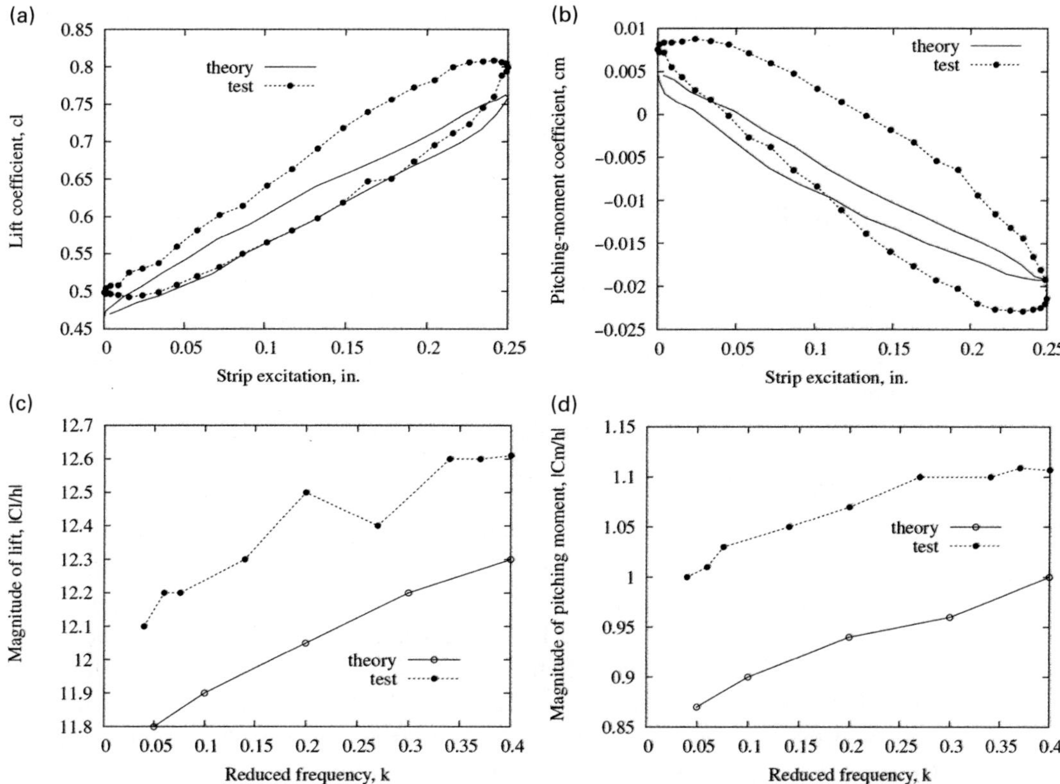

Figure 2.24 Lift coefficient (a, c) and pitching-moment coefficient (c, d) as functions of the Gurney flap height and the reduced frequency for $\alpha = 5°$ (Tang and Dowell 2007). Reproduced with permission, copyright © The American Institute of Aeronautics and Astronautics.

Gurney flap was given as $h = h_s(1+\sin(1.5\pi+2\pi ft))$, where f was the oscillatory frequency which varied from 1 to 10 Hz. A reduced frequency could be obtained by $k = 0.5fc/U_\infty$.

One typical result is shown in Figure 2.24(a) for the hysteresis loop of lift response and Figure 2.24(b) for the pitching-moment response. Both the lift coefficient and nose-down pitching momentum increase as the height of the Gurney flap becomes larger. The oscillating amplitude for the lift and pitching moment are 0.155 and −0.014, respectively. The averaged lift and pitching-moment values over one oscillation cycle are $C_L = 0.65$ and $C_M = -0.007$, which are larger than the static values of $C_L = 0.47$ and $C_M = 0.0055$ for the case without Gurney flap control. The influence of the oscillating frequency on the mean lift and pitching-moment coefficients is shown in Figure 2.24(c) and (d), respectively. It is indicated that the mean aerodynamic loadings increase as the oscillating frequency increases.

In addition, the dynamic flow control ability of the deployable Gurney flaps on the oscillating airfoils has also been validated, by works by Kinzel et al. (2010), Lee and Su (2012). Kinzel et al. (2010) found that the location of the Gurney flap greatly affected the unsteady aerodynamics. When the Gurney flap was positioned at the trailing edge, it appeared to remain effective through high deployment frequencies. Lee and Su (2012) focused on the impact of actuation start time and duration on the control effects, indicating that both earlier actuation and shorter duration, namely higher actuation frequency, were beneficial for obtaining the maximum lift coefficient.

2.5 Concluding Remarks

The Gurney flap is a simple device, such as a flat plate, which can be easily attached to the pressure surface of an airfoil. It can effectively increase the lift coefficient of airfoils, wings, and aircrafts, and shows great ability to shorten the takeoff/landing distance of the aircraft. However, the Gurney flap will also result in a drag penalty, and only under certain conditions can the Gurney flap increase the lift-to-drag ratio. Study of the parameters suggests that the Gurney flap should be vertically mounted onto the pressure surface at the trailing edge to provide the best lift enhancement, while the height of the Gurney flap should be within the local boundary layer thickness, but not too large.

The Gurney flap has not actually been applied to aircraft, though numerous studies in lab have validated its significant effects. One concern is that the Gurney flap may induce the drag penalty inevitably. Such a feature may influence the efficiency of the use of Gurney flaps, which may cause additional cruise drag and reduce economy. However, we have also proposed the strategy for engineering applications of the Gurney flaps. It is suggested that we can slightly decrease the cruising angle of the aircraft when the Gurney flap is used, in order to maintain the cruising lift condition but not increase the drag. In this way, it is expected that Gurney flaps could actually be used in the near future.

References

Albertani, R. Wind-tunnel study of Gurney flaps applied to micro aerial vehicle wing. *AIAA Journal*, 2008, 46(6): 1560–1562

Feng, L. H., Jukes, T. N., Choi, K. S., and Wang, J. J. Flow control over a NACA 0012 airfoil using dielectric-barrier-discharge plasma actuator with a Gurney flap. *Experiments in Fluids*, 2012, 52(6): 1533–1546

Kinzel, M. P., Maughmer, M. D., and Duque, E. P. N. Numerical investigation on the aerodynamics of oscillating airfoils with deployable Gurney flaps. *AIAA Journal*, 2010, 48(7): 1457–1469

Lee, T. and Gerontakos, P. Oscillating wing loadings with trailing-edge strips. *Journal of Aircraft*, 2006, 43(2): 428–436

Lee, T. PIV study of near-field tip vortex behind perforated Gurney flaps. *Experiments in Fluids*, 2011, 50(2): 351–361

Lee, T. and Ko, L. S. PIV investigation of flowfield behind perforated Gurney-type flaps. *Experiments in Fluids*, 2009, 46(6): 1005–1019

Lee, T. and Lee, L. Effect of Gurney flap on unsteady wake vortex. *Journal of Aircraft*, 2007, 44 (4): 1398–1402

Lee, T. and Su, Y. Y. Pitching airfoil with combined Gurney flap and unsteady trailing-edge flap deflection. *AIAA Journal*, 2012, 50(2): 503–507

Li, Y. C., Wang, J. J., and Zhang, P. F. Experimental investigation of lift enhancement on a NACA0012 airfoil using plate/serrated Gurney flaps. *Acta Aeronautica et Astronautica Sinica*, 2003b, 24(2): 119–123 (Chinese)

Li, Y. C., Wang, J. J., and Hua, J. Experimental investigations on the effects of divergent trailing edge and Gurney flaps on a supercritical airfoil. *Aerospace Science and Technology*, 2007, 11(2–3): 91–99

Li, Y. C., Wang, J. J., Tan, G. K., and Zhang, P. F. Effects of Gurney flaps on the lift enhancement of a cropped nonslender delta wing. *Experiments in Fluids*, 2002b, 32(1): 99–105

Li, Y. C., Wang, J. J., and Zhang, P. F. Effect of Gurney flaps on a NACA0012 airfoil. *Flow, Turbulence and Combustion*, 2002a, 68(1): 27–39

Li, Y. C., Wang, J. J., and Zhang, P. F. Influences of mounting angles and locations on the effects of Gurney flaps. *Journal of Aircraft*, 2003a, 40(3): 494–498

Liebeck, R. H. Design of subsonic airfoils for high lift. *Journal of Aircraft*, 1978, 15(9): 547–561

Liu, T. S. and Montefort, J. Thin-airfoil theoretical interpretation for Gurney flap lift enhancement. *Journal of Aircraft*, 2007, 44(2): 667–671

Nelson, J. and Koratkar, N. Effect of miniaturized Gurney flaps on aerodynamic performance of microscale rotors. *Journal of Aircraft*, 2005, 42(2): 557–561

Tang, D. and Dowell, E. H. Aerodynamic loading for an airfoil with an oscillating Gurney flap. *Journal of Aircraft*, 2007, 44(4): 1245–1257

Traub, L. W. Effects of Gurney flaps on an annular wing. *Journal of Aircraft*, 2009, 46(3): 1085–1088

Van Dam, C. P., Yen, D. T., and Vijgen, P. M. H. W. Gurney flap experiments of airfoil and wings. *Journal of Aircraft*, 1999, 36(2): 484–486

Wang, J. J., Li, Y. C., and Choi, K. S. Gurney flap – Lift enhancement, mechanisms and applications. *Progress in Aerospace Sciences*, 2008, 44(1): 22–47

Wang, J. J., Zhan, J. X., Zhang, W., and Wu, Z. Application of a Gurney flap on a simplified forward-swept aircraft model. *Journal of Aircraft*, 2006, 43(5): 1561–1564

Wang, J. J., Zhang, Z. J., and Feng, L. H. Influence of Gurney flap on longitudinal aerodynamic characteristics of multiple-control-surface UAV. *Journal of Beijing University of Aeronautics and Astronautics*, 2010, 36(6): 631–635 (in Chinese)

Wang, J. J. and Li, Y. C. The effects of Gurney flap on double delta wing aerodynamic performance in low speed wind-tunnel tests. *Acta Aerodynamica Sinica*, 2007, 25(2): 216–219, 225 (in Chinese)

Yu, T., Wang, J. J., and Zhang, P. F. Numerical simulation of Gurney flap on RAE-2822 supercritical airfoil. *Journal of Aircraft*, 2011, 48(5): 1565–1575

Zhao, Y. L., Zuo, L. X., Yu, D. S., and Wang, J. J. Effects of segmented Gurney flaps on the aerodynamics of a target drone model. *Acta Aerodynamica Sinica*, 2010, 28(5): 518–524 (in Chinese)

3 Vortex Generator

3.1 Background

A vortex generator is usually a small device placed onto a wall, that can induce a streamwise vortex. So this control technique is called the vortex generator. Actually, there are many devices that can be used as vortex generators, such as a rectangular plate, triangular plate, wishbone, doublet, airfoil, wedge, ramp, etc. When the vortex generators are used for flow control, the devices are usually arranged in one row along the spanwise direction to induce a set of vortices. For example, they can be placed in a co-rotating (CoR) manner, where each device is inclined in the same direction (Figure 3.1 (a)), or a counter-rotating (CtR) manner, where a pair of devices induces vortices with different rotation directions (Figure 3.1(b)).

Lin (2002) conducted a series of studies into vortex generators and their applications in low-speed flow. He presented a history of the development of vortex generators. The height of the vortex generators was originally one order of the boundary thickness δ. Brown et al. (1968) used rectangular-vane type vortex generators with 1.2δ to improve the performance of subsonic diffusers. Calarese et al. (1985) reduced the afterbody drag of a 1/72 scaled C-130 aircraft model with 0.21δ airfoil-type vortex generators. Nickerson (1986), Bragg and Gregorek (1987) have shown the ability of the vortex generators to increase the lift and reduce the drag of airfoils. Rao and Kariya (1988) indicated that vortex generators with height of 0.625δ had similar effects as those with $h = \delta$, while Lin et al. (1990) indicated that vortex generators with height of 0.1δ performed as well as larger ones with height of 0.4δ. However, the smaller-height devices had much lower device drag. Thus, vortex generators are usually referred to as micro-vortex generators nowadays, as their heights are usually of the order of 0.1 of the local boundary thickness. Vortex generators have already been applied in various fields for flow control, ranging from low-speed to high-speed fields, which has been recently reviewed by Lu et al. (2012).

The scale of vortex generators is small with light weight, and the devices are very simple, as only one flat plate can be used. Vortex generators can also be very easily attached to the object surface, with no need to change the object configuration. The induced streamwise vortex has great potential ability to enhance momentum mixing, and thus to achieve ideal effects. Thus, vortex generators are among the few control techniques that have been actually applied by engineers. For example, nearly all modern

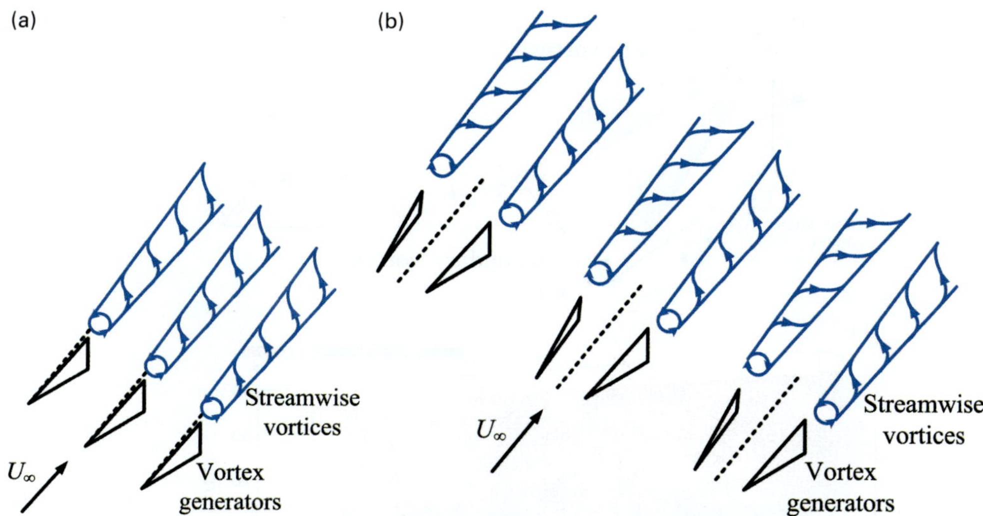

Figure 3.1 Schematic of vortex generators. (a) Co-rotating arrangement; and (b) counter-rotating arrangement.

airplanes have installed vortex generators in the suction surface of the main wings to improve the aerodynamic performance.

In this chapter, we will first present the fundamental characteristics of vortex generators and introduce the corresponding parameter influence. The applications of vortex generators in various fields for flow separation control and heat transfer enhancement are then introduced.

3.2 Fundamental Flow Characteristics

The vortex generator device is usually placed at an angle to the free-stream direction. When the fluids flow around the device, flow separates over the surface to form the shear layer, which rolls up into vortical structure (Figure 3.2(a)). The induced vortex can evolve in the downstream with the vortex axis nearly along the streamwise direction, known as the streamwise vortex (Figure 3.2(b)).

The basic characteristic of vortex generators is that the streamwise vortex can be induced, thus the vortex circulation is a quantitative measurement for the control ability of vortex generators. It is deduced that a larger vortex circulation could lead to a higher control effect. Ashilf et al. (2001) indicated that the vortex circulation increased with an increase in the vortex generator height. Then it reached a convergent value, which was then nearly independent of the vortex generator height. As the streamwise vortex formed from the vortex generators, it developed along the streamwise direction. Thus, the vortex strength varied as it evolved into downstream, as reported by Ashilf et al. (2002). There is a basic trend that the vortex circulation decreases as the vortex develops downstream.

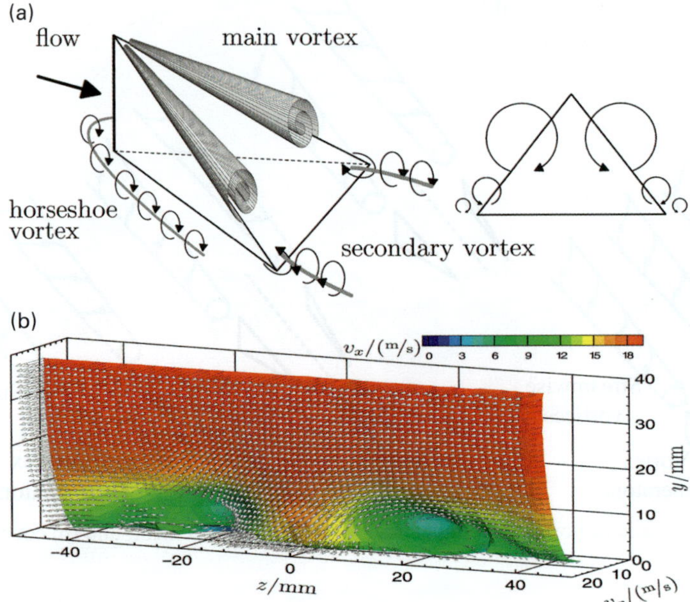

Figure 3.2 Formation (a) and downstream evolution (b) of the streamwise vortices induced by a backwards wedge vortex generator (Henze et al. 2011). Reproduced with permission, copyright © 2010 Elsevier Ltd. All rights reserved.

3.3   Boundary Layer Control

Godard and Stanislas (2006) studied the influence of different parameters of vortex generators on the boundary layer development. The parameters considered included vortex generator height h and length l, transverse spacing of the devices λ, the distance L between the trailing edges of two vanes, the skew angle β_{pd} of the vanes to the free stream, and the boundary layer thickness δ. The wall shear stresses at the minimum position $x_{Cfmin} = 18.58$ m and at two spanwise positions, namely in the symmetry plane of the device $z/\lambda = 0$ and at mid distance between two devices $z/\lambda = 0.5$, were measured with hot film probes. The measured wall shear stress τ was scaled by the value τ_0 for the natural case, namely $\Delta\tau = \tau - \tau_0$, which was used as a predictor for parameter influence. The main results are shown in Figure 3.3.

Figure 3.3(a) shows the effect of the skew angle β_{pd} and device spacing λ/h. We can see the significant variation of the shear stress in the spanwise direction, showing that the higher values are located on the symmetry plane of the devices. The $\Delta\tau/\tau_0$ increases with increasing device spacing, though the differences between the $\lambda/h = 8$ and 12 cases are small. On the other hand, $\Delta\tau/\tau_0$ decreases with increasing the skew angle for $\lambda/h = 4$. It increases at first and reaches the maximum at around $\beta_{\mathrm{pd}} = 18°$ and then decreases with skew angle for $\lambda/h = 8$ and 12, though the skew angle has little influence.

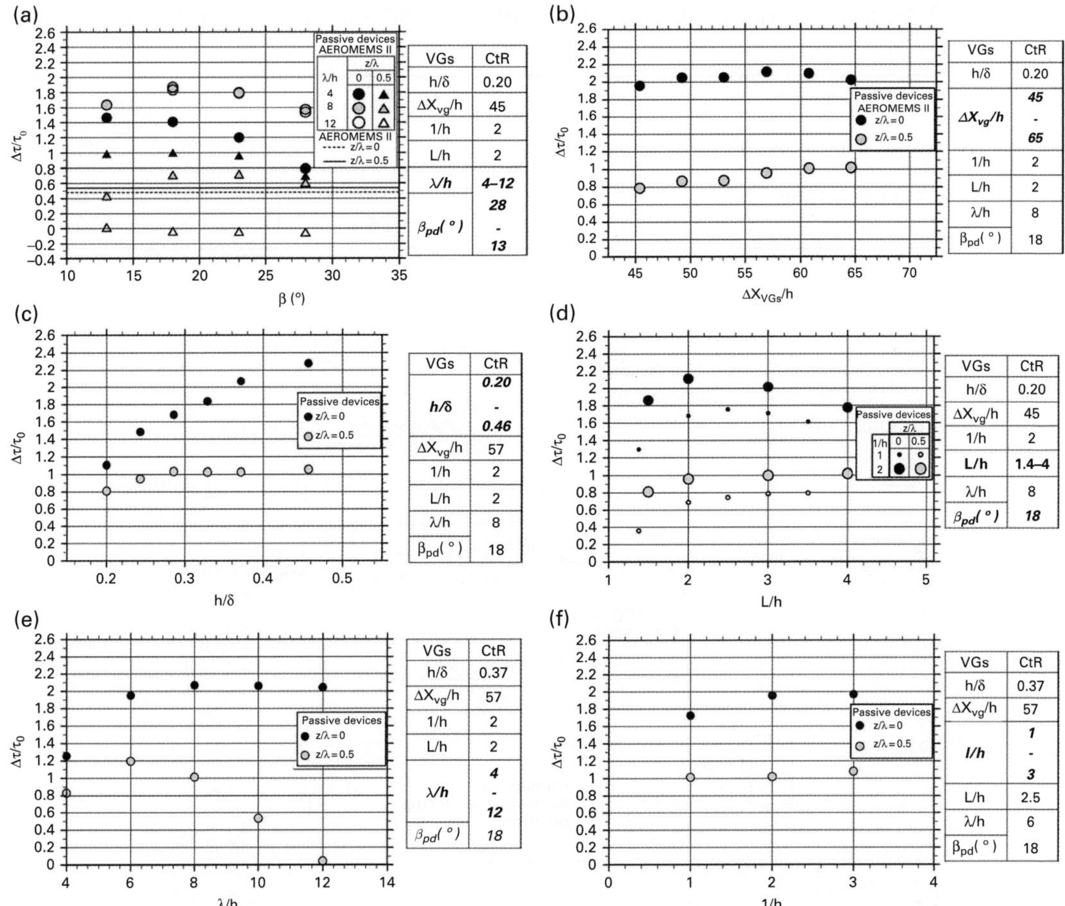

Figure 3.3 Influence of different parameters on the wall shear stress. (a) Skew angle; (b) device position; (c) device height; (d) distance between the trailing edges of two vanes; (e) transverse spacing the devices; and (f) device length (Godard and Stanislas 2006). Reproduced with permission, copyright © 2006 Elsevier Ltd. All rights reserved.

Figure 3.3(b) shows the influence of the device position Δx_{VGs}. There is no significant variation of $\Delta\tau/\tau_0$ at $z/\lambda = 0$ with a maximum of 10% and only a 20% variation at $z/\lambda = 0.5$ for the range investigated.

Figure 3.3(c) indicates that the $\Delta\tau/\tau_0$ increases with increasing the vortex generators height, especially at the symmetry plane $z/\lambda = 0$.

Figure 3.3(d) shows that $\Delta\tau/\tau_0$ for $l/h = 2$ is higher than that for $l/h = 1$. For both cases, the $\Delta\tau/\tau_0$ at $z/\lambda = 0$ reaches the maximum at around $L/h = 2.5$, while $\Delta\tau/\tau_0$ at $z/\lambda = 0.5$ is maximal for about $L/h > 3$.

Figure 3.3(e) shows that $\Delta\tau/\tau_0$ at the symmetry plane $z/\lambda = 0$ increases significantly from $\lambda/h = 4$ to $\lambda/h = 8$ and remains nearly constant for $\lambda/h > 8$, while $\Delta\tau/\tau_0$ at $z/\lambda = 0.5$ reaches an optimal value at $\lambda/h = 6$ and then decreases with increasing the device spacing λ/h.

Figure 3.3(f) shows that a significant variation of $\Delta\tau/\tau_0$ of about 20% is observed around $l/h = 2$ at the symmetry plane $z/\lambda = 0$, while it has nearly no variation with the device length at $z/\lambda = 0$.

Table 3.1 Optimal parameters for boundary layer control with vortex generators (Godard and Stanislas 2006)

							$\Delta\tau/\tau_0(\%)$	
Vortex generators	h/δ	$\Delta x_{VGs}/h$	l/h	L/h	λ/h	$\beta_{pd}(°)$	min	max
Counter-rotating	0.37	57	2	2.5	6	18	110	200
Co-rotating	0.37	57	2	–	6	18	55	105

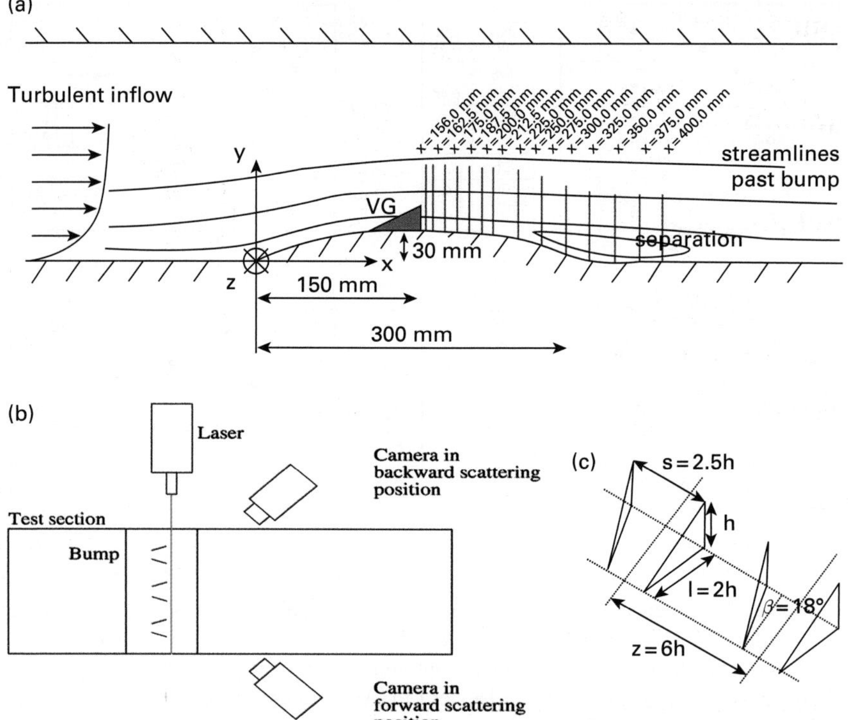

Figure 3.4 Schematic of the flow over a hump controlled by vortex generators. (a) Side view; (b) plane view; and (c) vortex generators (Velte et al. 2008). © IOP Publishing Ltd. Reproduced by permission of IOP Publishing.

Based on the above analysis, optimal parameters for both co-rotating and counter-rotating vortex generators were summarized by Godard and Stanislas (2006) for further study, which is listed in Table 3.1. The optimal parameters for the co-rotating vortex generators are the same as the counter-rotating ones. It also gives the spanwise minimum and maximum values of the shear stress variation. We can see that with the same geometries, the co-rotating configuration is twice as efficient as the counter-rotating one.

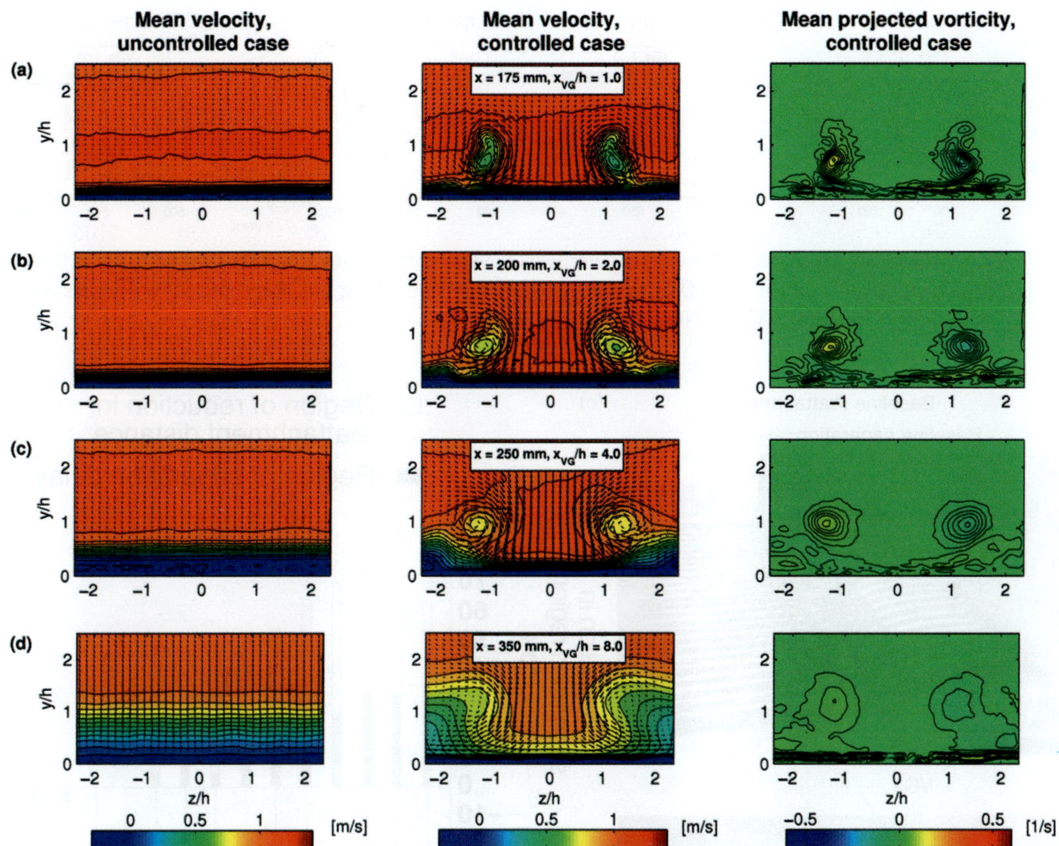

Figure 3.5 Mean velocity for the natural (left column) and control (middle column) cases and mean vorticity for the control case (right column) at different streamwise positions. (a) $x_{VG}/h = 1$; (b) $x_{VG}/h = 2$; (c) $x_{VG}/h = 4$; (d) $x_{VG}/h = 8$ (Velte et al. 2008). © IOP Publishing Ltd. Reproduced by permission of IOP Publishing.

3.4   Flow Separation Control

The streamwise vortex induced by the vortex generators can transfer momentum from the outer flow to the near wall region, which can be used for delaying flow separation. For the flow over a hump or back-facing ramp, a large separation region occurs due to serious adverse pressure gradient. There have been many studies to confirm the effect of the vortex generators in this field.

Figure 3.4 shows a sketch for the control of the separated flow around a hump. The hump height was 30 mm, chord length was 300 mm, and the width was 600 mm. Triangular-vane counter-rotating vortex generators were used for flow control. The vortex generators were positioned at 50% bump chord length. Stereoscopic particle image velocimetry (PIV) was used to measure the crossflow field at different streamwise positions. The experiments were conducted at a free-stream velocity of $U_\infty = 1$ m/s, corresponding to Reynolds number of $Re = 20\,000$ based on the bump chord length.

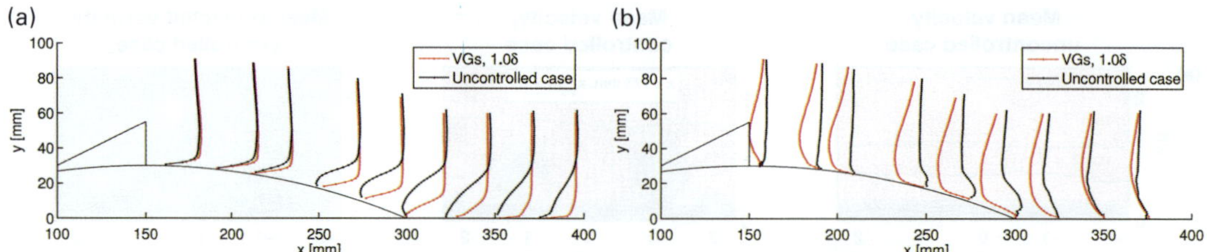

Figure 3.6 Mean streamwise (a) and vertical (b) velocities profiles along the symmetry plane for the natural and control cases (Velte et al. 2008). © IOP Publishing Ltd. Reproduced by permission of IOP Publishing. CC BY-NC-SA.

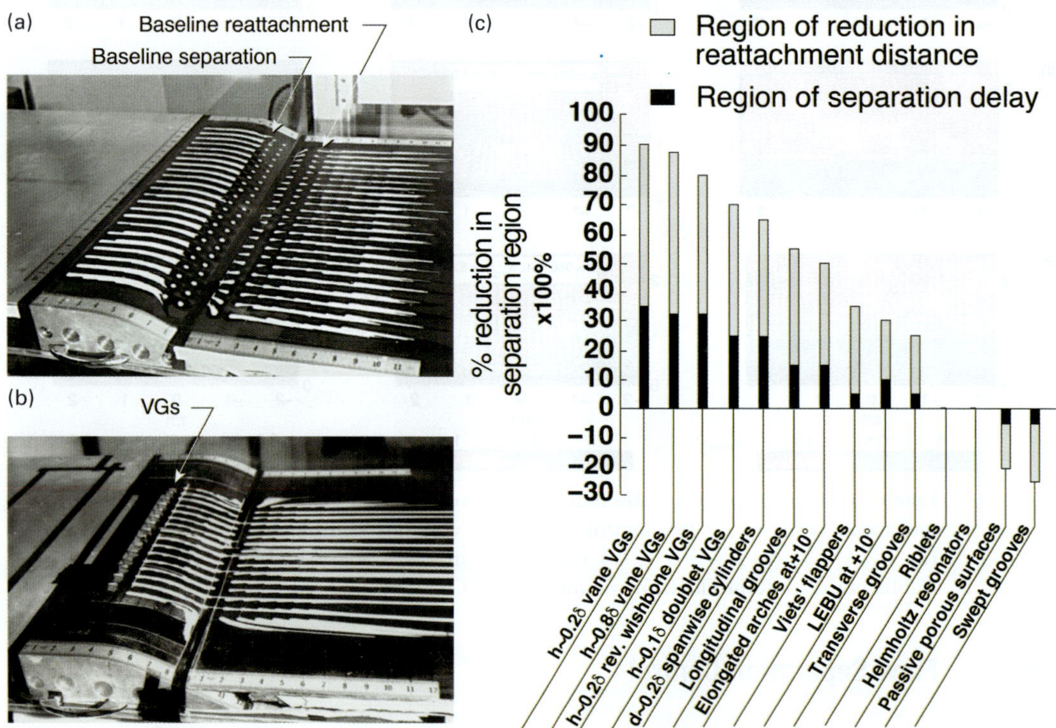

Figure 3.7 Oil flow visualization over a back-facing ramp for the natural case (a) and control case with 0.2δ high vane-type counter-rotating vortex generators (b). (c) Effectiveness in flow separation control for various devices (Lin 1999). Reproduced with permission, copyright © The American Institute of Aeronautics and Astronautics.

The mean velocity and streamwise vorticity are shown in Figure 3.5 for the natural and control cases. For the natural case, the near wall velocity decreases along the streamwise position, while the reverse flow gradually occurs near the wall surface, showing the appearance of the separation region. The vortex generators cause the flow to rotate, creating counter-rotating streamwise vortices, which are obviously seen from the middle and right columns. The streamwise vortices distort the flow greatly. The high-momentum fluid is transported from the outer flow into the near wall region around the symmetry plane of the devices, while the low momentum fluid is transported upwards between two pairs of the devices. Thus, the reverse flow region is reduced with control, in particular in the symmetry plane region.

The variation in the velocity field induced by the vortex generators is more clearly shown in Figure 3.6 by exhibiting the streamwise and vertical velocity profiles along the symmetry plane, namely $z = 0$. The streamwise vortices increase the streamwise velocity and the downwash flow, which benefits separation elimination.

Other applications are to control flow over a back-facing ramp, such as the works by Lin (1999), Gardarin et al. (2008), Serakawi and Ahmad (2012). One example is shown in Figure 3.7(a) and (b) by Lin (1999). From the flow visualization results, we can see the separation line and the reattachment line, which denote the streamwise scale of the separation region. With 0.2δ high vane-type counter-rotating vortex generators, it is indicated that the separation line is delayed while the reattachment line is advanced, which contributes to a decrease in the separation region. Gardarin et al. (2008) have also shown similar effects. Figure 3.7(b) only shows the control for one type of vortex generator. However, considerable effects may also be achieved for other types.

The control effect is scaled by the delay in the separation line and the reduction in the distance between the separation line and the reattachment line, which sums the reduction in the extent of the separation region. This is summarized in Figure 3.7(c) for various passive flow control techniques. It is indicated that the most effective group of flow control devices are vortex generators, such as vane, wishbone, and doublet types, which induce the streamwise vortex. Note that the 0.2δ height vane vortex generator may be even more effective in reducing the flow separation region than the 0.8δ height example. The second most effective group of separation control is the devices that can induce the spanwise vortex, such as spanwise cylinders, grooves, arches, flappers, and large-eddy breakup devices. The third group of devices has no influence on the flow separation, such as riblets and Helmholtz resonators, or even enhances flow separation, such as passive porous surfaces and swept grooves. Note that this summary actually only gives suggestions on control effects for different devices under specific conditions, which can provide a reference for researchers. However, the control effects under other conditions may be different.

3.5   Lift Enhancement and Drag Reduction

3.5.1   Circular Cylinder

Vortex generators were used by Ünal and Atlar (2010) for flow control around a circular cylinder, as shown in Figure 3.8. The diameter of the circular cylinder was 70 mm, and the length was 810 mm. Rectangular-vane vortex generators were used with the height of 1.6 mm, length of 3.2 mm, and spanwise space of 6.4 mm. The vortex generators were symmetrically placed on two rows on each side of the cylinder. The angular positions of the vortex generators varied as 50°, 60°, 65°, and 70°, denoted as VG50, VG60, VG65, and VG70, respectively. The free-stream velocity was 0.588 m/s, corresponding to the

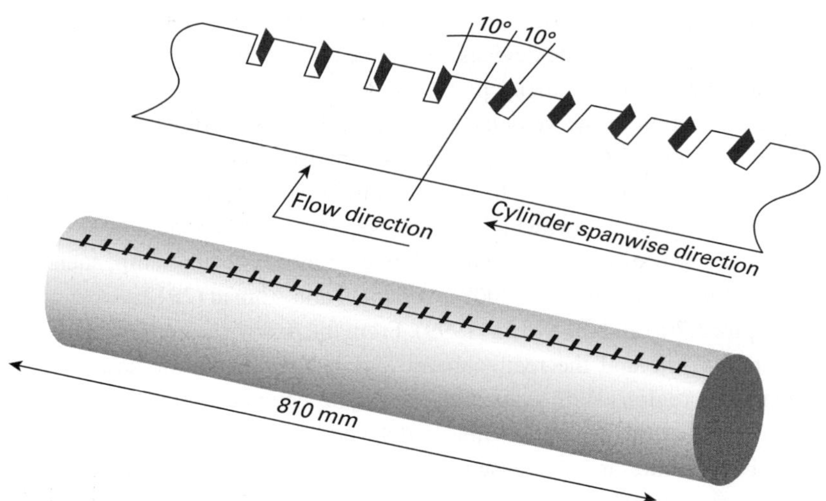

Figure 3.8 Schematic of the flow over a circular cylinder controlled by vortex generators (Ünal and Atlar 2010). Reproduced with permission, copyright © Springer-Verlag 2009.

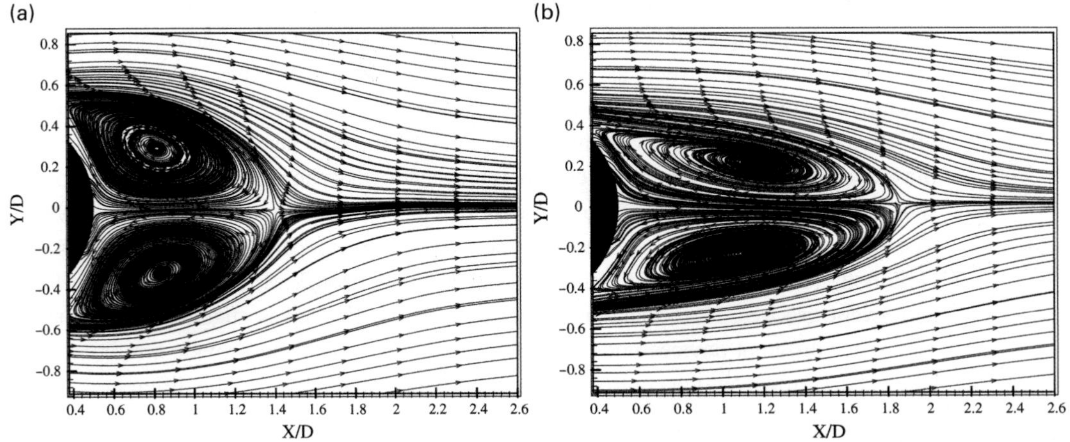

Figure 3.9 Time-averaged streamline of the circular cylinder for the natural (a) and VG70 control (b) cases (Ünal and Atlar 2010). Reproduced with permission, copyright © Springer-Verlag 2009.

Reynolds number based on the cylinder diameter of Re = 41300. PIV measurements were conducted in the flow field of the middle span plane.

Figure 3.9 shows the mean streamlines for the natural case and one typical control case with VG70. The control effect obviously indicates that the recirculation region behind the cylinder becomes narrower and more elongated than that in the natural case. This is also observed for other control cases.

A summary of the characteristic parameters of the near wake is listed in Table 3.2. The recirculation bubble length l_c is defined as the distance between the cylinder centroid and the location of the zero streamwise velocity along the wake centerline, while the vortex

Table 3.2 Characteristic values along the wake centerline of the circular cylinder (Ünal and Atlar 2010)

Test cases	l_c/D	l_f/D	$(U_{RMS}/U_\infty)_{max}$	$(V_{RMS}/U_\infty)_{max}$	C_D
Natural case	1.39	1.40	0.35	0.69	1.14
VG50	1.58	1.66	0.28	0.52	0.89
VG60	1.65	1.69	0.28	0.50	0.85
VG65	1.69	1.72	0.27	0.46	0.84
VG70	1.85	2.03	0.24	0.40	0.82

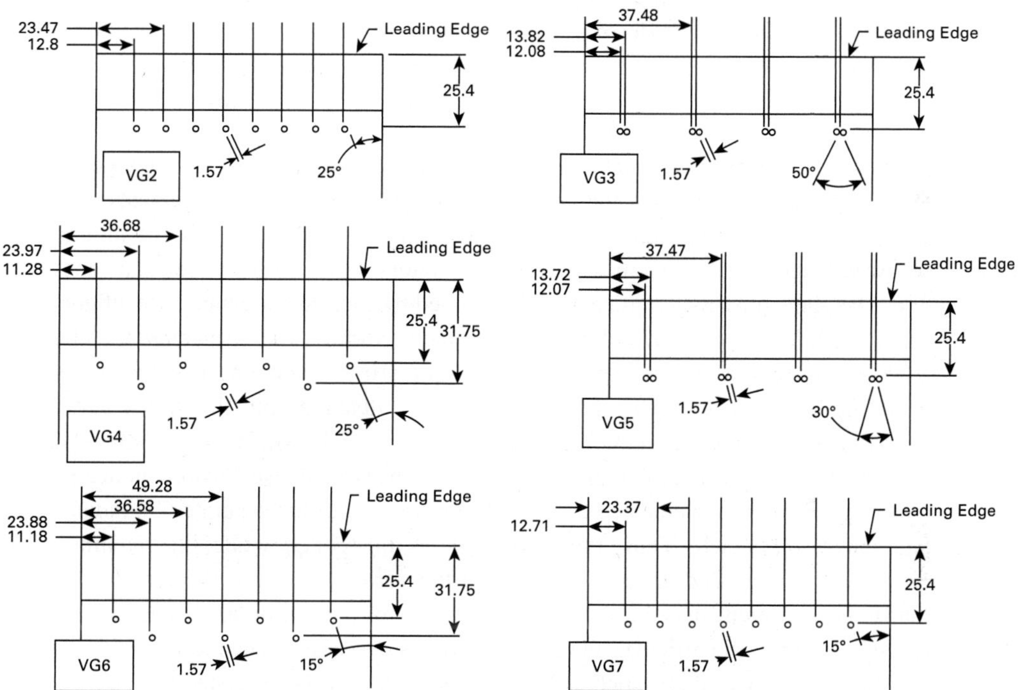

Figure 3.10 Different arrangements of vortex generators for airfoil control (Seshagiri et al. 2009). Reproduced with permission, copyright © 2008 by Lance W. Traub.

formation length l_f is defined as the distance between the cylinder centroid and the location of the maximum root mean square (RMS) streamwise velocity along the wake centerline. We can see that vortex generators increase the length scale of the near wake, while reducing the maximum RMS streamwise and vertical velocities. It is suggested that the upper and lower wake vortices are formed at a further downstream position because of the delay of separation, but with a weakened mutual interaction. The effect increases with an increase in the vortex generator position angle. As expected, the mean drag on the circular cylinder is reduced, with the maximum reduction about 28% for the VG70 case.

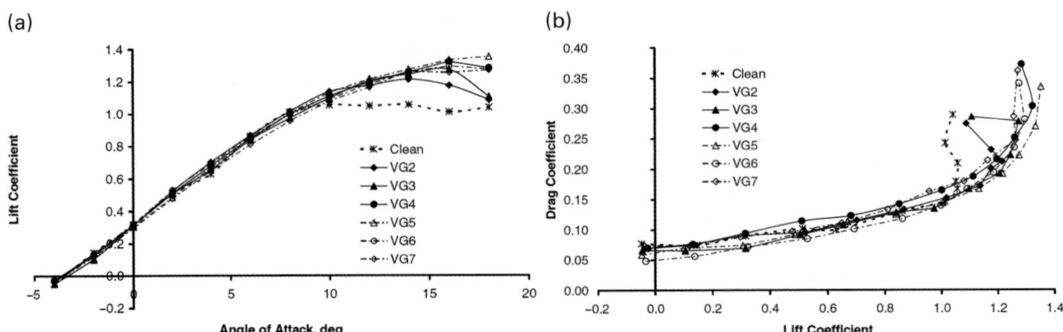

Figure 3.11 Lift coefficient (a) and drag coefficient (b) of the airfoil for the natural and control cases (Seshagiri et al. 2009). Reproduced with permission, copyright © 2008 by Lance W. Traub.

3.5.2   Airfoil

There have been many researches on the flow control of airfoils with vortex generators. Seshagiri et al. (2009) tested the control effects of six different arrangements of vortex generators, as shown in Figure 3.10. The chord length of the airfoil was 165 mm and the spanwise length was 101.6 mm. The vortex generators were made of rectangular vanes with 2 mm height. Co-rotating vortices were generated by four vortex generator configurations VG2, VG4, VG6, and VG7, while counter-rotating vortices were generated by VG3 and VG5. The experiments were conducted at $Re = 1.6 \times 10^5$ based on the airfoil chord length.

Before stall, the flow over the airfoil is nearly attached, thus the vortex generators nearly have no influence on the lift coefficient (Figure 3.11(a)). However, after stall, the flow over the airfoil is completely separated, and the streamwise vortices induced by the vortex generators may enhance the momentum mixing of the separated boundary layer and make it reattach. Thus, the stall angle is postponed while the maximum lift coefficient is increased under the control of the vortex generator. We can also see that the drag coefficient may be reduced with the flow control at small angles of attack (Figure 3.11(b)). In particular, the zero lift drag is decreased. The reduction in drag has also been found by others, such as Kerho et al. (1993). It is suggested that this result is because the vortex generators reduce the scale of the laminar separation bubble over the suction surface. However, the drag may increase at higher angles of attack with vortex generators. Comparing the effects among different configurations, the 15° incidence offers better post stall performance than that of 25°, in particular for the VG5 case.

Besides the applications in the main airfoils, vortex generators could also be applied in multi-element airfoils, such as in the studies by Lin et al. (1994) and Klausmeyer et al. (1996). During the takeoff/landing stages, the leading-edge slat and trailing-edge flap will be deflected to provide a high lift coefficient. However, the flow over the suction surface of the trailing-edge flap will encounter serious flow separation due to its high angle of attack. Thus, the vortex generators can be installed on the suction surface of the flap near the leading edge (Figure 3.12(a)). Since the height of the vortex generators is small, it can be stowed inside the flap during the cruise stage (Figure 3.12(b)).

Figure 3.13 presents the pressure distributions over the whole airfoil without and with control. The trapezoid-vane counter-rotating vortex generators are mounted at 25% of

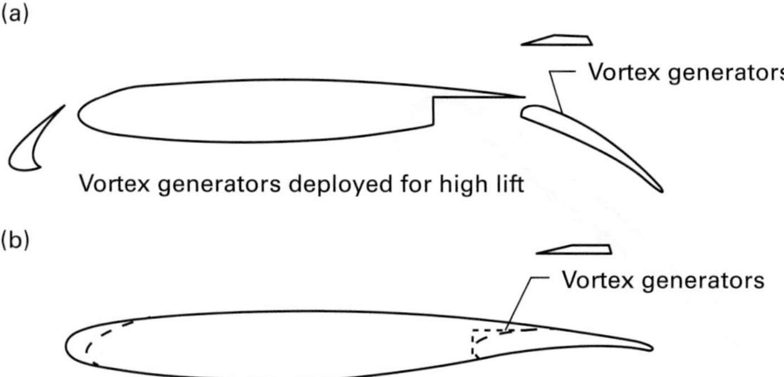

Figure 3.12 Schematic of the application of vortex generators in the multi-element airfoil. (a) Takeoff/ landing stages; (b) cruise stage (Lin et al. 1994). Reproduced with permission, copyright © The American Institute of Aeronautics and Astronautics.

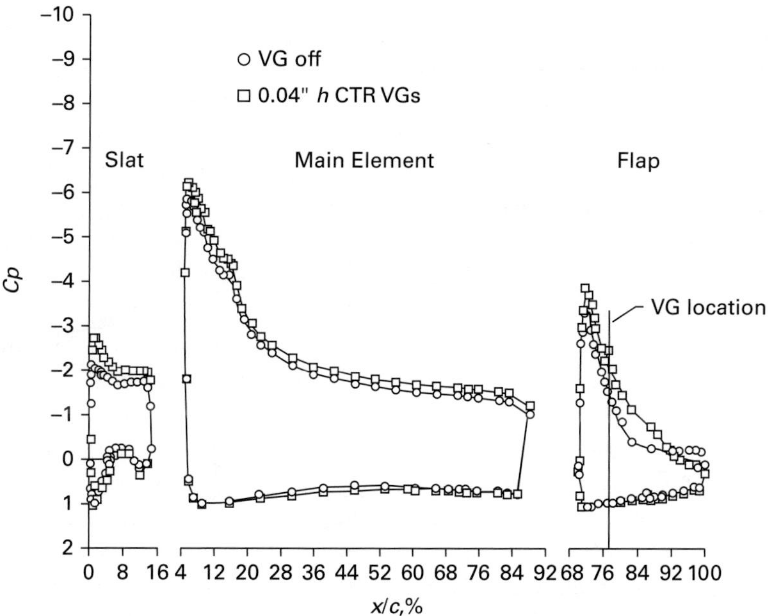

Figure 3.13 Pressure distribution over the entire airfoil for the natural and control cases (Lin et al. 1994). Reproduced with permission, copyright © The American Institute of Aeronautics and Astronautics.

the flap chord. The global effects of the vortex generators on the multi-element airfoil are that the suction over all elements is increased. The increase of suction pressure results in a significant lift enhancement before stall, which is shown in Figure 3.14(a) for two Reynolds numbers. We can also see that the Reynolds number effects on the lift coefficient for both natural and control cases are negligible. The benefits of the vortex generators can also be concluded from the polar curves, showing that the drag coefficient can be greatly reduced by vortex generators at the same lift coefficient as the natural case

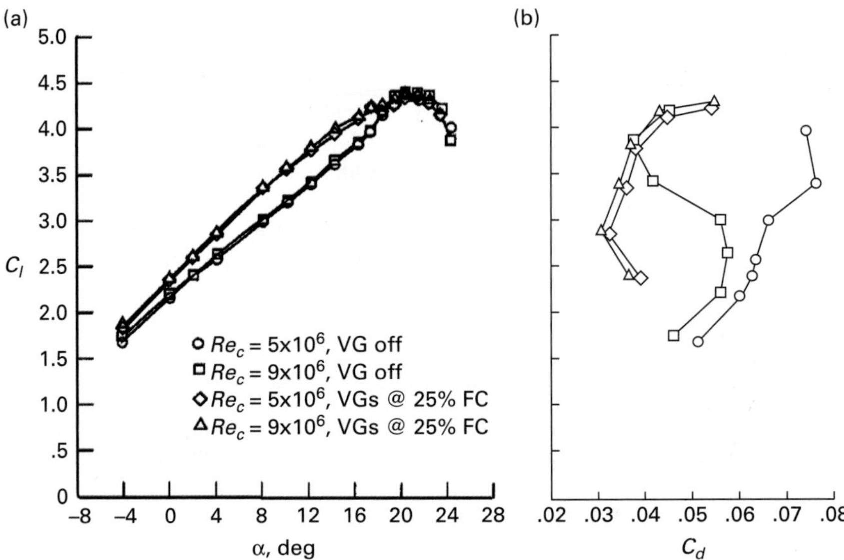

Figure 3.14 Lift coefficient versus angle of attack (a) and lift coefficient versus drag coefficient (b) of the airfoil for the natural and control cases (Lin et al. 1994). Reproduced with permission, copyright © The American Institute of Aeronautics and Astronautics.

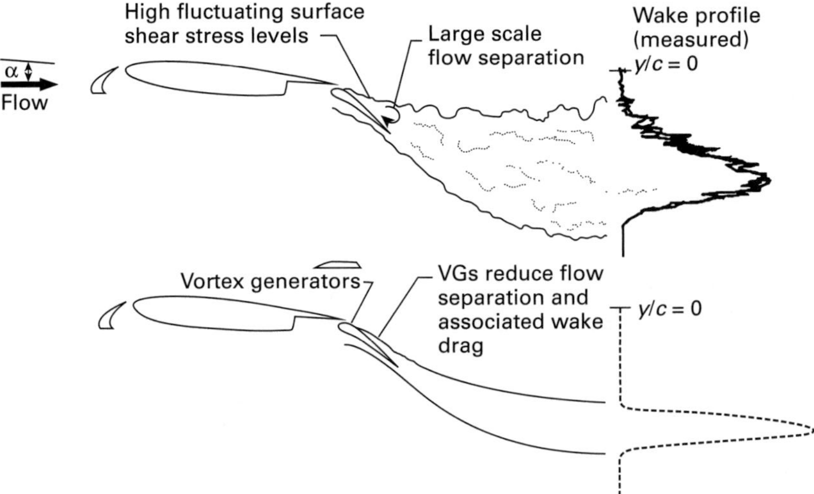

Figure 3.15 Near-wake development of the multi-element airfoil for the natural (upper row) and control (lower row) cases (Lin et al. 1994). Reproduced with permission, copyright © The American Institute of Aeronautics and Astronautics.

(Figure 3.14(b)). This can be clearly interpreted by the variations in the near wake profiles, as shown in Figure 3.15. The scale of the separation region over the flap is reduced, while the near wake profile becomes narrower with flow control than that in natural case. Such variations benefit the drag reduction.

3.5.3   Aircraft

Vortex generators are among the most useful techniques and have been widely used in modern passenger aircraft. Control effects have also been validated in military aircrafts. For example, Calarese et al. (1985) reduced the afterbody drag of a 1/72 scaled C-130 aircraft model by vortex generators; Tai (2003) validated that vortex generators positioned at the right position of the overwing fairing surface of a V22 model could reduce the flow separation region and thus increase the lift-to-drag ratio of the whole aircraft.

A recent study has been conducted by Ito et al. (2015), who examined the effects of co-rotating rectangular vortex generators on transonic swept wings. The NASA CRM was used as the basic configuration of the aircraft for flow control. The simulation was performed at $Ma = 0.85$ and $Re = 2\,270\,000$. The height of the vortex generators was 0.8 mm and the incidence angle to the free stream was 32.6°. These vortex generators were placed on the wing from the leading edge of 20% chord length and with three different spanwise spacings. As expected, the lift coefficient was increased while the drag coefficient at the same lift coefficient was reduced by flow control. However, it was found that the spanwise spacing of the vortex generators had little influence on the aerodynamic forces in their experiments.

Ito et al. (2015) also studied the skin friction coefficient over the wing suction surface for the natural and control cases at the cruise angle of attack $\alpha = 4°$. It was indicated that the vortex generators could make the separation region smaller, though the flow became more three-dimensional. In the downstream of some vortex generators, there were areas with larger friction than that in the natural case.

3.6   Heat Transfer Enhancement

The subject of heat transfer enhancement has significant impact on the development of efficient heat exchangers. Accompanied by enhanced momentum mixing, vortices also benefit the heat transfer enhancement. Thus, vortex generators have been widely used to improve the performance of thermal systems, such as heat exchangers and channels. Ahmed et al. (2012) presented a recent review of this field.

The control of heat exchangers with vortex generators have been conducted by many researchers, and one example is shown in Figure 3.16 by Leu et al. (2004). There were three-row tubes connected with fins for the heat exchangers. Three different configurations of the counter-rotating rectangular vortex generators were investigated, with height of 5 mm, length of 10 mm, and thickness of 2 mm. The incidence angle is 30°, 45°, and 60° relative to the main flow direction for these three configurations.

Figure 3.17 compares the natural and control cases. Figure 3.17(a) shows the isotherms on the fin surface for four cases, which are determined by infrared thermovision. It is indicated that the temperature gradient between the tube rows becomes larger as the incidence angle of vortex generators is increased. Figure 3.17(b) shows that the local Nusselt number between the regions of vortex generators and tubes is increased. The maximum improvement occurs for the case with 60° incidence angle. The thermal variations induced by the vortex generators can be well explained by the flow visualization shown in Figure 3.17(c). The color for the dye in front of the first row tube is black, and the

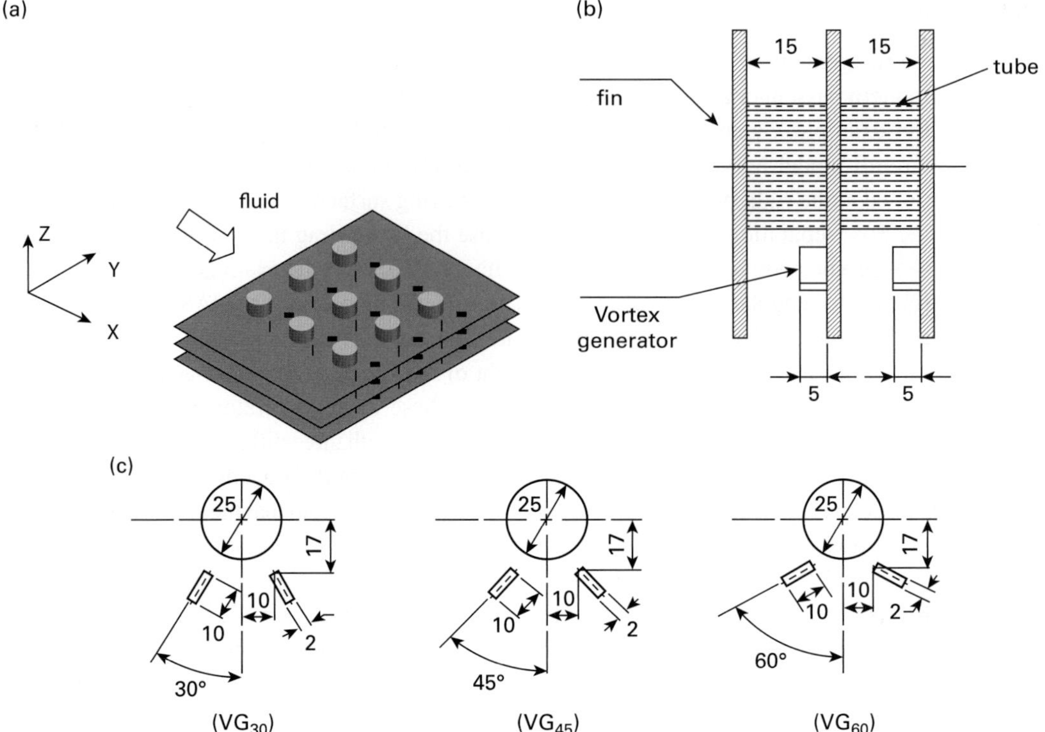

Figure 3.16 Schematic of the heat exchanger controlled by vortex generators. (a) Top view; (b) side view; (c) three configurations of vortex generators (Leu et al. 2004). Reproduced with permission, copyright © 2004 Elsevier Ltd. All rights reserved.

color behind the tube is red. A pair of streamwise vortices are formed downstream of the vortex generators, which enhance the flow mixing in the downstream. As the span angle is increased, the strength of streamwise vortices is enhanced.

3.7   Concluding Remarks

We have introduced the fundamental characteristics and applications of vortex generators. One principal advantage is simplicity. The device is small and simple, and is easily attached to the control body. The flow control mechanism is also very clear. The streamwise vortices induced by the vortex generators can enhance the momentum mixing, and thus delay flow separation and enhance heat transfer. Thus, vortex generators have actually been applied in a few industrial fields, such as modern passenger aircraft. This shows the advantages of vortex generators, since most of the other control devices are still at the lab-investigation stage, and face some difficulties in engineering applications.

However, there is also one issue in using vortex generators for flow control. There are many different kinds of vortex generators, and many different geometric parameters even for the same type. There are also various arrangements on the control body.

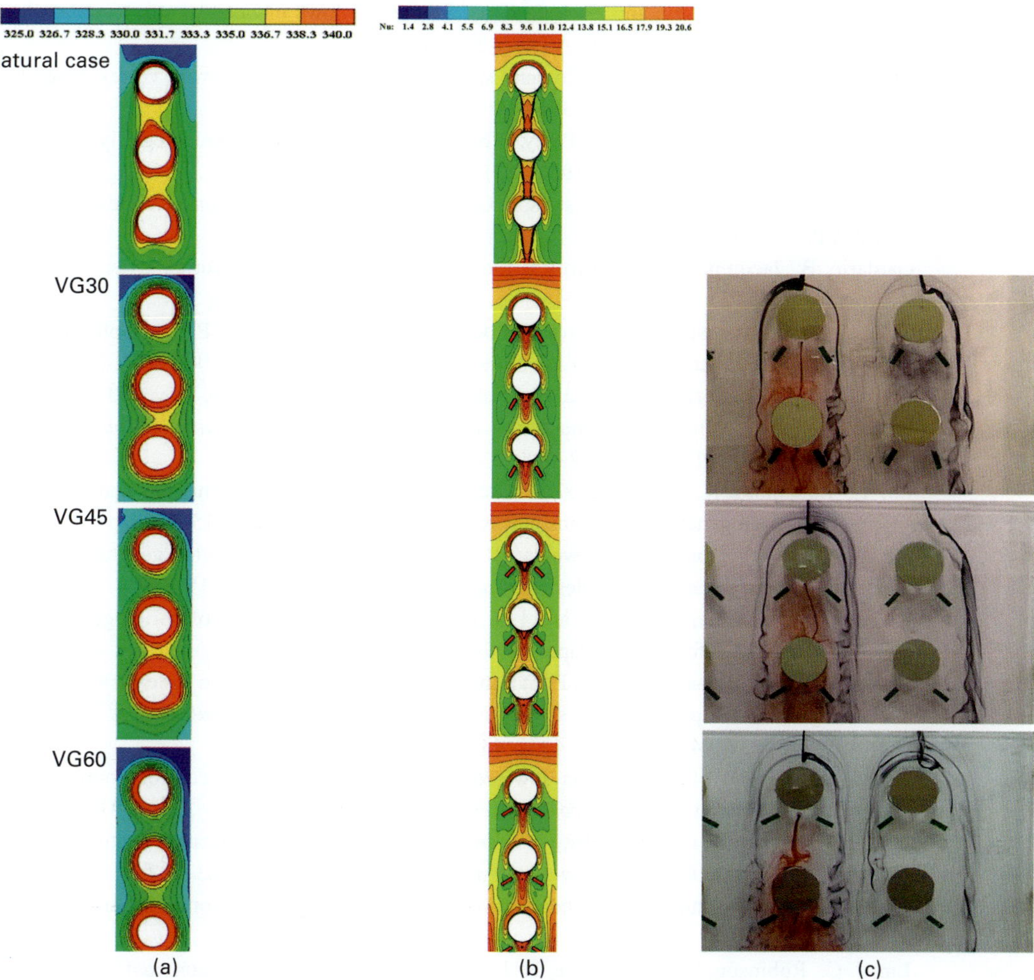

Figure 3.17 Isotherms (a), local *Nu* distributions (b), and flow visualization (c) on the fin surfaces of the heat exchanger for the natural and control cases (Leu et al. 2004). Reproduced with permission, copyright © 2004 Elsevier Ltd. All rights reserved.

In addition, the flow condition may also determine the control results. It makes the control parameters become very complex for the ideal effect. Thus, system studies must be made before vortex generators are finally applied in a certain field.

References

Ahmed, H.E., Mohammed, H. A., and Yusoff, M. Z. An overview on heat transfer augmentation using vortex generators and nanofluids: approaches and applications. *Renewable and Sustainable Energy Reviews*, 2012, 16(8): 5951–5993

Ashilf, P. R., Fulker, J. L., and Hackett, K. C. Research at DERA on sub boundary layer vortex generators (SBVGs). AIAA Paper 2001–0887

Ashilf, P. R., Fulker, J. L., and Hackett, K. C. Studies of flows induced by sub boundary layer vortex generators (SBVGs). AIAA Paper 2002–0968

Bragg, M. and Gregorek, G. Experimental study of airfoil performance with vortex generators. *Journal of Aircraft*, 1987, 24(5): 305–309

Brown, A. C., Nawrocki, H. F., and Paley, P. N. Subsonic diffusers designed integrally with vortex generators. *Journal of Aircraft*, 1968, 5(3): 221–229

Calarese, W., Crisler, W. P., and Gustafson, G. L. Afterbody drag reduction by vortex generators. AIAA Paper 1985–0354

Gardarin, B., Jacquin, L., and Geffroy P. Flow separation control with vortex generators. AIAA Paper 2008–3773

Godard, G. and Stanislas, M. Control of a decelerating boundary layer. Part 1: Optimization of passive vortex generators. *Aerospace Science and Technology*, 2006, 10(3): 181–191

Henze, M., Wolfersdorf, J., Weigand, B., Dietz, C. F., and Neumann, S. O. Flow and heat transfer characteristics behind vortex generators – a benchmark dataset. *International Journal of Heat and Fluid Flow*, 2011, 32(1): 318–328

Ito, Y., Yamamoto, K., Kusunose, K., Koike, S., Nakakita, K., Murayama, M., and Tanaka, K. Effect of vortex generators on transonic swept wings. AIAA Paper 2015–1238

Kerho, M., Hutcherson, S., Blackwelder, R. F., and Liebeck, R. H. Vortex generators used to control laminar separation bubbles. *Journal of Aircraft*, 1993, 30(3): 315–319

Klausmeyer, S. M, Papadakis, M., and Lin, J. C. A flow physics study of vortex generators on a multi-element airfoil. AIAA Paper 1996–0548

Leu, J. S., Wu, Y. H., and Jang, J. Y. Heat transfer and fluid flow analysis in plate-fin and tube heat exchangers with a pair of block shape vortex generators. *International Journal of Heat and Mass Transfer*, 2004, 47(19–20): 4327–4338

Lin, J. C. Review of research on low-profile vortex generators to control boundary-layer separation. *Progress in Aerospace Sciences*, 2002, 38(4–5): 389–420

Lin, J. C., Howard, F. G., and Selby, G. V. Small submerged vortex generators for turbulent flow separation control. *Journal of Spacecraft and Rockets*, 1990, 27(5): 503–507

Lin, J. C. Control of turbulent boundary-layer separation using micro-vortex generators. AIAA Paper 1999–3404

Lin, J. C., Robinson, S. K., McGhee, R. J., and Valarezo, W. O. Separation control on high-lift airfoils via micro-vortex generators. *Journal of Aircraft*, 1994, 31(6): 1317–1323

Lu, F. K., Li, Q., and Liu, C. Microvortex generators in high-speed flow. *Progress in Aerospace Sciences*, 2012, 53: 30–45

Nickerson, J. Jr. A study of vortex generators at low Reynolds numbers. AIAA Paper 1986–0155

Rao, D. M. and Kariya, T. T. Boundary-layer submerged vortex generators for separation control– an exploratory study. AIAA Paper 1988–3546

Serakawi, A. R. and Ahmad, K. A. Experimental study of half-delta wing vortex generator for flow separation control. *Journal of Aircraft*, 2012, 49(1): 76–81

Seshagiri, A., Cooper, E., and Traub, L. W. Effects of vortex generators on an airfoil at low Reynolds numbers. *Journal of Aircraft*, 2009, 46(1): 116–122

Tai, T. C. Effect of midwing vortex generators on V-22 aircraft forward-flight aerodynamics. *Journal of Aircraft*, 2003, 40(4): 623–630

Ünal, U. O. and Atlar, M. An experimental investigation into the effect of vortex generators on the near-wake flow of a circular cylinder. *Experiments in Fluids*, 2010, 48(6): 1059–1079

Velte, C. M., Hansen, M. O. L, and Cavar, D. Flow analysis of vortex generators on wing sections by stereoscopic particle image velocimetry measurements. *Environmental Research Letters*, 2008, 3(1): 015006

4 Roughness

4.1 Background

Since the famous pipe flow experiment by Reynolds in 1883, it is known that laminar flow in a pipe undergoes transition to turbulence above the threshold Reynolds number of about $Re = 2100$–2300. Laminar and turbulent flows are quite different. For example, the friction drag coefficient in turbulent flow can be as much as 20 times higher than that in laminar flow. For modern large aircraft, the friction drag may account for about 50% of the total drag. Thus, in order to be more cost-effective and environmentally friendly, it is always required that the friction drag should be as low as possible. Flow control plays an important role in achieving this goal, by either controlling the flow in the laminar or turbulent states or delaying the laminar to turbulence transition. It is suggested that readers should first gain a basic understanding of the turbulent boundary layer and the influence of roughness height on the laminar, transitional and turbulent flows. Some of the main conclusions are presented in the following referring to the works of Schlichting and Gersten (2000), Jiménez (2004), Chwang and Dong (2011), and Kundu et al. (2013).

The turbulent boundary layer can be divided into the inner layer and outer layer, as shown in Figure 4.1. There are also three sublayers within the inner layer, namely the viscous sublayer region, the logarithmic region, and the buffer region. The velocity distributions in the different regions are presented as follows:

(1) Viscous sublayer region: $\dfrac{u}{u_e} = \dfrac{yu_e}{v}$ or $u^+ = y^+$, $0 \leq y^+ < (5 \sim 10)$.

(2) Logarithmic region: $u^+ = 5.85\lg y^+ + 5.56$, $y^+ \geq (30 \sim 70)$, $y/\delta \leq 0.2$.

(3) Buffer region: there is no specific function to describe this region, which is located between the viscous sublayer and logarithmic region, $(5 \sim 10) \leq y^+ < (30 \sim 70)$.

(4) Outer layer: $u^+ = \dfrac{1}{\kappa}\ln y^+ + C + W\!\left(\dfrac{y}{\delta}\right)$, where $W\!\left(\dfrac{y}{\delta}\right) = \dfrac{2\Pi}{\kappa}\sin^2\!\left(\dfrac{\pi y}{2\delta}\right)$ is the law of the wake, $0.2 \leq y/\delta \leq 1$.

There have been numerous studies in the literature which reviewed the influence of the roughness on the friction drag of different flow states. The drag coefficient for a smooth pipe flow is shown in Figure 4.2 according to the following equations. There are three typical curves from 1 to 3. Curve 1 stands for the drag coefficient for the laminar flow, namely

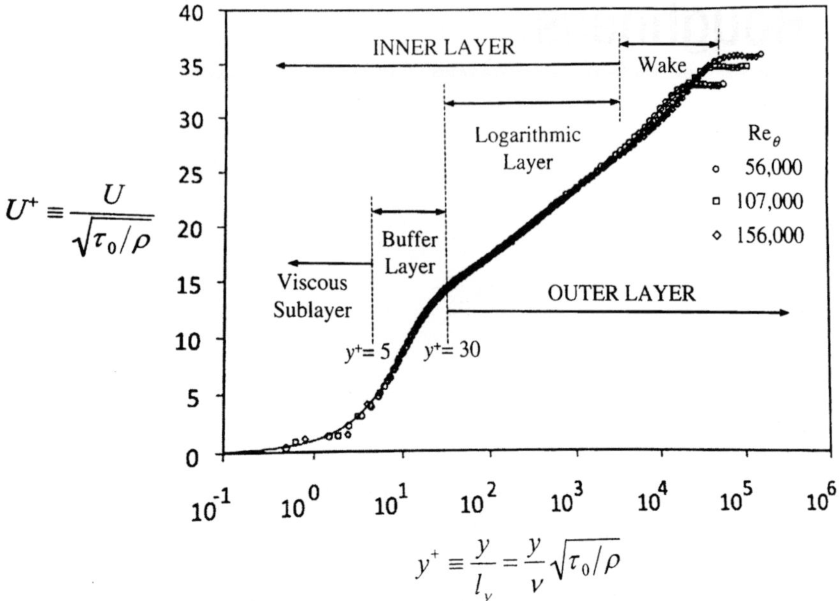

Figure 4.1 Velocity profile of a turbulent boundary layer (Kundu et al. 2013). Reproduced with permission, copyright © 2008 Elsevier Ltd. All rights reserved.

$$\lambda = \frac{64}{Re}.$$
(4.1)

Curve 2 stands for the Blasius solution,

$$\lambda = 0.3164\left(\frac{u_m d}{\nu}\right)^{-1/4},$$
(4.2)

where u_m is the mean velocity in the cross-section. However, the Blasius solution may depart from the experimental data at high Reynolds numbers, $Re > 10^5$. Thus, curve 3 gives the best fitting for the experimental data, namely

$$\frac{1}{\sqrt{\lambda}} = 2.0\lg\frac{u_m d}{\nu}\sqrt{\lambda} - 0.8.$$
(4.3)

This formula was first proposed by Ludwig Prandtl, and was validated by the experiments of Nikuradse for Reynolds number up to 3.4×10^6.

The effects of roughness on the drag coefficient of a turbulent pipe flow is shown in Figure 4.3. The transition from laminar flow to turbulent flow is mainly influenced by the Reynolds number rather than the roughness. The critical Reynolds number is usually $Re = 2300$, as denoted by the point A in the figure. The region between points A and B indicate the transition zone.

When the pipe flow is turbulent, the flow can be divided into three different regions.

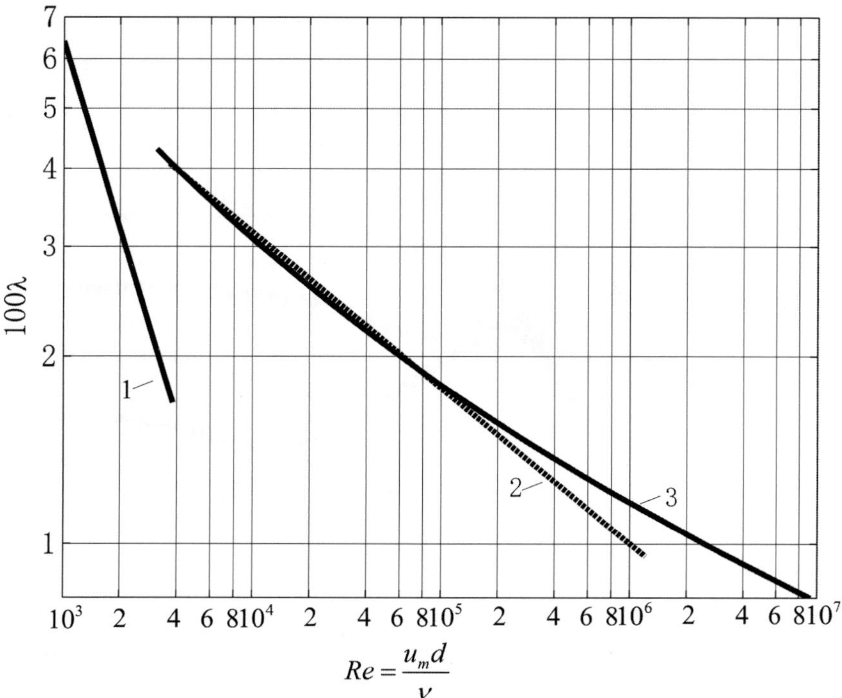

Figure 4.2 Drag coefficient of a smooth pipe.

(1) Hydraulically smooth. All data with different roughnesses fall into the Blasius curve or the Prandtl curve. However, for the various roughnesses, the hydraulically smooth only occurs in a certain range of Reynolds number. A higher roughness (smaller relative smoothness) deviates earlier from the hydraulically smooth region to the transition region. It has been indicated that when the height of the roughness is in the range

$$0 \le \ \frac{k_s u_e}{\nu} < 5, \tag{4.4}$$

the flow belongs to hydraulically smooth. In that case, the roughness is within the viscous sublayer, and does not influence the flow.

(2) Completely rough. This can be seen from the right part of the dashed line DE in Figure 4.3. In this region, the drag coefficient is completed determined by the roughness, but nearly has no relationship with the Reynolds number. The threshold for the roughness height is

$$\frac{k_s u_e}{\nu} > 70. \tag{4.5}$$

It is suggested that the roughness is in the logarithmic region.

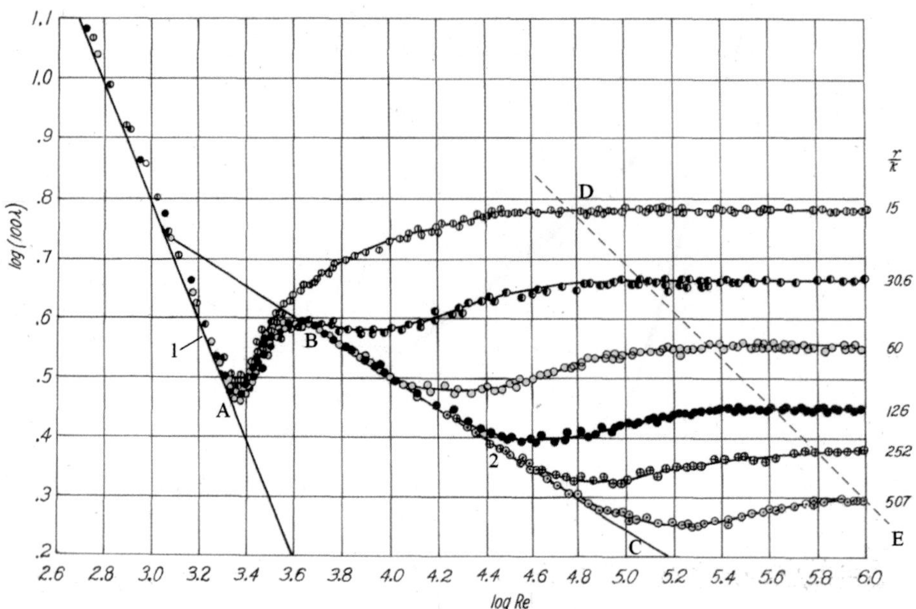

Figure 4.3 Effect of roughness on the drag coefficient of a turbulent pipe (Adapted from Nikuradse 1933).

(3) Transition region. This describes the transition region from hydraulically smooth to completely rough, where the drag coefficient is determined by both Reynolds number and the roughness height. The range of the roughness height is

$$5 \leq \frac{k_s u_e}{\nu} \leq 70, \tag{4.6}$$

which is also around the buffer region of the boundary layer.

Besides the influence on the turbulent boundary layer, the roughness also influences the boundary layer transition. In general, the critical Reynolds number for boundary layer transition is $Re = 2100$–2300 for pipe flow. The traditional literature indicate that the roughness may introduce disturbance to the flow field, thus promoting the flow transition. However, the height of the roughness plays an important role. When the roughness height is smaller than the threshold, the roughness may not influence the flow transition. Thus, some of the criteria for the roughness height are defined as follows.

(1) The critical roughness height. Numerous experiments indicate that the critical roughness height is

$$\frac{k_{crit} u_e}{\nu} = 7. \tag{4.7}$$

When the roughness height is smaller than the critical height, the roughness will not affect the flow transition.

(2) The effective roughness height. The flow transition may just take place directly at the position of the roughness when its height reaches the effective height

$$\frac{k_{effect}u_e}{v} = 15 - 20. \tag{4.8}$$

(3) The transitional roughness height. When the height of the roughness is between the critical height and effective height, the roughness can influence the flow transition. However, the transition location may change with the different roughness heights.

In this section, some of the traditional knowledge from the literature has been presented, followed by detailed introduction of recent findings of the influence of roughness on transition and the turbulent flow.

4.2 Roughness on Transition

4.2.1 Evolution of Vortical Structure

For the applications of the roughness (or roughness element) for flow control, one also needs to know the basic characteristics of flow over the roughness, as shown in Figure 4.4. A schematic of vortical structure around a roughness is shown in

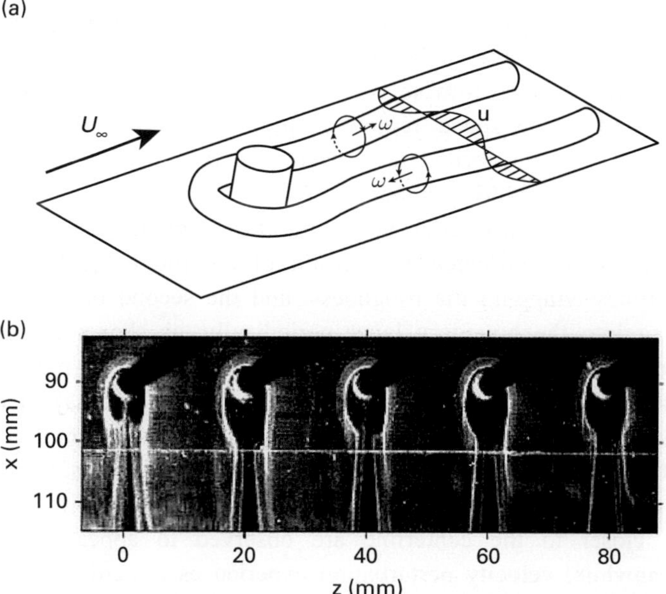

Figure 4.4 (a) Schematic of vortical structure around a roughness and the lift-up effect; and (b) hydrogen bubble flow visualization around a row of roughnesses at $Re_k = 108$ (Zhang et al. 2011). Reproduced with permission, copyright © Springer-Verlag 2011.

Figure 4.4(a), which can also be observed from the flow visualizations around a set of roughnesses shown in Figure 4.4(b). Here, the Reynolds number based on the roughness height is, $Re_k = u_k k/v = 108$, where u_k is the velocity at the top of the roughness when the roughness is absent. The fluids upstream of the roughness separate from the wall due to the adverse pressure gradient. The separated fluids will roll into a standing vortex that wraps around the windward side of the roughness and a pair of counter-rotating streamwise vortices which are stretched downstream and located on both sides of the roughness. Further downstream, the streamwise vortices entrain high-momentum fluids towards the wall near the centerline of the roughness. Meanwhile, the low-speed fluids in the regions between the roughnesses are pushed upwards. This is called the lift-up effect, which transforms a velocity deficit region close to the roughness into a high-speed region downstream.

The streamwise evolution of the mean streamwise velocity and the RMS value at a fixed wall-normal height $y = 3$ mm is shown in Figure 4.5 for three roughness Reynolds numbers. There is a velocity deficit immediately downstream of the roughness. Due to the lift up effect, high-speed fluids are transported into this low-speed region by streamwise vortices and then it gradually becomes the high-speed region. However, as the flow evolves downstream, the difference of the streamwise velocity in the spanwise direction gradually diminishes. The mean velocity for three Reynolds numbers shows similar variations. However, the evolution of the RMS velocity seems to be different for different Reynolds number. In the near wake region, a single peak of the RMS velocity occurs in each low-speed region for $Re_k = 227$, while two peaks appear in the local high-speed regions for the other two higher Reynolds number cases. Zhang et al. (2011) pointed out that this was because the strength of the streamwise vortices increased with the roughness Reynolds number. Each streamwise vortex was strong enough to produce a high-speed region and a low-speed region at high Reynolds numbers, while two streamwise vortices were needed to convert one low-speed region into a high-speed region at low Reynolds numbers.

In terms of flow pattern, Acarlar and Smith (1987) pointed out that two types of vortical structures are presented as a result of the roughness. The first one is stationary vortices wrapping the roughness, and the second one is hairpin structures shedding into the boundary layer periodically, as shown in Figure 4.6(a). Corresponding to the observed characteristic flow structures, the mean and fluctuation streamwise velocities are typically shown in Figure 4.6(b). As a result of the vortex legs located at either side of the roughness, fluids from different wall normal position exchange, creating high-speed streaks at both sides of roughness elements in the near wake. The distribution pattern fades off in the far wake, and new streaks closer to the centerline are observed to appear with increasing strength. Meanwhile, velocity perturbation experiences a significant augmentation in the far wake, where two lobes are present corresponding to the location of local maxima of the spanwise shear layer.

It was found that hairpin vortices were formed and shed downstream periodically from the roughness, which can be clearly seen from the hydrogen bubble flow

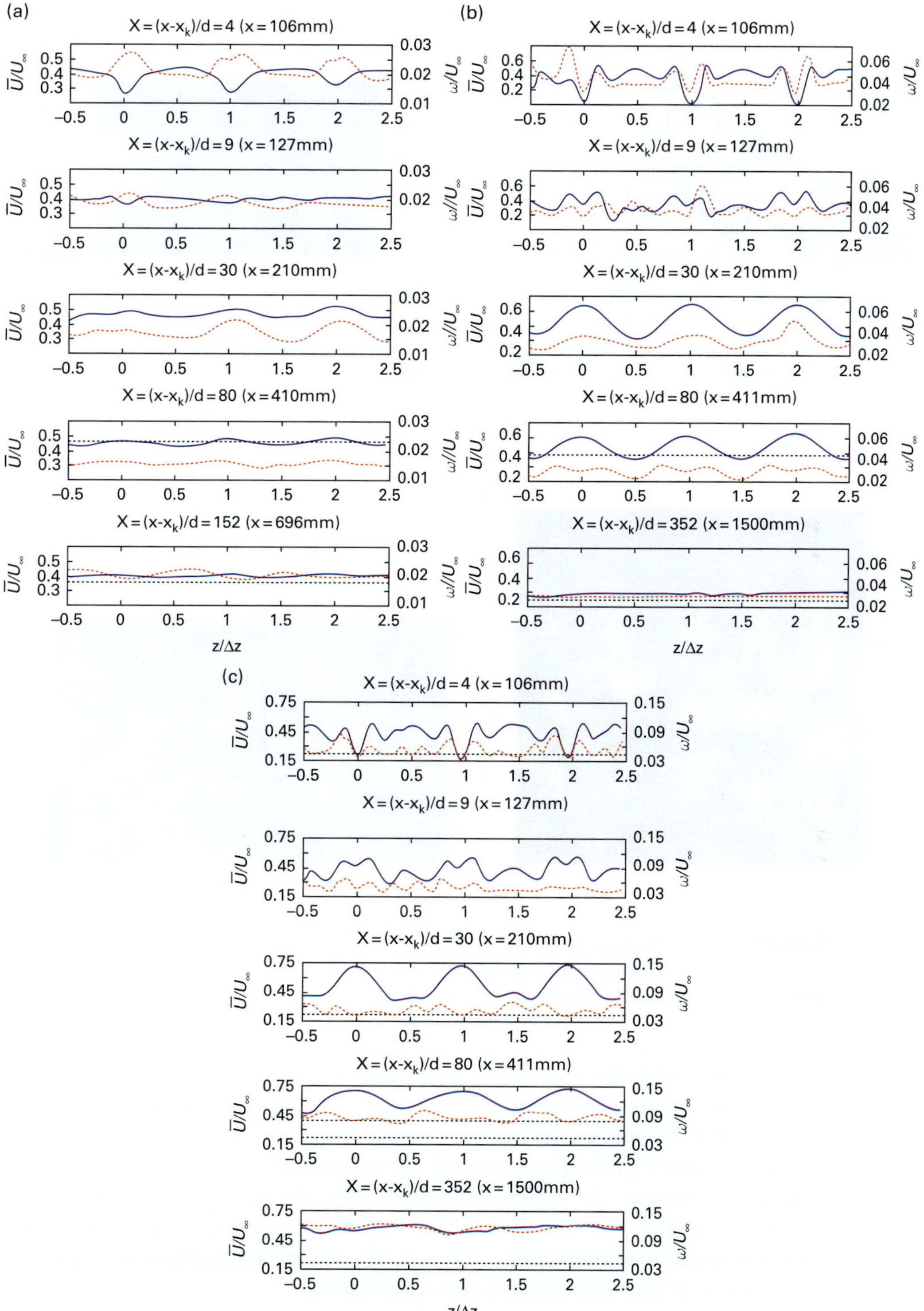

Figure 4.5 Streamwise evolution of the mean streamwise velocity (blue solid line) and its root mean square (RMS) value (red dash dotted line) at $y = 3$ mm. (a) $Re_k = 227$, (b) $Re_k = 339$, (c) $Re_k = 443$ (Zhang et al. 2011). Reproduced with permission, copyright © Springer-Verlag 2011.

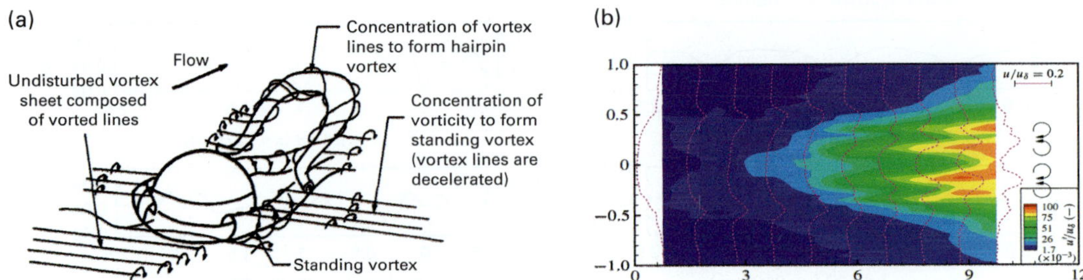

Figure 4.6 (a) Flow structures around a half-sphere roughness (Acarlar and Smith 1987). Reproduced with permission, copyright © Cambridge University Press 1987. (b) Distribution of mean (dashed lines) and fluctuation (contours) streamwise velocities (Plogmann et al. 2014). Reproduced with permission, copyright © Cambridge University Press 2014.

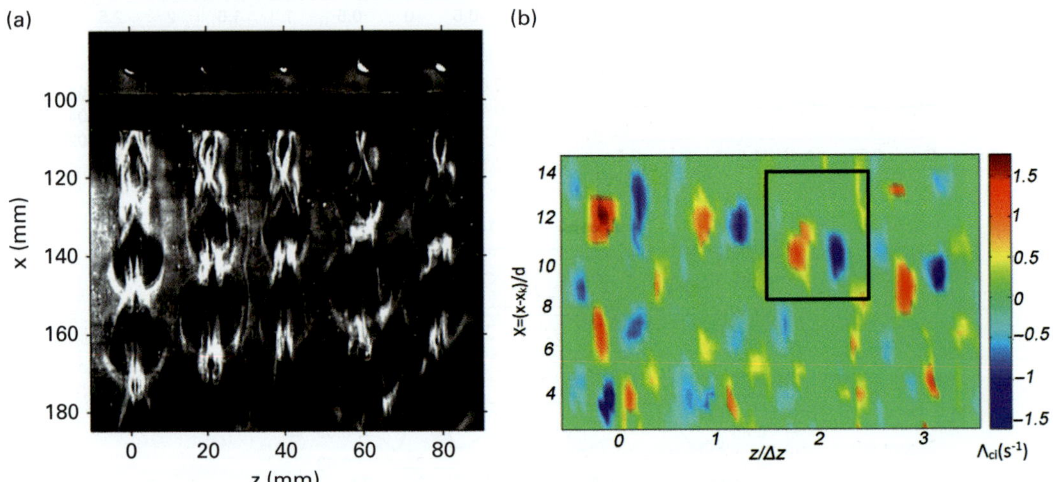

Figure 4.7 Periodic shedding of hairpin vortices for $Re_k = 443$. (a) Hydrogen bubble flow visualization at $y = 2$ mm; and (b) Λ_{ci} field at $y = 3$ mm (Zhang et al. 2011). Reproduced with permission, copyright © Springer-Verlag 2011.

visualization shown in Figure 4.7(a). The shedding frequency was $f = 2.9$ Hz, corresponding to a Strouhal number $St = fk/U_\infty = 0.161$. The hairpin vortices were also quantitatively identified by the swirling strength Λ_{ci} method, as shown in Figure 4.7(b). It presents several pairs of concentrated Λ_{ci} regions downstream of each roughness. It actually represents the wall-normal vortex pairs that are convected downstream. The history of the Λ_{ci} is shown in Figure 4.8, which jumps to some finite positive value when a positive vortex passes through. The dominant frequency from Figure 4.8 agrees well with that observed from flow visualization.

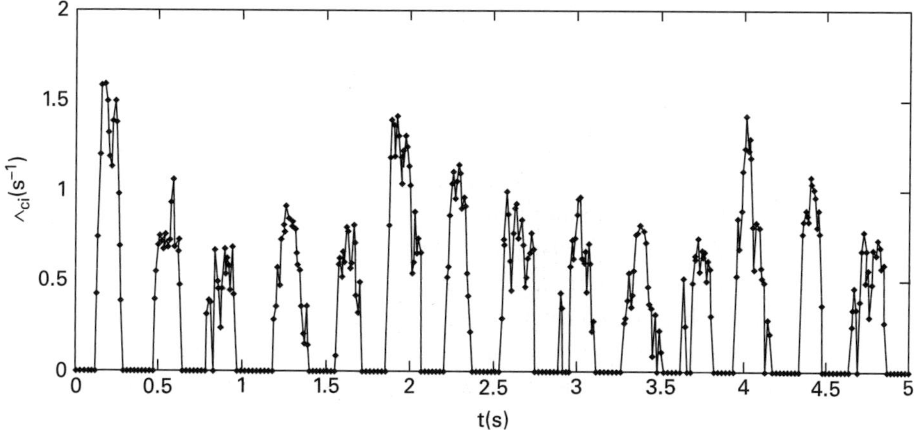

Figure 4.8 History of Λ_{ci} at point A to show the shedding of hairpin vortices (Zhang et al. 2011). Reproduced with permission, copyright © Springer-Verlag 2011.

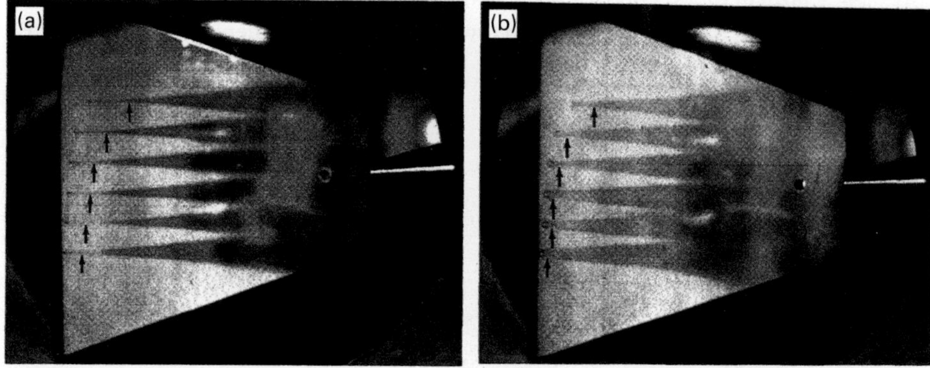

Figure 4.9 Fluorene sublimation of the boundary layer over a flat plate controlled by spherical (a) and triangular (b) roughnesses. The arrow marks the onset of transition (Hicks and Harper 1970).

4.2.2 Promoting Transition

Traditionally, the roughness is considered as a method of promoting transition. Schneider (2008) presents a review of this field, and one typical experiment was conducted by Hicks and Harper (1970). Figure 4.9 shows typical results for spherical and triangular roughness on the boundary layer developed in a flat plate at $Ma = 2.91$ and $Re_k =5500$. A turbulent wedge is typically present after the roughness and the arrow marks the onset of transition. In addition, flow transition begins further upstream for the triangular roughness than for the spherical roughness at all stations.

Although a small roughness element would not trigger transition in advance, a higher one may have an impact on the transition process. The minimum height that has an impact on the onset of the transition process, is usually referred to as a critical height, and the roughness Reynolds number is thus called critical roughness Reynolds number.

A further increase in the roughness height results in a forward movement of the transition position until it stagnates somewhere near the roughness. Hence, an effective value, namely roughness Reynolds number above which transition starts right after the roughness, can be defined. The value is usually referred to as effective roughness Reynolds number. While effective roughness Reynolds number is found to be nearly constant, critical roughness Reynolds number is a function of roughness shape, Mach number, free-stream turbulence, etc., as shown in Figure 4.10.

Depending on the height of roughness elements, a different tripping mechanism dominates the impact on the transition process (Plogmann et al. 2014). For low roughness heights, roughness induced small-scale velocity variation, which could result in energy transfer from large-scale free-stream turbulence to Tollmien–Schlichting (TS) waves. For medium roughness heights, roughness elements induced velocity streaks, which turns two-dimensional TS waves into their three-dimensional counterparts. Consequently, the growth rate of the disturbance is observed to increase, resulting in the appearance of wave interaction and disturbance that does not grow on a smooth surface. Energy transfer from the large-scale spanwise structures to smaller ones and the growth of a secondary wave contribute to the increase of velocity fluctuation and mean profile, marking the last step towards the fully turbulent boundary layer. Besides, in the

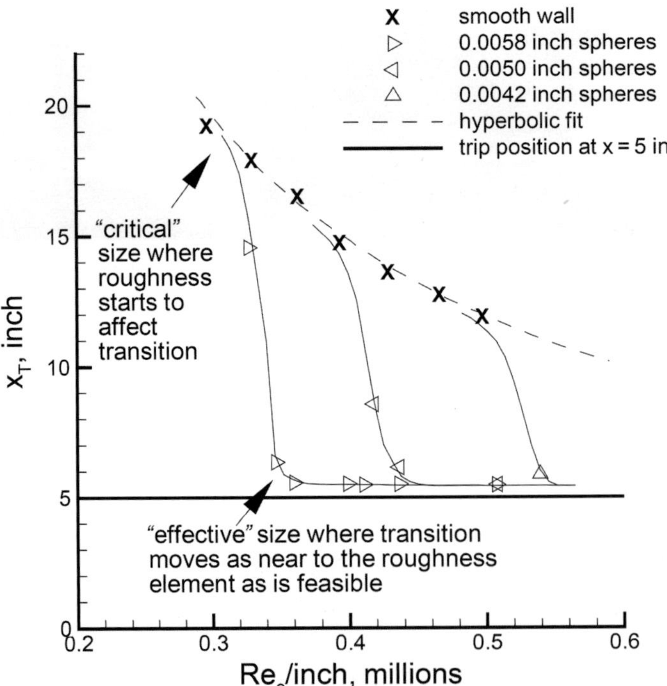

Figure 4.10 Definition of critical and effective roughness and their influence factors (Schneider 2008). Reproduced with permission, copyright © 2007 by Steven P. Schneider.

Table 4.1 Parameters and effects of roughness bump size (Zhou and Wang 2012)

Case	x_e	l_{AC}	l_{BD}	l_{EF}	C_L	C_D	C_{Dp}	C_{Df}	C_L/C_D
AOA_4	N/A	N/A	N/A	N/A	0.600	2.34e-2	1.38e-2	0.97e-2	25.6
AOA_4c	0.05	0.045	0.045	0.0035	0.593	2.05e-2	1.00e-2	1.05e-2	28.9
AOA_4c.w	0.05	0.090	0.045	0.0035	0.579	2.10e-2	0.94e-2	1.16e-2	27.6
AOA_4c.h	0.05	0.045	0.045	0.005	0.579	2.05e-2	0.95e-2	1.12e-2	28.0

(a)   (b)

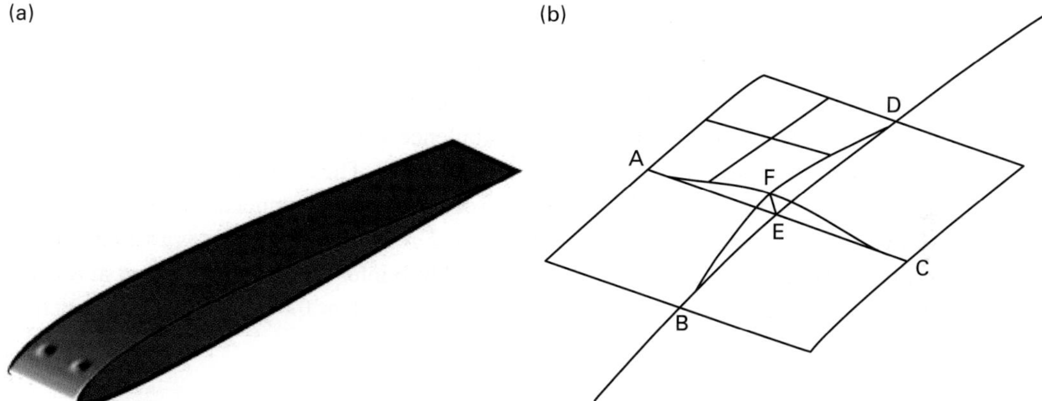

Figure 4.11 (a) Roughness bump on the wing surface and (b) roughness geometries (Zhou and Wang 2012). Reproduced with permission, copyright © 2011 by Y. Zhou and Z. J. Wang.

case of a spanwise periodic roughness elements array, alternating high and low speed streamwise streaks result in the formation of shear layers between adjacent streaks. When the velocity gradient in the shear layer is substantially large, inviscid instability may lead to growth of disturbances and finally transition to the fully developed turbulence (Andersson et al. 2001). For high roughness heights, K–H instability might be responsible for the roughness related bypass transition.

Similar to their influence on the boundary layer over a flat plate, roughness elements are also able to promote transition over an airfoil. Zhou and Wang (2012) numerically simulated the impact of leading-edge roughness on flow around a SD7003 airfoil, as shown in Figure 4.11(a). A periodic boundary condition was used in the spanwise direction and the span width of the wing is chosen to be 20% of the chord. The angle of attack was set to 4° and the Reynolds number was $Re = 6\times10^4$. The main flow parameters for the roughness shown in Figure 4.11(b) are listed in Table 4.1. There were two elements arranged in the spanwise direction. Here, the l_{EF}, l_{BD}, and l_{AC} indicate height, length, and width of the roughness bump, while x_E shows the location of the roughness relative to the leading edge.

The distributions of mean pressure coefficient and friction coefficient on the suction surface are shown in Figure 4.12 for the natural and control cases. Results in Figure 4.12 indicate that the transition position of the airfoil is shifted forward dramatically with the roughness control. Thus the size of the separation bubble over the suction surface is reduced in comparison with the natural case, while the friction coefficient is increased before about $x/c = 0.7$. As a result, the pressure drag over the entire airfoil is reduced while the friction drag is increased with roughness. However, the total mean drag is reduced and the lift is also increased, leading to an increase in the lift-to-drag ratio.

4.2.3   Delaying Transition

It is not always the case that the roughness will promote flow transition and increase the friction drag. One significant work was conducted by Fransson et al. (2006) to show the effect of delaying flow transition to turbulence with roughness. A schematic of flat plate under roughness control is shown in Figure 4.13. The boundary layer developed on a flat plate horizontally mounted in the test section of a wind tunnel at a fixed free-stream velocity of $U_\infty = 5$ m/s. Forty-two cylindrical roughness elements with diameter of 4.2 mm, height of 1.4 mm, and spaced at 14.7 mm intervals were positioned at 80 mm from the leading edge of the plate. Two slots with width of 0.8 mm and span of 330 mm

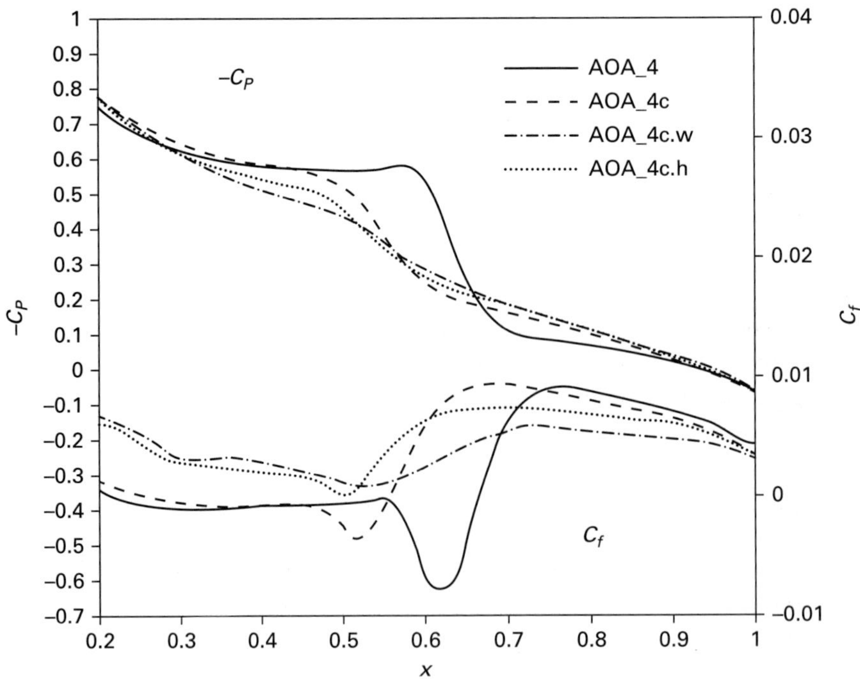

Figure 4.12 Mean pressure coefficient and mean skin-friction coefficient on the wing surface (Zhou and Wang 2012). Reproduced with permission, copyright © 2011 by Y. Zhou and Z. J. Wang.

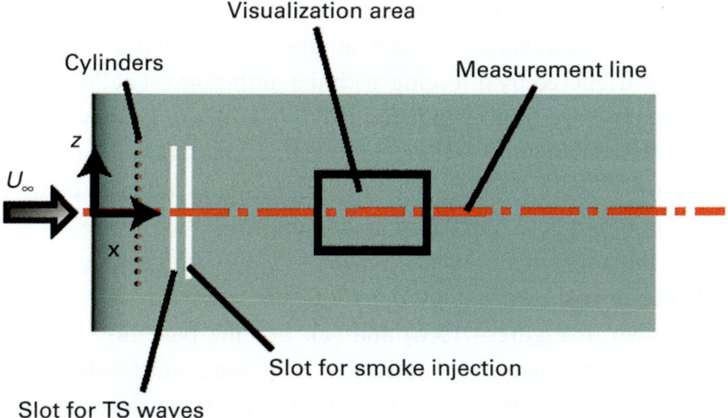

Figure 4.13 Schematic of flat plate under roughness control (Fransson et al. 2006). Reprinted figure with permission. Copyright 2006 American Physical Society.

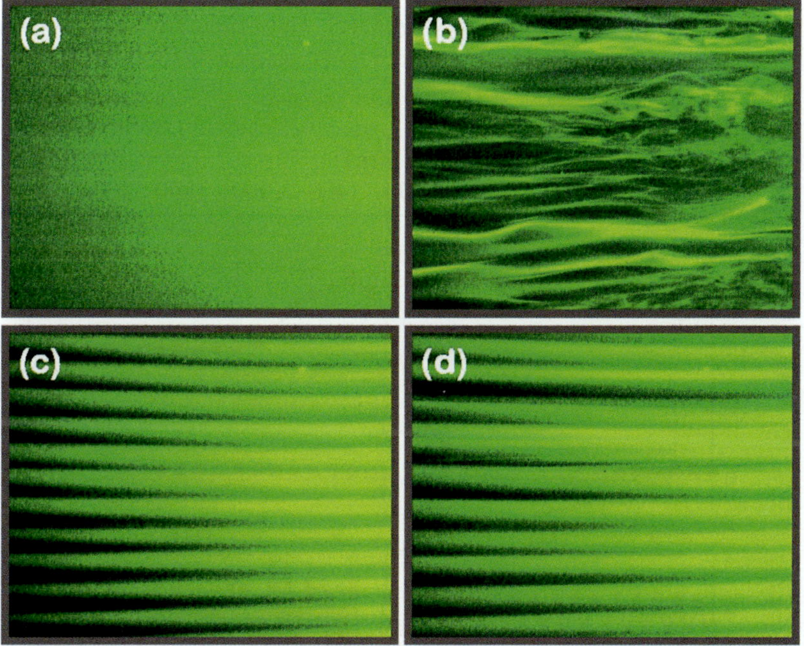

Figure 4.14 Smoke flow visualization of boundary layer over the flat plate. (a) Boundary layer without excitation, laminar flow; (b) boundary layer with excitation of 201 mV, turbulent flow; (c) boundary layer without excitation but with roughness control, laminar flow; and (d) boundary layer with excitation of 450 mV and roughness control, laminar flow (Fransson et al. 2006). Reprinted figure with permission. Copyright 2006 American Physical Society.

were positioned at $x = 190$ mm and 213 mm, which were used as slots for TS wave generation and smoke injection, respectively.

Typical flow visualization for different states is shown in Figure 4.14. Figure 4.14(a) shows the two-dimensional laminar flow without excitation, while Figure 4.14(b) shows the turbulent flow triggered by a forcing with the amplifier set at 201 mV. With roughness control, streaky base flow is induced (Figure 4.14(c)). Even when the excitation amplitude is increased to 450 mV, when the base flow should become turbulent, the flow is still kept laminar (Figure 4.14(d)). This finding provides valuable evidence to exhibit the effect of the roughness to delay flow transition.

Evolution of the root mean square (RMS) streamwise velocity fluctuation in the streamwise direction is shown in Figure 4.15(a), while the time histories at two typical locations are shown in Figure 4.15(b) and (c). For the base flow, the RMS velocity fluctuation increases gradually from about $x = 1050$ mm, and reaches the maximum at about $x = 1350$ mm where the flow is transitional. It finally decreases to a plateau at around $x = 1650$ mm, where transition is completed and a fully turbulent signal is observed. However, the flow is still laminar with roughness control, as it is indicated that the RMS velocity fluctuation remains at a small amplitude and sinusoidal all along the measurement region. This can be seen by comparing the signals without and with roughness at $x = 1650$ mm in Figure 4.15(c). Thus, Fransson et al. (2006) showed that it was possible to use the streaks induced by the roughness to suppress T-S waves at low

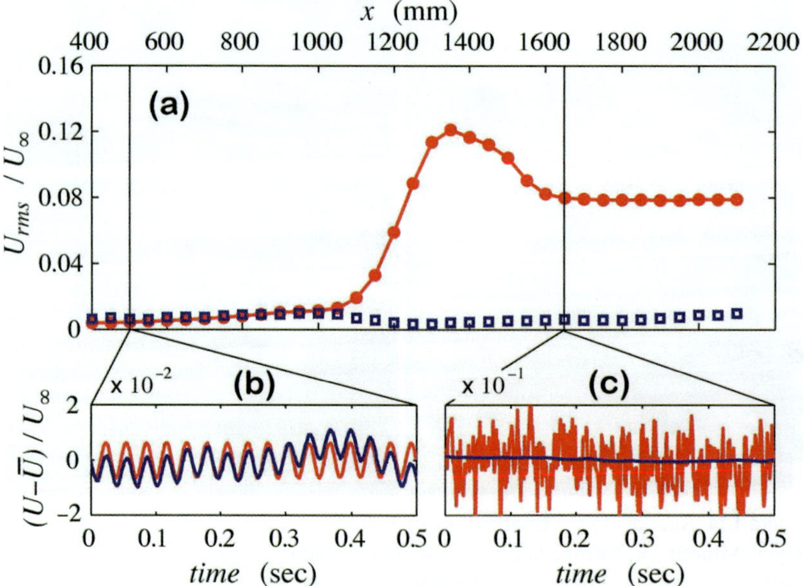

Figure 4.15 (a) Evolution of the RMS streamwise velocity fluctuation in the streamwise direction when T-S waves are excited at $f = 32$ Hz and 1000 mV; (b) time signal at $x = 400$ mm; and (c) time signal at $x = 1650$ mm. The red lines and solid symbols correspond to the two-dimensional boundary layer without roughness control, whiles blue lines and open symbols correspond to the boundary layer with roughness control (Fransson et al. 2006). Reprinted figure with permission. Copyright 2006 American Physical Society.

amplitudes in a boundary layer which otherwise would be transitional in the absence of the streaks. The results were essentially unchanged when random noise was added to the forcing.

Besides the applications in the boundary layer developed on a flat plate, spanwise periodically arranged roughness elements were also proved to be effective in delaying transition on a swept wing. Saric et al. (2011) examined the effect of roughness elements on a swept wing hung on a Cessna O-2 during a flight experiment, and the experimental configuration is shown in Figure 4.16. The swept wing had a 30° leading-edge sweep, and the angle of attack was −4°. The experiment was conducted under a chord Reynolds number of Re = 7.5×10^6. An infrared (IR) image was adopted as a method to capture flow transition, which can be determined by the colour over the wing surface. In general, the colder area denoted by the dark orange colour indicates laminar flow, while the lighter area denotes turbulent flow. The flow in all figures is from right to left.

The roughness array was placed near the leading edge around x/c = 1% to 1.3%. The roughness height was 12 μm and the space between neighboring roughness elements was set to be 2.25 mm. Infrared images of the suction surface of swept wing without and with leading-edge roughness element array are presented in Figure 4.17 to show the effects. It is suggested that the roughness elements array has a significant impact on transition process. The transition position obtained from the IR image is delayed from around $0.3c$ for the natural case to around $0.6c$ for the control case.

Similar results on a swept wing at $Re = 2.4 \times 10^6$ have also been reported by Saric et al. (1998), who postponed the transition position from x/c = 0.71 to 0.8. However, Hosseini et al. (2013) achieved a delay of 20%c. These achievements were based on the observation of Reibert et al. (1996) when roughness elements were installed, only unsteady waves with wavelengths equal to $1/n$ times the distance between neighboring roughnesses were activated; here n denotes an arbitrary positive integer. According to the observation, properly chosen distances between adjacent roughness elements can avoid

Figure 4.16 (a) Photograph to show the swept wing hung on a Cessna O-2; and (b) details over the swept wing surface (Saric et al. 2011). By permission of the Royal Society.

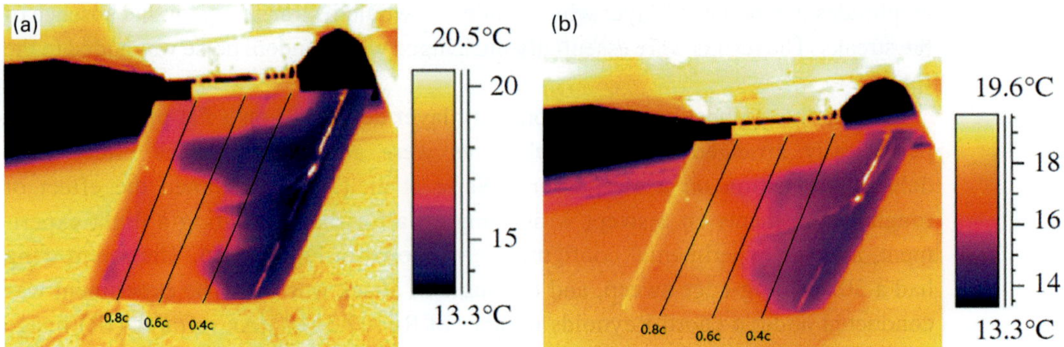

Figure 4.17 IR image of the suction surface of swept wing without (a) and with (b) leading-edge roughness element array. The flow direction is from right to left in the figure (Saric et al. 2011). By permission of the Royal Society.

the growth of most unstable disturbances and thus decrease the growth rate of velocity perturbation. Consequently, the flow transition could be postponed.

4.2.4   Effects on Bypass Transition

Besides the regular transition, where the flow disturbances are small, there is also another kind of transition, namely the bypass transition, which happens when strong disturbances exist. The wake/boundary layer interaction is an important type of bypass transition and it also frequently occurs in engineering. Pan et al. (2008) revealed the general path of the bypass transition induced by a two-dimensional cylinder wake, as shown in Figure 4.18. Periodic Kármán vortex shed from the circular cylinder, caused the generation of the secondary vortices near the wall with the same shedding frequency as the Kármán vortex. The instability of the secondary vortices led to the formation of hairpin vortices. In the following evolution, the hairpin vortices congregated to form the hairpin packets, as shown in Figure 4.19.

Roughness elements were further used by Wang et al. (2011) to control the bypass transition of a flat plate boundary layer induced by a circular cylinder wake. A sketch of the experimental setup is shown in Figure 4.20. The free-stream velocity was $U_\infty = 0.084$ m/s. A plate with length of 2000 mm, width of 500 mm, thickness of 10 mm and a 4:1 elliptical leading edge, was mounted parallel to the bottom of the water channel. The origin of the coordinate system was placed on the leading edge of the plate. A circular cylinder with diameter of $D = 20$ mm and length of 550 mm was placed spanwise above the plate at $x_c = -25$ mm and $y_c = 65$ mm.

The side view of the flow visualization is shown in Figure 4.21. Shear layer forms over the wall surfaces due to flow separation induced by the Kármán vortices shedding from the circular cylinder. The shear layer develops into the secondary spanwise vortex, which forms and convects downstream periodically. The secondary vortices can also be

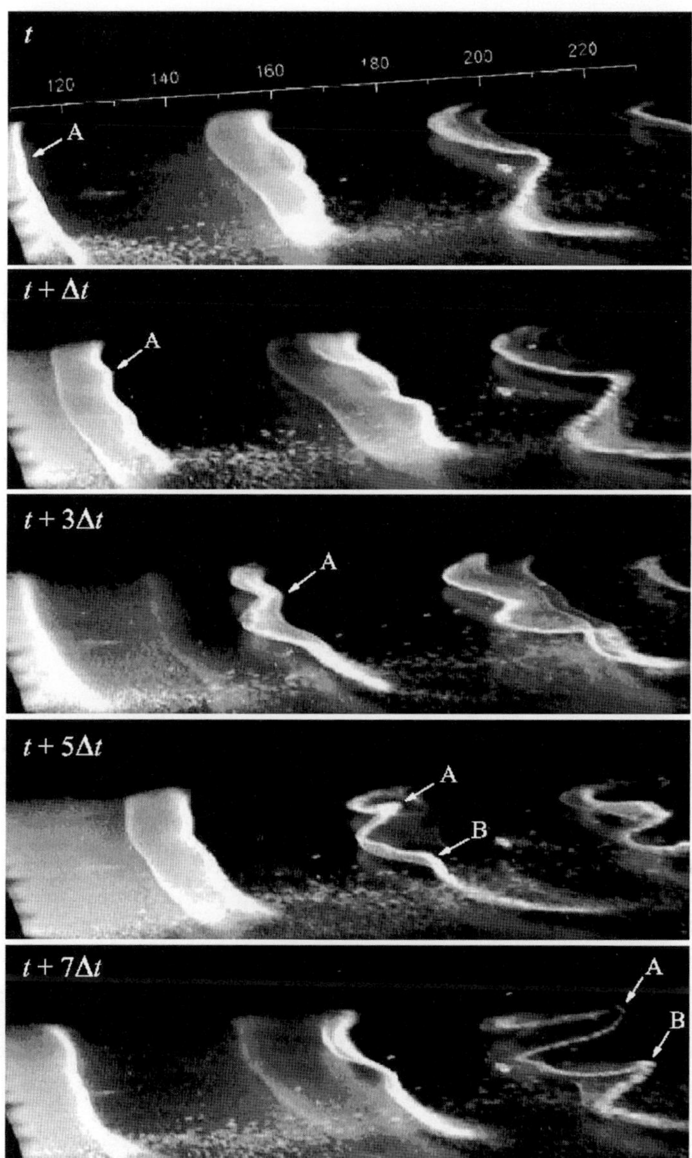

Figure 4.18 Illustration of the bypass transition process induced by a two-dimensional cylinder wake (Pan et al. 2008). Reproduced with permission, copyright © Cambridge University Press 2008.

clearly seen from the plane view, as shown in Figure 4.22. However, the shear layer is cut by the roughness, resulting in a row of discrete vortex segments.

The segmented secondary vortices between the two neighboring roughness elements are unstable. As they are convected downstream, two types of instabilities are observed, namely large-scale instability and small-scale instability, which are shown

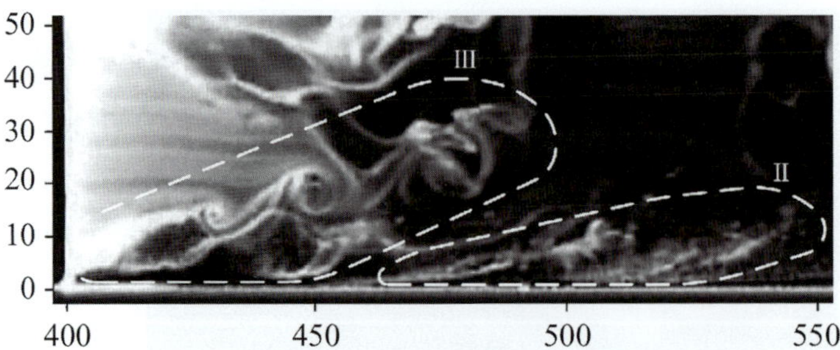

Figure 4.19 Illustration of the hairpin packets (Pan et al. 2008). Reproduced with permission, copyright © Cambridge University Press 2008.

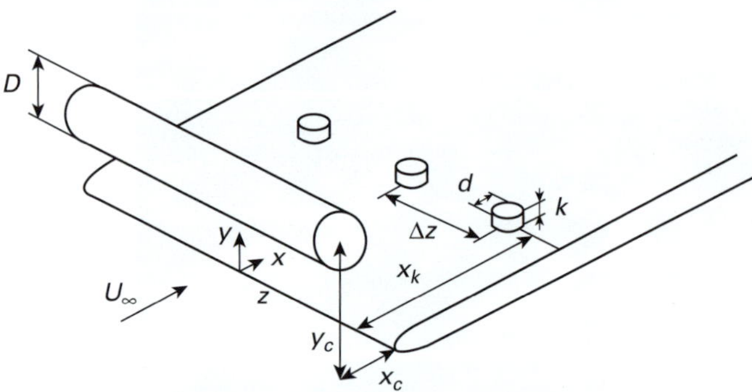

Figure 4.20 Sketch of the roughness control for flat plate boundary layer bypass transition induced by a circular cylinder wake (Wang et al. 2011). Reproduced with permission, copyright © The Visualization Society of Japan 2010.

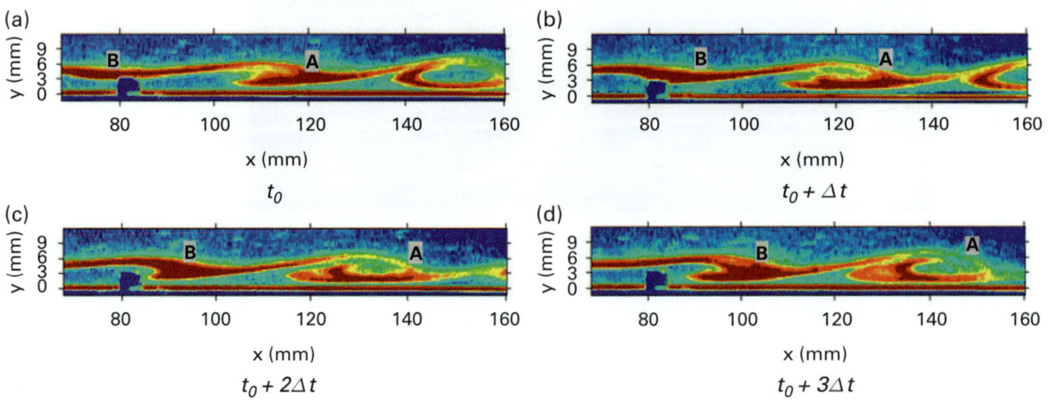

Figure 4.21 Generation of secondary vortex. Side view, $\Delta t = 0.24$ s, hydrogen bubble wire position: $x_w = 20$ mm, $y_w = 2$ mm (Wang et al. 2011). Reproduced with permission, copyright © The Visualization Society of Japan 2010.

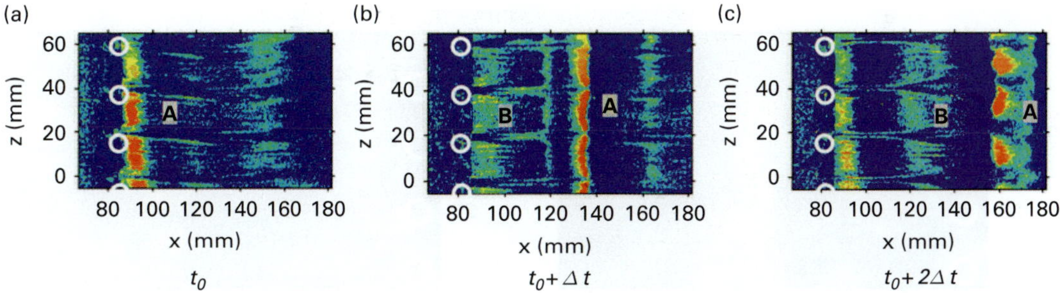

Figure 4.22 Generation of secondary vortex. Plane view, $\Delta t = 0.24$ s, $x_w = 20$ mm, $y_w = 2$ mm (Wang et al. 2011). Reproduced with permission, copyright © Springer 2011.

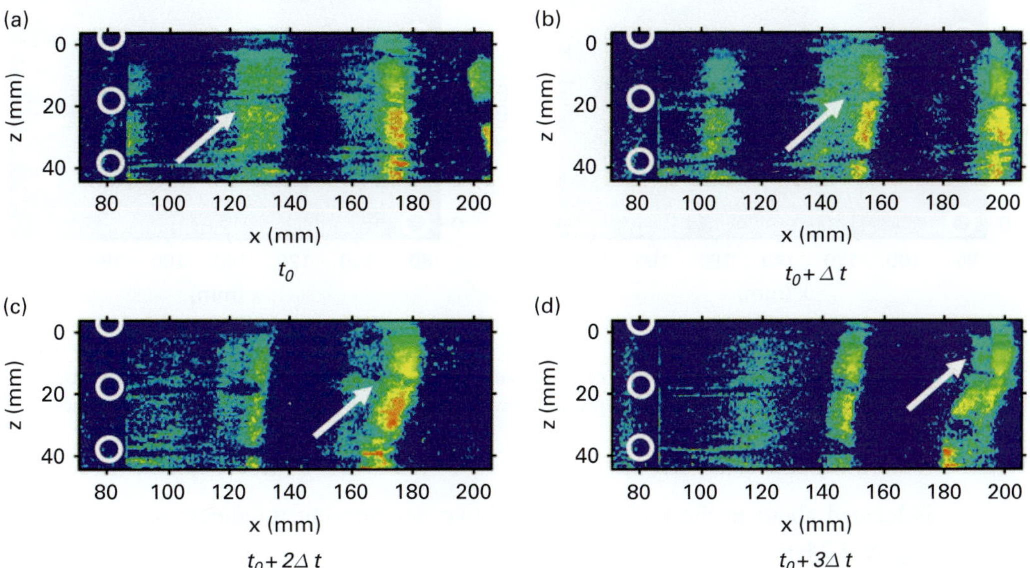

Figure 4.23 Large-scale instability of secondary vortex. Plane view, $\Delta t = 0.52$ s, $x_w = 60$ mm, $y_w = 3$ mm (Wang et al. 2011). Reproduced with permission, copyright © The Visualization Society of Japan 2010.

in Figures 4.23 and 4.24, respectively. Wang et al. (2011) suggested that the large-scale instability occurred when the high-shear layer was cut shallowly by the roughness. Thus, the segmented vortices could re-connect themselves and act as one large vortex, and then evolved into a large-scale hairpin vortex. In comparison, the small-scale instability occurred when the roughnesses deeply cut the shear layer. Then each segmented vortex evolved separately, forming an array of hairpin vortices with nearly the same size and spacing. Note that the size of the hairpins is roughly equal to the spacing between roughnesses and the head of the hairpin vortex

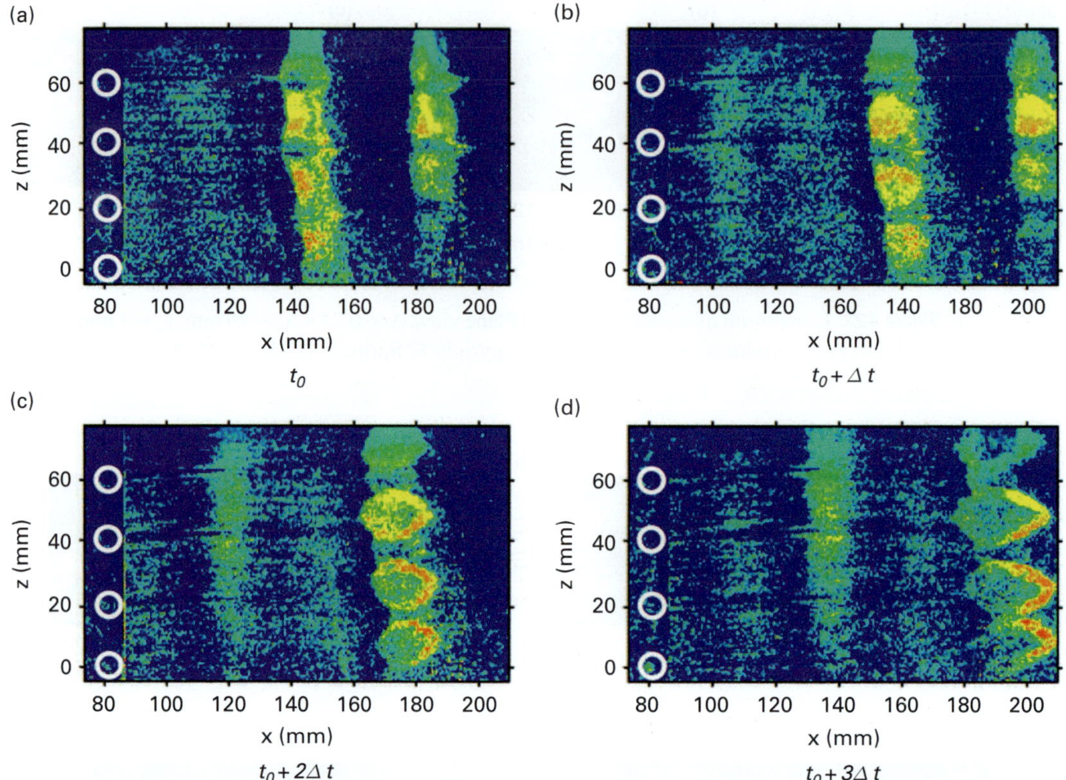

Figure 4.24 Small-scale instability of secondary vortex. Plane view, $\Delta t = 0.4$ s, $x_w = 60$ mm, $y_w = 2$ mm (Wang et al. 2011). Reproduced with permission, copyright © The Visualization Society of Japan 2010.

is located about in the middle plane of two neighboring roughnesses, as shown in Figure 4.24.

When the hairpin vortex is formed and convected downstream, there are also usually two scenarios for its evolution. The first kind is that a row of hairpin vortices is convected downstream side by side, and there exists weak mutual interaction between them, as shown in Figure 4.24. On the other hand, The small-scale hairpin vortices generated during small-scale instability may present some complex behavior, such as vortex merging, as shown in Figure 4.25. It shows four legs (A1, A2, B1 and B2) of two hairpin vortices at time t_0. Two adjacent legs (A2 and B1) are attracting and annihilating each other at $t_0 + \Delta t$, and they become weaker and finally disappear. Then, two outer legs (A1 and B2) are stretched and the heads of both vortices begin to join together at $t_0 + 2\Delta t$, merging into one larger hairpin vortex.

The general roadmap of the bypass transition influenced by roughness is summarized by Wang et al. (2011) in Figure 4.26. The periodic wake vortices shedding from the cylinder first induce a separated high-shear layer upstream of the roughness. Then the roughness incises the high-shear layer, forming segmented spanwise vortices. Two kinds of instability

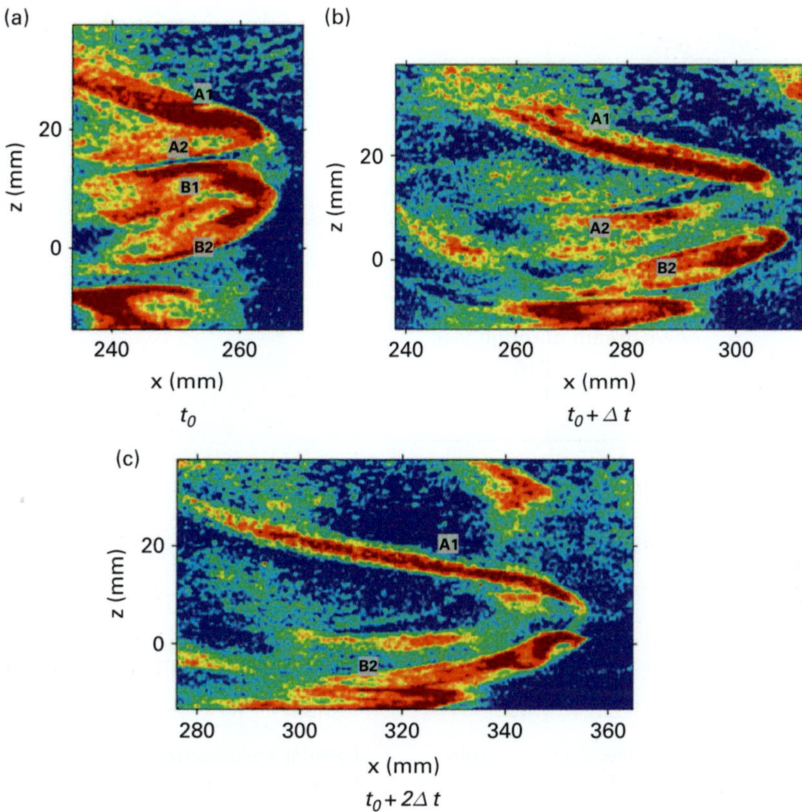

Figure 4.25 Merging of secondary vortices. Plane view, $\Delta t = 0.6$ s, $x_w = 75$ mm, $y_w = 2$ mm (Wang et al. 2011). Reproduced with permission, copyright © The Visualization Society of Japan 2010.

may occur according to the different depth of incision, denoted as large- and small-scale instabilities. A large-scale hairpin vortex may be induced for the former case, while for the latter case each segmented vortex evolves into one small-scale hairpin vortex. The small-scale hairpin vortex may convect downstream side by side or merge with each other.

4.3 Roughness on Turbulence

4.3.1 Effects on Friction Drag

Roughness was used by many researchers to control the turbulent flow. Frohnapfel et al. (2007) conducted an interesting study about the influence of surface roughness on the development of turbulent flow. The purpose here is to show recent developments on the influence of roughness height on the turbulent drag reduction. Though the configuration is somewhat similar to the riblet, it is still called roughness for the sake of consistency.

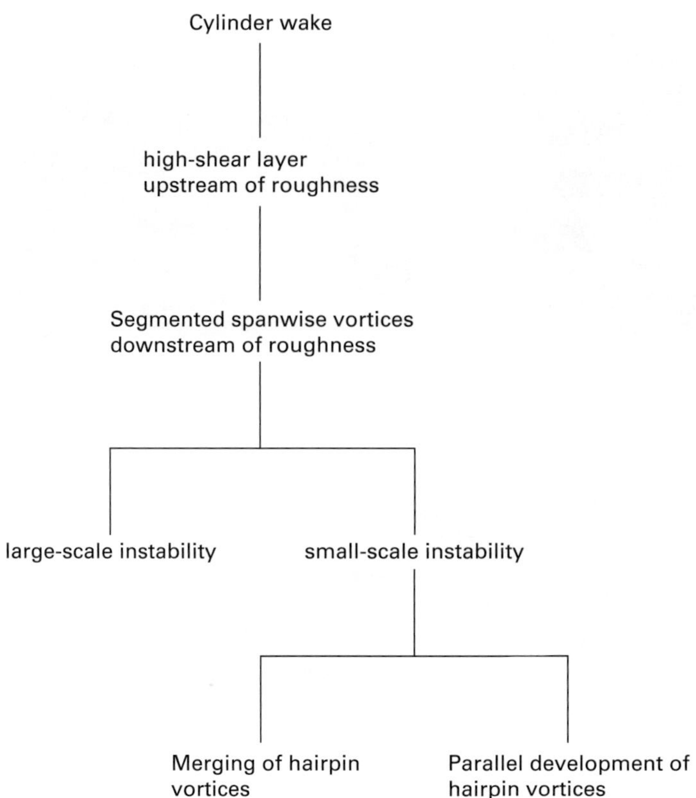

Figure 4.26 A general roadmap of the bypass transition influenced by roughness (Wang et al. 2011). Reproduced with permission, copyright © The Visualization Society of Japan 2010.

The experiment was conducted in a wind tunnel. The test section was 3.5 m long and 0.3 m wide and could be adjusted to heights of $H = 25$ and 35 mm, as shown in Figure 4.27(a). A trip was installed at the channel inlet to ensure turbulent flow conditions. Roughness was inserted on the wall in the free stream direction, as shown in Figure 4.27(b). The test section was divided into two parts of equal length, the upstream roughness section and the downstream smooth section. Fourteen pressure taps of 300 μm diameter were installed on both of the channel sidewalls with an intervals of 0.2 m. The drag reduction could be estimated based on the pressure drops $\Delta p/\Delta x$ over the smooth and the grooved surface parts, namely $\Delta \tau = 1 - (\Delta p/\Delta x)_{\text{grooved}}/(\Delta p/\Delta x)_{\text{smooth}}$. The experimental Reynolds number based on the channel height was ranging from 1000 to 64 000.

The pressure drop measurements at different Reynolds numbers over the entire test section are shown in Figure 4.28. The straight lines closely follow the trend in the pressure drop for the grooved part (solid line) and the smooth part (dashed line). The slope of the fitting line can reflect the friction drag coefficient, as a smaller slope stands for a smaller friction drag. Thus it is indicated that drag reduction of about 0.55%, 21.8%, and 1.8% could be obtained for the control cases at $Re = 2450$, 2870, and 26 830,

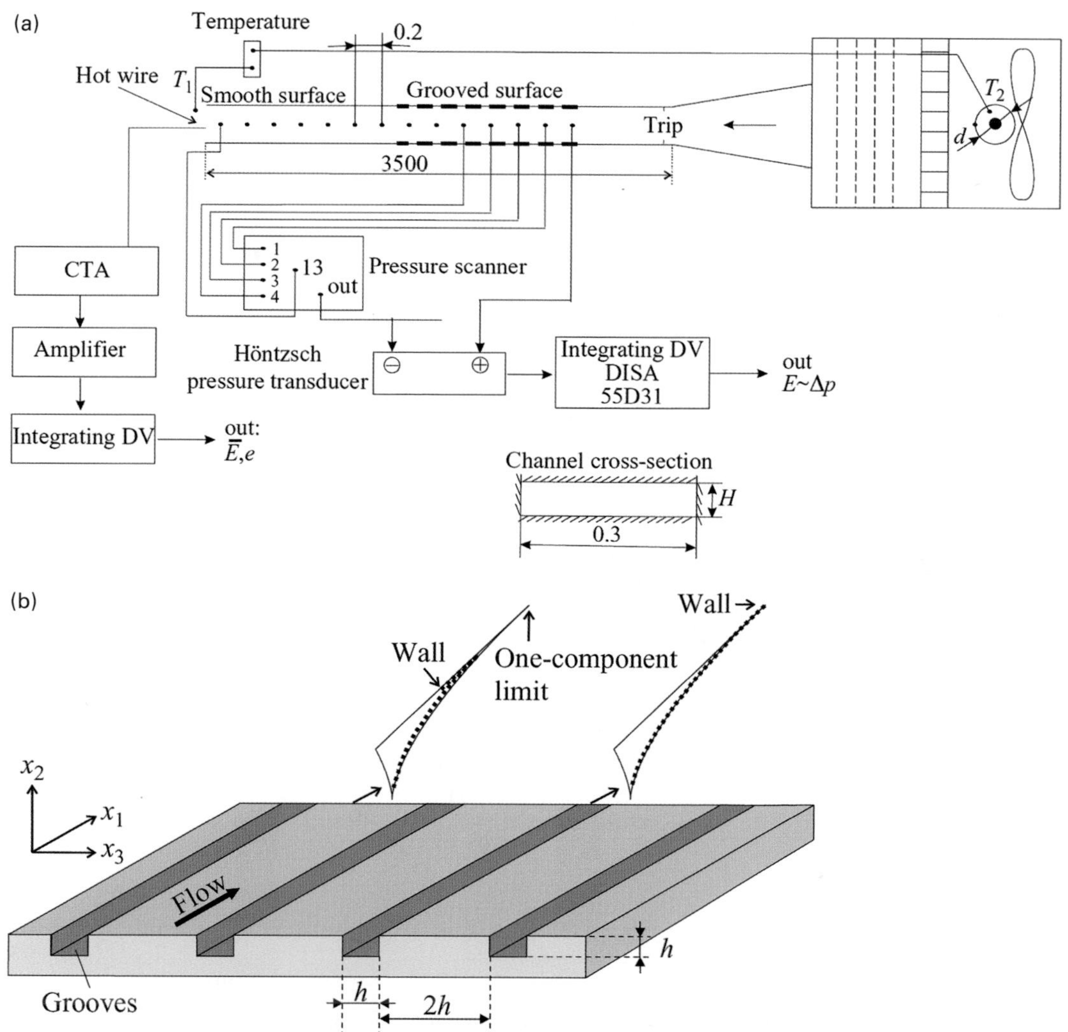

Figure 4.27 (a) Experimental setup and (b) details of the roughness on wall (Frohnapfel et al. 2007). Reproduced with permission, copyright © Cambridge University Press 2007.

respectively. However, roughness control may increase the drag coefficient by about 1.4% at $Re = 3620$.

The drag reduction obtained with channel heights of 25 mm and 35 mm is shown in Figure 4.29(a) and (b) as a function of the Reynolds number. For the 25 mm cases, high drag reduction up to 14% is obtained at low Reynolds numbers around $Re = 2300$–2900. However, the measured drag reduction deduced from pressure drop measurements is within the limit of the measurement accuracy at higher Reynolds numbers around $Re = 5000$–45000. For the 35 mm cases, significant drag reduction is achieved in the range of $Re = 2700$–3200, where drag reduction up to 25% is measured.

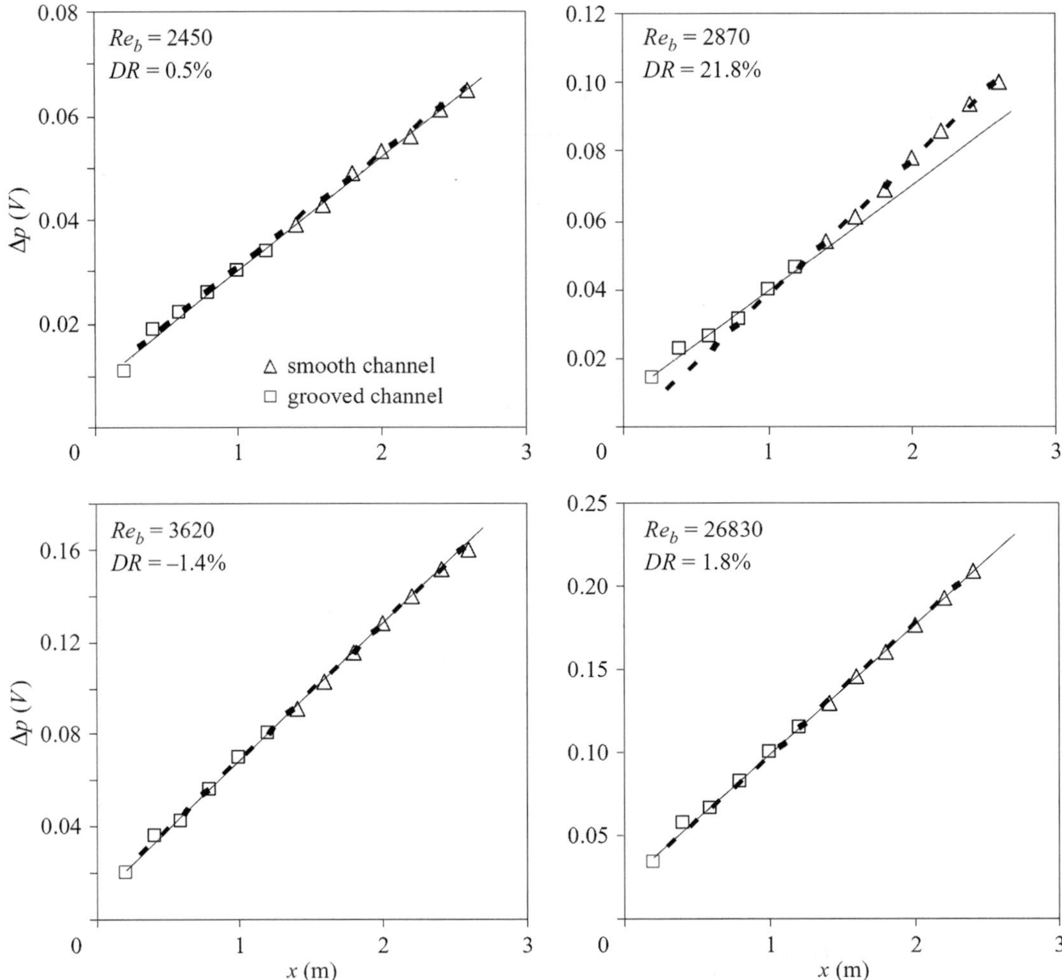

Figure 4.28 Pressure drop over the entire test section for different Reynolds numbers. The first half has roughness and the second half with smooth walls. The pressure gradient is determined by a best-fit line for each part (Frohnapfel et al. 2007). Reproduced with permission, copyright © Cambridge University Press 2007.

All data are summarized in Figure 4.29(c) as a function of the dimensionless size of the roughness. It is indicated that significant drag reduction is obtained at around $h^+ = 0.8$, which is approximately half of the Kolmogorov scale. This finding seems surprising since the wall roughness is within the viscous sublayer, which is usually considered with a dimensionless height of $y^+ = 5$. However, the traditional literature showed that the roughness height below the viscous sublayer was hydraulically smooth and had no influence on the pressure drop in channel or pipe flow. Frohnapfel et al. (2007) provided a new and interesting result about the effects of roughness on turbulence, which of course needs to be further validated by more evidence.

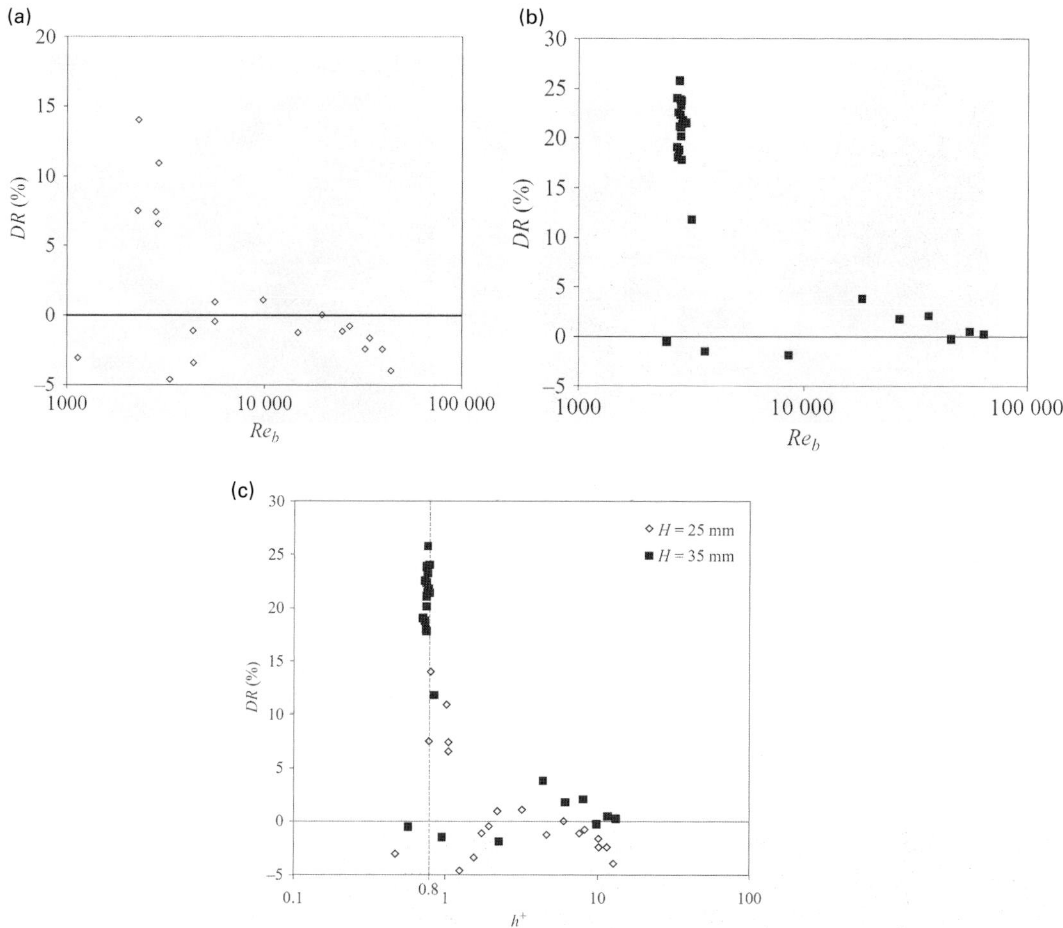

Figure 4.29 Drag reduction obtained with a channel height $H = 25$ mm (a) and 35 mm (b) plotted versus Reynolds number and (c) dimensionless size of the roughness (Frohnapfel et al. 2007). Reproduced with permission, copyright © Cambridge University Press 2007.

4.3.2   Effects on Coherent Structures

Kline et al. (1967) found the existence of low and high-speed streaky structures in the turbulent boundary layer in the 1960s, indicating that the turbulence is not totally random, but contains coherent structures. These streaky structures usually exist near the sublayer region. During downstream convection, the head of the low-speed streak may arise gradually, resulting in the generation of the spanwise vortex. The streak evolution is usually accompanied by the burst event, including the ejection and sweep, which contributes to the generation of turbulent kinetic energy. Thus, it suggests a strategy for drag reduction of the turbulent boundary layer by manipulating the streak with roughness.

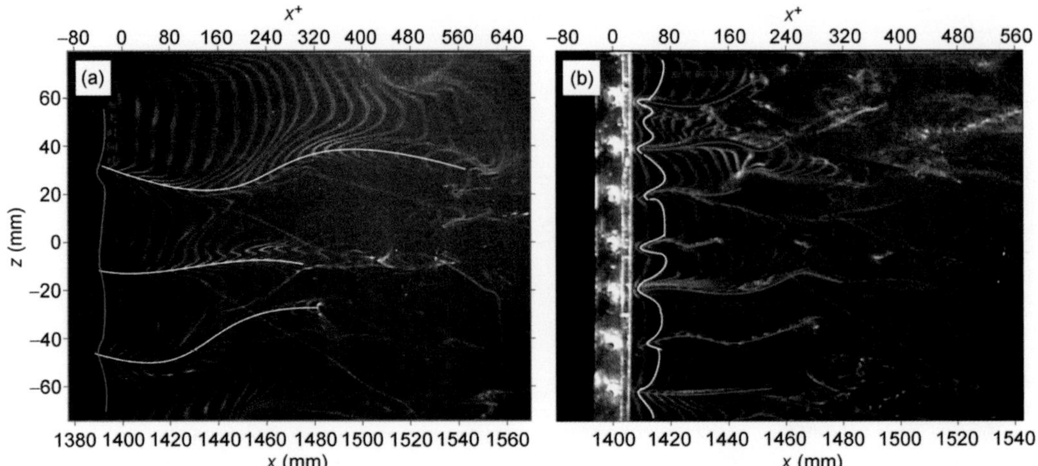

Figure 4.30 Instantaneous near-wall streak pattern for the natural (a) and roughness control (b) cases. The platinum wire was placed at $x_w = 1410$ mm and $y_w = 8$ mm (Zhang et al. 2015). Reproduced with permission, copyright © Science China Press and Springer-Verlag Berlin Heidelberg 2015.

Zhang et al. (2015) studied the effects of roughness on the streak evolution. The turbulent boundary layer developed on a 2000 mm-long flat plate with a 4:1 elliptical leading edge at $U_\infty = 86.1$ mm/s. A 6 mm-diameter rod was transversely stuck to the wall at $x = 70$ mm from the leading edge, resulting in a fully developed turbulent boundary layer after $x = 700$ mm. Cylindrical roughness elements with diameter of 4 mm and height of $k = 8$ mm ($k/\delta = 0.2$) were attached on the wall surface at $x = 1400$ mm. The roughness spanwise spacing was chosen to be $\Delta z = 20$ mm and 40 mm. The low-speed streaks for the natural case can be clearly seen from Figure 4.30(a). With roughness control, the streak is induced downstream of each roughness (Figure 4.30(b)). Thus, more low-speed streaks are observed than in the natural case. However, the oscillation amplitude of the streaks for the control case is significantly smaller than that of the natural case.

The probability density function (PDF) was calculated to determine the mean streak spanwise spacing at typical streamwise and wall-normal positions from 20 000 snapshots of PIV measurement. Figure 4.31 shows the variation in the streamwise position for the natural and roughness control cases. For the natural case, the spacing increases with the wall normal height, however it has nearly no change with the streamwise location below the log layer. The roughness control decreases the spacing of the streak, though the spacing also gradually increases with the streamwise position to reach the value of the natural case. The control effects also seem to become more significant at higher wall-normal locations, and the most obvious effects occur at $y^+ = 44$.

The distribution of RMS streamwise velocity fluctuation for all cases is shown in Figure 4.32. For the control case of $\Delta z = 20$ mm, the streamwise velocity fluctuation can be reduced up to 25% below $y^+ = 12$ in the vicinity of the roughness elements, while

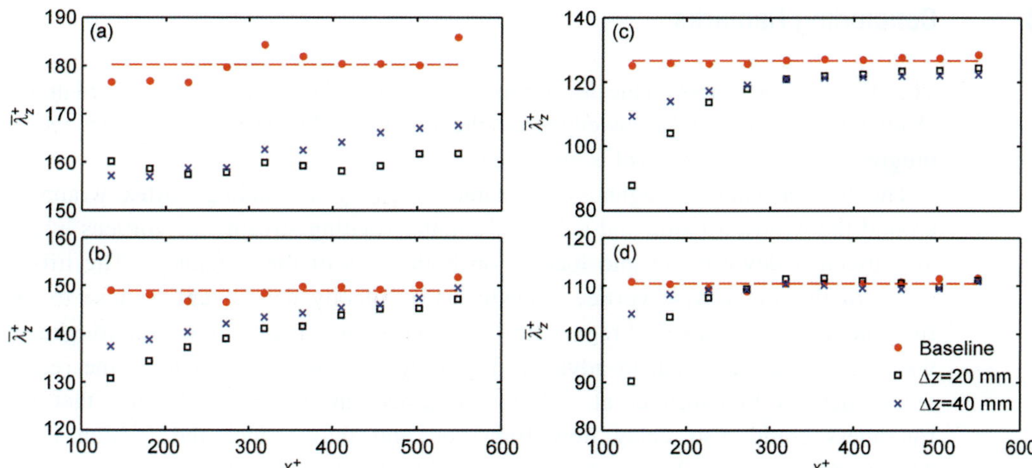

Figure 4.31 Variation of mean streak spacing along the streamwise direction for the natural and control cases. (a) $y^+ = 44$; (b) $y^+ = 32$; (c) $y^+ = 20$; (d) $y^+ = 12$ (Zhang et al. 2015). Reproduced with permission, copyright © Science China Press and Springer-Verlag Berlin Heidelberg 2015.

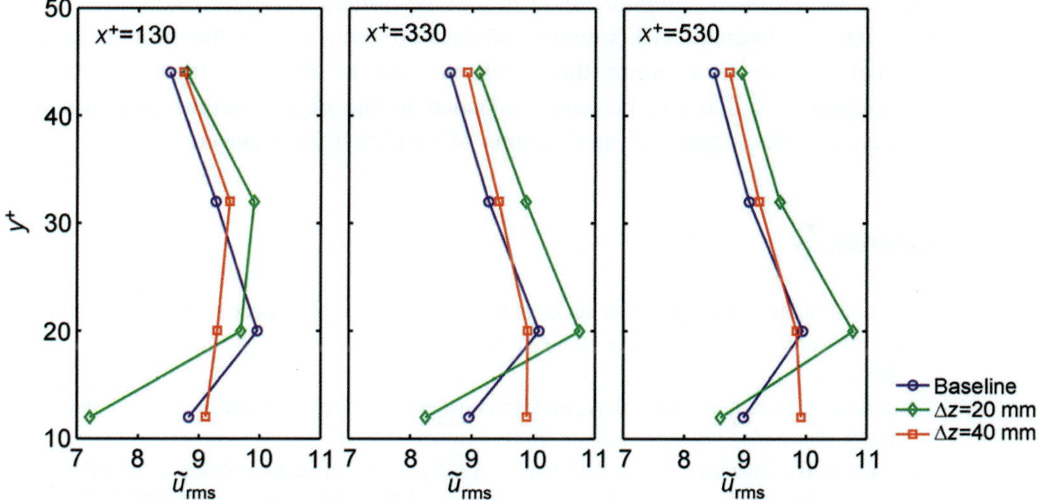

Figure 4.32 Distribution of RMS streamwise velocity fluctuation for the natural and control cases. The unit is mm/s (Zhang et al. 2015). Reproduced with permission, copyright © Science China Press and Springer-Verlag Berlin Heidelberg 2015.

roughness with spacing of $\Delta z = 40$ mm may increase it at the same wall normal position. However, for the control case of $\Delta z = 20$ mm, the reduction gradually diminishes along the streamwise and wall-normal directions. On the other hand, the roughness control may increase the velocity fluctuation for $y^+ \geq 20$. Thus, Zhang et al. (2015) suggested that $y^+ = 20$ might be the interface between different effects.

4.4   Concluding Remarks

The effects of roughness on the development of the boundary layer are introduced in this chapter. Some traditional knowledge from the literature is first presented and then recent progress on roughness control is presented.

The flow around a roughness is characterized by a standing vortex wrapping around the windward side and a pair of counter-rotating streamwise vortices which are stretched downstream and located on both sides of the roughness. The lift-up effect of the streamwise vortices transforms a velocity deficit region close to the roughness into a high-speed region further downstream. Traditionally, the roughness was used as an approach to advance flow transition, however some of the recent work such as Fransson et al. (2006) suggested an interesting finding that the roughness could also delay flow transition and thus reduce the friction drag. The roadmap of the bypass transition could also be altered by the roughness, which is largely influenced by the depth of incision on the spanwise vortex. The roughness also influences the coherent structures in the turbulent boundary layer. Another interesting work was conducted by Frohnapfel et al. (2007), who obtained significant turbulent drag reduction by up to 25% at around $h^+ = 0.8$, though the traditional literature showed that the roughness height below the viscous sublayer was hydraulically smooth and had no influence on the friction drag in channel or pipe flow. Since the roughness can be made to form very simple configurations and it can be easily attached to the object surface, such findings suggest significant potential applications of roughness in engineering.

References

Acarlar, M. S. and Smith, C. R. A study of hairpin vortices in a laminar boundary layer. Part 1. Hairpin vortices generated by a hemisphere protuberance. *Journal of Fluid Mechanics*, 1987, 175: 1–41

Andersson, P., Brandt, L., Bottaro, A., and Henningson, D. S. On the breakdown of boundary layer streaks. *Journal of Fluid Mechanics*, 2001, 428: 29–60

Frohnapfel, B., Jovanovic, J., and Delgado, A. Experimental investigations of turbulent drag reduction by surface-embedded grooves. *Journal of Fluid Mechanics*, 2007, 590: 107–116

Chwang, A. L. and Dong, Z. N. *Viscous Fluid Mechanics* (Second Edition). Tsinghua University Press, 2011 (in Chinese)

Fransson, J. H. M., Talamelli, A., Brandt, L., and Cossu, C. Delaying transition to turbulence by a passive mechanism. *Physical Review Letters*, 2006, 96(6): 064501

Jiménez, J. Turbulent flows over rough walls. *Annual Review of Fluid Mechanics*, 2004, 36: 173–196

Hicks, R. M. and Harper, W. R. Jr. A comparison of spherical and triangular boundary-layer trips on a flat plate at supersonic speeds. Technical Memorandum, 1970, NASA-TM-X-2146

Hosseini, S. M., Tempelmann, D., Hanifi, A., and Henningson, D. S. Stabilization of a swept-wing boundary layer by distributed roughness elements. *Journal of Fluid Mechanics*, 2013, 718: R1

Kline, S. J., Reynolds, W. C., Schraub, F. A., and Runstadler, P. W. The structure of turbulent boundary layers. *Journal of Fluid Mechanics*, 1967, 30(4): 741–773

Kundu, P. K., Cohen, I. M., and Dowling, D. R. *Fluid Mechanics* (Fifth Edition). Elsevier, 2013

Nikuradse, J. Strömungsgesetze in rauhen Rohren. VDI Forschumgsheft 361, 1933

Pan, C., Wang, J. J., Zhang, P. F., and Feng, L. H. Coherent structures in bypass transition induced by a cylinder wake. *Journal of Fluid Mechanics*, 2008, 603: 367–389

Plogmann, B., Würz, W., and Krämer, E. On the disturbance evolution downstream of a cylindrical roughness element. *Journal of Fluid Mechanics*, 2014, 758: 238–286

Reibert, M. S., Saric, W. S., Jr., Carrillo, R. B., and Chapman, K. L. Experiments in nonlinear saturation of stationary crossflow vortices in a swept-wing boundary layer. AIAA Paper 1996–0184

Saric, W. S., Carpenter, A. L., and Reed, H. L. Passive control of transition in three-dimensional boundary layers, with emphasis on discrete roughness elements. *Philosophical Transactions of The Royal Society A-Mathematical Physical and Engineering Sciences*, 2011, 369(1940): 1352–1364

Saric, W. S., Jr., Carrillo, R. B., and Reibert, M. S. Leading-edge roughness as a transition control mechanism. AIAA Paper 1998–0781

Schlichting, H. and Gersten, K. *Boundary Layer Theory*. Springer-Verlag, 2000

Schneider, S. P. Effects of roughness on hypersonic boundary-layer transition. *Journal of Spacecraft and Rockets*, 2008, 45(2): 193–209

Wang, J. J., Zhang, C., and Pan, C. Effects of roughness elements on bypass transition induced by a circular cylinder wake. *Journal of Visualization*, 2011, 14(1): 53–61

Zhang, C., Pan, C., and Wang, J. J. Evolution of vortex structure in boundary layer transition induced by roughness elements. *Experiments of Fluids*, 2011, 51(5): 1343–1352

Zhang, X., Pan, C., Shen, J. Q., and Wang, J. J. Effect of surface roughness element on near wall turbulence with zero-pressure gradient. *Science China-Physics Mechanics & Astronomy*, 2015, 58(6): 064702

Zhou, Y. and Wang, Z. J. Effects of surface roughness on separated and transitional flows over a wing. *AIAA Journal*, 2012, 50(3): 593–609

5 Polymer

5.1 Background

Polymer is a traditional passive control technique for turbulence drag reduction. It was found that minute concentrations of polymers could reduce the drag in turbulent flows by up to 80% (Procaccia et al. 2008). Thus, the study of drag reduction by polymer is important for the potential applications in engineering. For example, the drag reduction could result in a dramatic reduction of energy consumption in oil transportation. In heating or cooling systems, it could also significantly reduce the loss of heat transfer.

In particular, polymer shows some different characteristics compared with other techniques of drag reduction. The drag reduction with polymer additives is observed to be non-effective in a laminar pipe flow. Other characteristics of polymer involved in drag reduction are a linear molecular structure, extensibility, solubility, and viscoelasticity (White and Mungal 2008). It was generally thought that Toms (1948) first discovered the effect of drag reduction from polymers. After that, numerous studies of drag reduction of pipe and channel flows using polymer have been conducted. In early studies, Virk et al. (1967) found an asymmetric value for the maximum drag reduction (MDR) in a pipe flow, which was independent of polymers and pipe diameter. With advances in experimental techniques and numerical simulations, we can gain more detailed insights into the drag reduction of polymer.

In this chapter, we will mainly introduce the applications of polymer for drag reduction in pipe and channel turbulent flow, and then the variations in the coherent structures will also be discussed to reveal the control mechanism.

5.2 Polymer for Pipe Flow

5.2.1 Main Parameters

For a fully developed turbulent flow through a straight pipe with diameter d (radius of $r = 0.5d$), the mean shear stress τ_w at the wall can be determined based on the pressure drop ΔP over a distance Δx,

$$\tau_w = \frac{d}{4}\frac{\Delta P}{\Delta x}. \tag{5.1}$$

The wall shear stress is usually expressed in terms of the friction factor f defined as

$$f = \frac{\tau_w}{0.5\rho U_m^2}, \tag{5.2}$$

where U_m is the mean velocity in the pipe and ρ is the fluid density.

The expressions of f for laminar and fully turbulent pipe flow of a Newtonian fluid are respectively given as

$$f = \frac{64}{\mathrm{Re}}, \tag{5.3}$$

$$\frac{1}{\sqrt{f}} = 2.0\lg\mathrm{Re}\sqrt{f} - 0.8. \tag{5.4}$$

where $Re = \rho U_m d/\mu$ and μ is the viscosity of the fluid.

For a fluid with polymer additives, the viscosity is usually dependent on the shear rate, so the usual definition of the Reynolds number could not be used. Thus, the Reynolds number Re_w could be calculated based on the viscosity μ_w at the pipe wall (Ptasinski et al. 2001), namely

$$\mu_w = \frac{\tau_w}{\gamma_w}, \tag{5.5}$$

where γ_w is the local shear rate at the wall. Then the wall Reynolds number is defined as

$$\mathrm{Re}_w = \frac{\rho U_m d}{\mu_w}. \tag{5.6}$$

Following Equation (5.1), the drag reduction defined as the reduction percentage of pressure drop due to the addition of polymers could be calculated by

$$DR = \frac{(\Delta P/\Delta x)_0 - (\Delta P/\Delta x)}{(\Delta P/\Delta x)_0} \times 100\%, \tag{5.7}$$

where the subscript "0" denotes the baseline Newtonian fluid.

5.2.2   Velocity Statistics

One traditional application for polymer is to control the pipe flow, as it has been recognized that up to 100% of the total pipe drag is coming from the friction drag. One typical work was conducted by Ptasinski et al. (2001), and the experimental setup is shown in Figure 5.1. The main part of the setup consisted of a smooth straight pipe with a length of 34 m and an inner diameter of $d = 40.37$ mm. The fluid was pumped from an open reservoir by a disk pump. A ring was used just after the entrance of the pipe to force the flow to become turbulence.

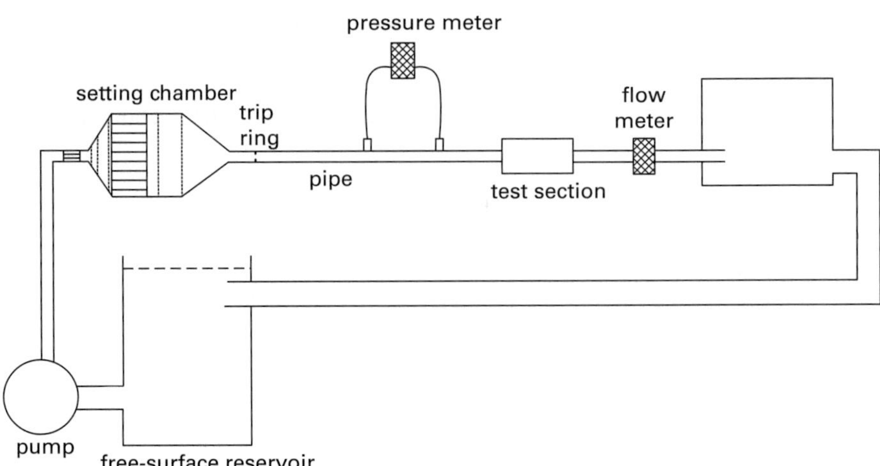

Figure 5.1 Schematic of the experimental setup for drag reduction control of pipe flow by using polymer additives (Ptasinski et al. 2001) Reproduced with permission, copyright © 2001 Kluwer Academic Publishers.

The flow returned from the exit of the pipe through a return pipe to the reservoir. The test section was 26 m after the entrance of the pipe, and the flow rate through the pipe was measured with a flow meter just downstream of the test section. The velocity was measured with the two-component LDV system. The pressure drop over a segment of the pipe was measured with a membrane differential pressure transducer. The segment was 16 m downstream of the pipe entrance and had a length of 8 m.

The polymer solutions consisted of Superfloc A110, namely a hydrolyzed polyacrylamide. The molecular weight of this polymer which consisted of approximately 10^5 monomers, was 6–8×10^6 g/mol. The polymer solutions were mixed slowly in the free surface reservoir until it became homogeneous. The weights of the polymer were chosen as 20, 103, 175, and 435 wppm (weight parts per million) Superfloc A110.

The profiles of the mean streamwise velocity and RMS velocity fluctuations are shown in Figure 5.2 for the basic Newtonian flow and polymer control flow. It is indicated that the mean streamwise velocity of the polymer control cases is almost the same as the baseline case in the viscous sublayer, while that in the logarithmic region is shifted upwards with polymer additives, and also increases with the polymer concentration (Figure 5.2(a)). The slope of the logarithmic profile for 20 wppm polymer additives is very similar to that of the baseline case, while that for other control cases becomes larger. It is indicated that the streamwise RMS velocity fluctuation with polymer is reduced near the wall, while the value is increased in the logarithmic region (Figure 5.2(b)). On the other hand, the radial RMS velocity fluctuations are suppressed as a result of polymer addition for all regions in the pipe except in the viscous sublayer (Figure 5.2(c)).

The drag reduction could be calculated based on the mean streamwise velocity profiles, and one conclusion is summarized by Yang and Dou (2010) for both smooth

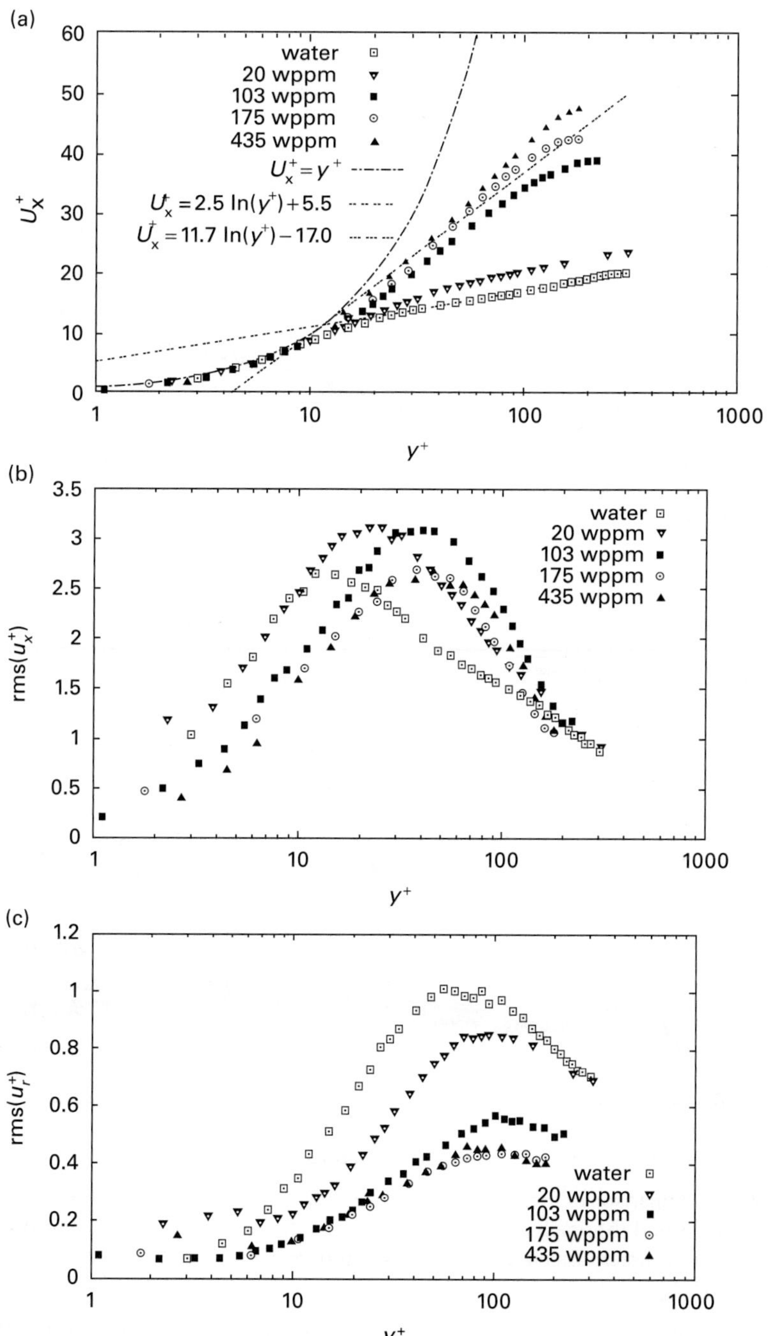

Figure 5.2 Profiles of mean streamwise velocity (a), streamwise RMS velocity fluctuation (b) and radial RMS velocity fluctuation (c) for the basic Newtonian flow and polymer control flow (Ptasinski et al. 2001). Reproduced with permission, copyright © 2001 Kluwer Academic Publishers.

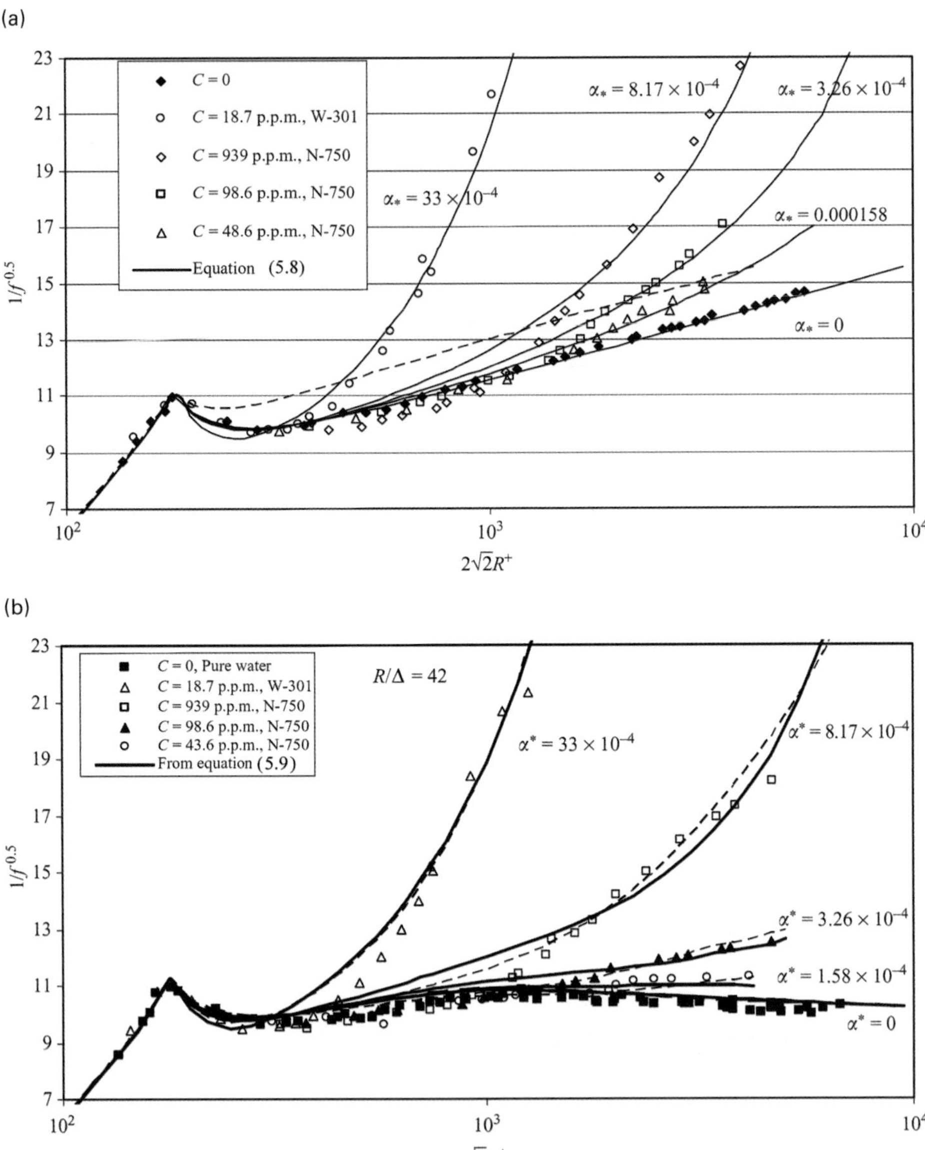

Figure 5.3 Comparison of friction factor for the basic Newtonian flow and polymer control flow. (a) Smooth pipe; and (b) rough pipe with relative smoothness $r/k_s = 42$ (Yang and Dou 2010). Reproduced with permission, copyright © Cambridge University Press 2010.

and rough pipe flows. Figure 5.3 shows the friction factor for the basic Newtonian flow and polymer control flow. Here, the Reynolds number R^+ is defined as $R^+ = u_w r/v$. Both the Reynolds number and the polymer concentration determine the friction factor. For the smooth pipe flow, there is a formula which represents such a relationship, namely

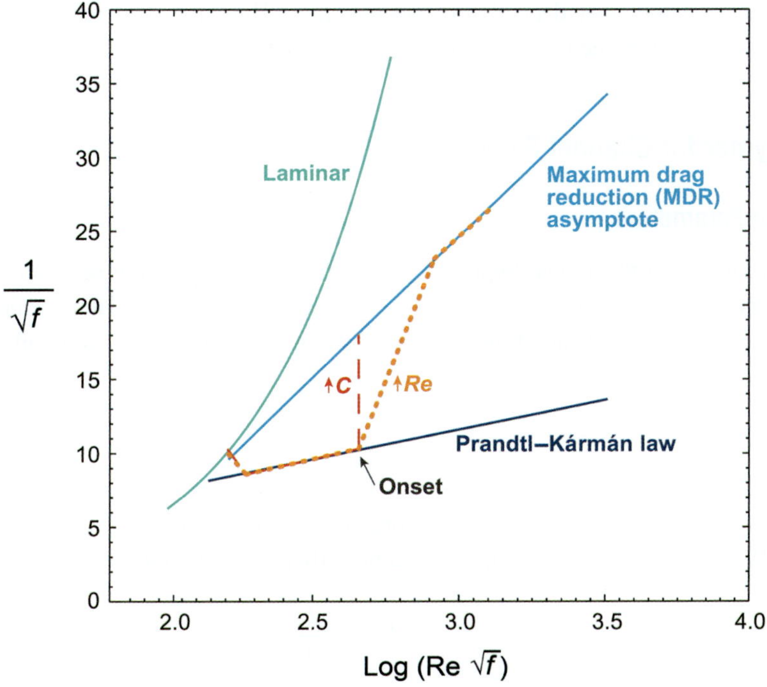

Figure 5.4 Schematic of friction factor with polymer injection (White and Mungal 2008). Reproduced with permission of Annual Review of Fluid Mechanics, Volume © by Annual Reviews.

$$\frac{U_o}{u_\tau} = \frac{1-p_t}{4}\frac{R^+}{D_f} + p_t\left(2.5\ln\frac{R^+}{D_f} - 66.69(\frac{R^+}{D_f^{3.5}})^{-0.72} + 5.8D_f^2 - 4\right), \qquad (5.8)$$

where U_o is the overall averaged velocity, p_t is the probability of turbulent occurrence, $D_f = 1 + \alpha_f p_t^2\, R^+$ is the drag reduction parameter, and α_f is an elastic factor depending on the polymer type and concentration. For the rough pipe flow, the friction factor can be obtained by

$$\frac{U_o}{u_\tau} = \frac{1-p_t}{4}\frac{R^+}{D_f} + p_t\left(2.5\ln\frac{R^+}{D_f} - 66.69(\frac{R^+}{D_f^{3.5}})^{-0.72} + 5.8D_f^2 - 4 - C\right), \quad (5.9)$$

where C is an integral constant based on boundary condition.

Figure 5.3 only gives an example for the change of friction factor induced by the polymer additives. The schematic of variations of the friction factor with Reynolds number and the polymer concentration is further proposed by White and Mungal (2008), as shown in Figure 5.4. The onset of drag reduction is determined as the point of departure from the Prandtl–Kármán law. For the same Reynolds number, the drag reduction initially increases with polymer concentration but saturates beyond a certain value. The borderline on the drag reduction is the maximum drag reduction asymptote. Similarly, for the same polymer concentration, the drag reduction initially increases with the Reynolds number along a unique trajectory that depends on concentration. Thus, the drag reduction of polymer additives is determined by both the Reynolds number and the polymer concentration.

5.3   Polymer for Channel Flow

5.3.1   Main Parameters

There has been theoretical hypothesis about the onset of the drag reduction by polymer additives, as has been introduced by Min et al. (2003). It is suggested that the drag reduction occurs when the relaxation time is longer than the timescale of the near-wall turbulence,

$$\lambda > \frac{\nu}{u_\tau^2}, \tag{5.10}$$

where λ is the relaxation time.

On the other hand, there is another prediction of the onset of drag reduction based on the Weissenberg number by applying a perturbation method to viscoelastic models,

$$We = \frac{\lambda U}{\delta}, \tag{5.11}$$

where U is the centerline velocity of the fully developed flow and δ is the channel half-height $0.5h$. The drag reduction occurs when

$$We_\tau = \frac{\lambda u_\tau^2}{\nu} > \alpha, \tag{5.12}$$

where We_τ is the Weissenberg number normalized by u_τ and ν, and α depends on the viscoelastic model.

5.3.2   Velocity Statistics

One work of control of turbulent channel flow by polymer additives was conducted by Min et al. (2003) with direct numerical simulation. The calculation domain was $7\delta \times 2\delta \times 3.5\delta$ in the streamwise (x), wall-normal (y), and spanwise (z) directions, respectively. The Reynolds number was $Re = U\delta/\nu = 3000$.

Time history of the mean pressure gradient normalized by that of Newtonian flow is shown in Figure 5.5, where the drag reduction by polymer additives could also be reflected. It is indicated that the pressure is nearly the same as the baseline case for $We \leq 1$. But drag reduction occurs for $We > 1$, with a mean drag reduction by about 12%, 20%, and 27% achieved for $We = 2, 3$, and 4, respectively. It is indicated that there exists a threshold for the Weissenberg number to obtain drag reduction. Min et al. (2003) suggested that the onset Weissenberg number We_τ was about 6 based on the experimental data. However, the value might differ for different situations. For example, Berman (1977) suggested the onset of drag reduction for We_τ was ranged from 1 to 8 depending on the properties of polymers and solvents.

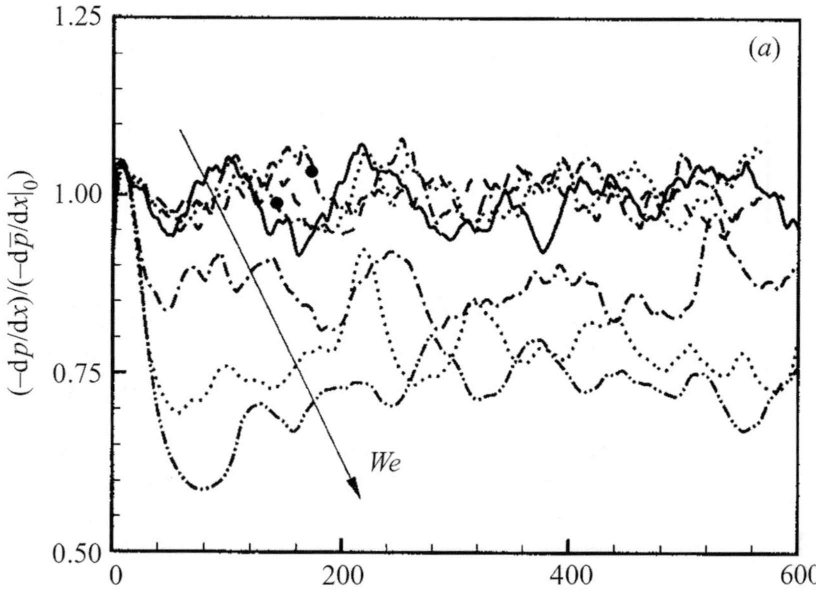

Figure 5.5 Time history of the mean pressure gradient normalized by that of Newtonian fluid flow to show the drag reduction at $Re = 3000$. ——, Newtonian; ·········, ●, $We = 0.1$; $- \cdot - \cdot -$, ●, $We = 0.5$; $- - -$, $We = 1$; $- \cdot - \cdot -$, $We = 2$; ·········, $We = 3$; $- \cdot \cdot - \cdot \cdot -$, $We = 4$ (Min et al. 2003). Reproduced with permission, copyright © Cambridge University Press 2003.

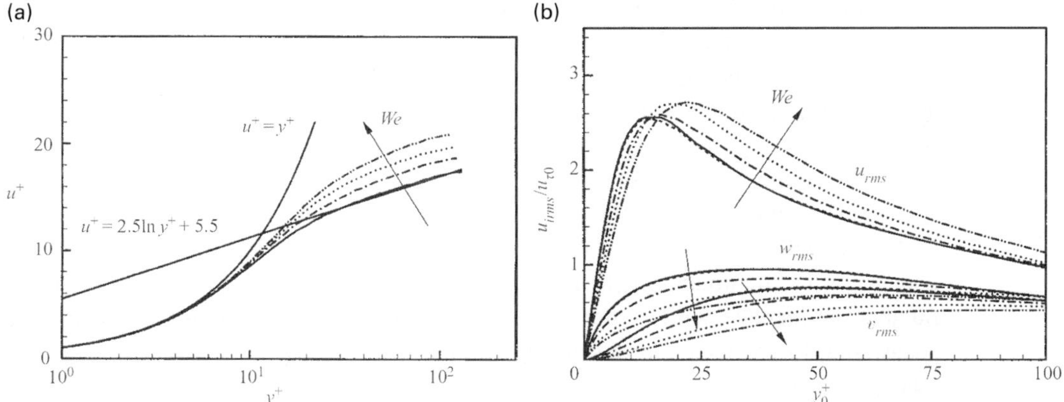

Figure 5.6 (a) Mean streamwise velocity normalized by the wall-shear velocity u_τ. (b) RMS velocity fluctuation normalized by the wall-shear velocity $u_{\tau 0}$ for Newtonian flow, where $y_0^+ = yu_{\tau 0}/\nu$. Lines and symbols are the same as those in Figure 5.5 (Min et al. 2003). Reproduced with permission, copyright © Cambridge University Press 2003.

The profiles of mean streamwise velocity are shown in Figure 5.6. It is indicated that the mean velocity for $We \leq 1$ agrees well with those for the Newtonian flow without polymer additives. However, the drag reduction cases for $We > 1$ show regular variations. An upward shift of the mean streamwise velocity in the logarithmic region is

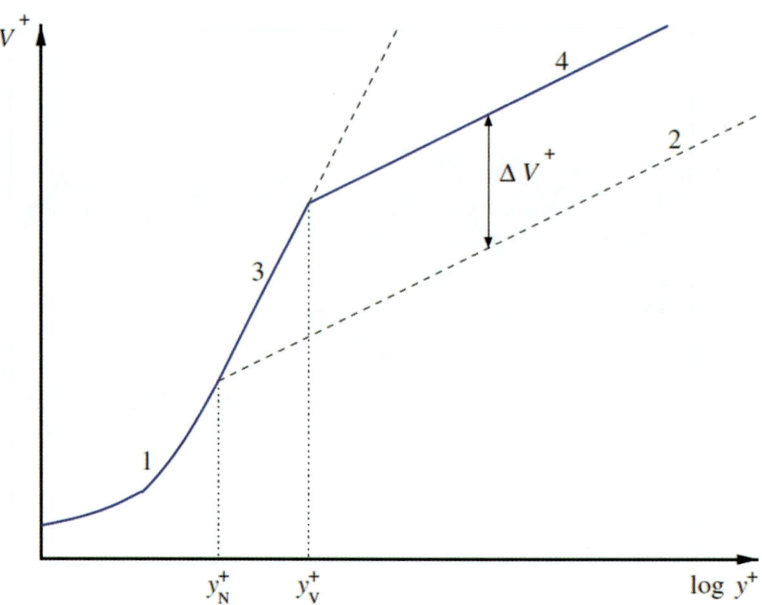

Figure 5.7 Schematic of mean velocity profile. Region 1, viscous sublayer. Region 2, logarithmic layer for the turbulent Newtonian flow. Region 3, maximum drag reduction asymptotic profile in the viscoelastic flow. Region 4, Newtonian plug in the viscoelastic flow (Procaccia et al. 2008). Reproduced with permission, copyright © 2008 The American Physical Society.

observed. Meanwhile, the viscous sublayer thickness is also increased. This is a common observation for the flow characteristics of polymer drag reduction, where a schematic is further shown in Figure 5.7. Three different velocity regions exist, namely viscous sublayer, logarithmic layer, and maximum drag reduction asymptotic region. The velocity in the logarithmic layer of the drag reduction case follows a log law with the Newtonian slope, but with some velocity increment.

The distributions of the RMS velocity fluctuations are shown in Figure 5.6(b). Variations with the Newtonian flow case occur for $We > 1$. In the near wall region about $y_0^+ < 15$, the streamwise RMS velocity fluctuations decrease with the Weissenberg number but increase with it afterwards. On the other hand, the RMS wall-normal and spanwise velocity fluctuations decrease with the Weissenberg number for $We > 1$ in the whole channel.

One can also compare the variations of the velocity fluctuations between the baseline and polymer control cases from the scatter plot shown in Figure 5.7. For the basic case (Figure 5.8(a)), the scatter plot shows a symmetrical elliptic shape with the major axis inclined to the coordinate axes at a certain angle. There are more scatter points in the second and fourth quadrants than in the first and third. It is suggested that the Reynolds shear stress is mainly produced by ejection (second quadrant event) and sweep (fourth quadrant event). For the polymer control case (Figure 5.8(b)), the distribution region becomes smaller and narrower than that for the baseline case, with the major axis nearly parallel to the abscissa. The absolute value of the velocity fluctuation also decreases in

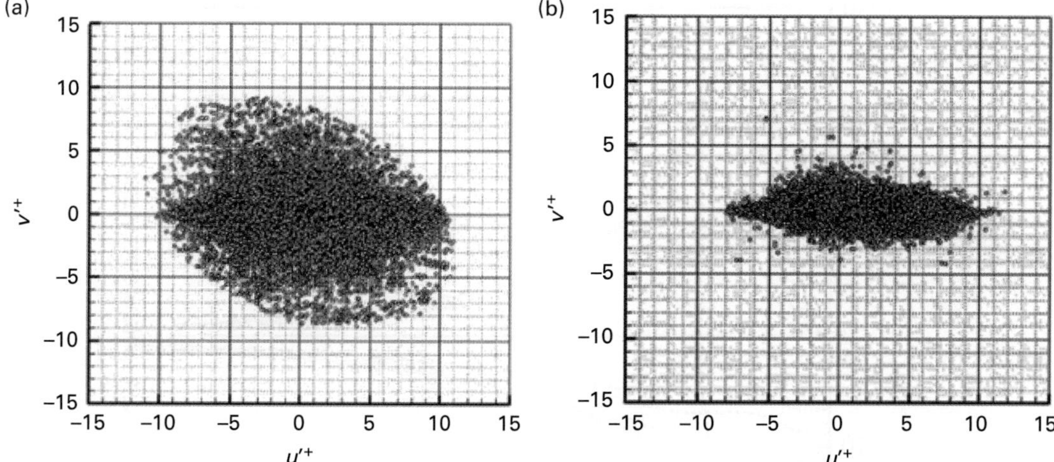

Figure 5.8 Scatter plot of streamwise and wall-normal velocity fluctuations. (a) Basic Newtonian flow at $y^+ = 20$; and (b) flow with polymer additives at $y^+ = 22$ (Guan et al. 2013). Reproduced with permission, copyright © Springer Nature 2013.

the second and fourth quadrants, indicating that the ejection and sweep events are weakened by polymer additives.

The production P_k and dissipation ε_k of the turbulent kinetic energy could also be calculated based on the statistic parameters in the flow field, namely

$$P_k = -\overline{u'v'}\frac{d\overline{u}}{dy}, \tag{5.13}$$

$$\varepsilon_k = \frac{1}{\mathrm{Re}}\overline{\frac{\partial u'_i}{\partial x_j}\frac{\partial u'_i}{\partial x_j}}. \tag{5.14}$$

The results are shown in Figure 5.9 with different Weissenberg numbers. It is indicated that the production and dissipation of the turbulent kinetic energy decrease with the polymer additives for $We > 1$ throughout the channel. In general, the variations of statistic parameters shown in Figure 5.5 to Figure 5.9 reflect the typical flow features associated with drag reduction by polymer additives.

5.4 Polymer for Coherent Structures

The variations of the coherent structures in the turbulent boundary layer could be used to clearly interpret the drag reduction with polymers. One typical snapshot of the near-wall vortical structures for the basic Newtonian flow and 60% drag reduction case is shown in Figure 5.10. It is indicated that the polymer additives alter the nature and strength of the vortical structures, resulting in a significant modification of the near-wall structures of the boundary layer. It is considered to be coupled with the variations in the flow

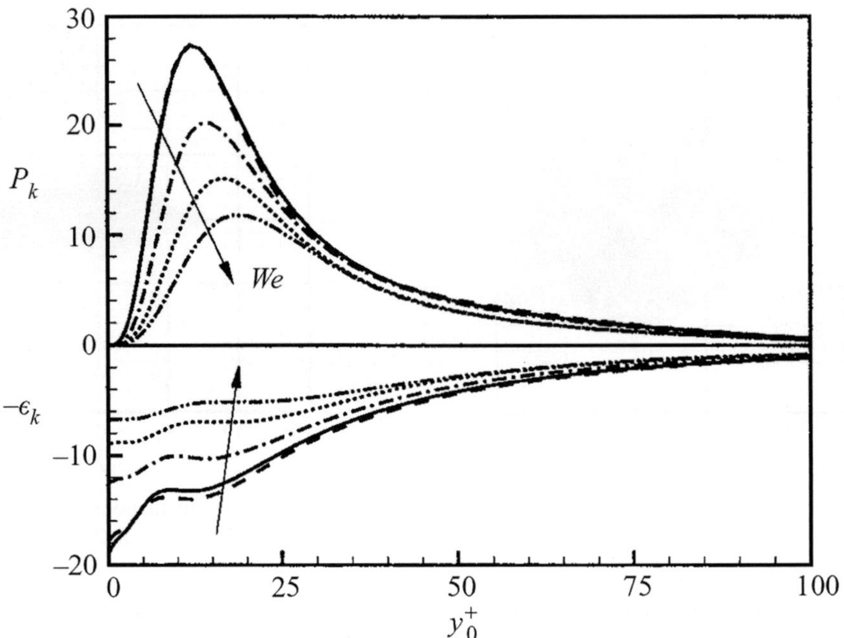

Figure 5.9 Profiles of the production P_k and dissipation ε_k of the turbulent kinetic energy normalized by u_{e0}^3/δ (Min et al. 2003). Reproduced with permission, copyright © Cambridge University Press 2003.

statistics, thus the mean velocity and its fluctuations are also modified and the shear in the boundary layer is redistributed.

Another typical snapshot of the flow field for the high drag reduction case is shown in Figure 5.11. The vortices are identified by the second invariant of the velocity gradient tensor Q. The polymer stretch C_{kk} is also exhibited which is related to the viscoelastic extra-stress tensor τ_{ij} by,

$$\tau_{ij} = \frac{1}{1 - C_{KK}/L^2} C_{ij} - \delta_{ij}, \tag{5.15}$$

where C_{ij} is the conformation tensor defined as the ensemble averaged value of the end-to-end distance of the polymer chains, L is the maximum admissible polymer chain extension, and δ_{ij} is the Kronecker delta. It is indicated that the development of polymer stretch and streamwise vortices is coupled together. Numerous vortical structures are observed close to the inflow region, followed by a high polymer stretch region which produces large stresses that suppress the vortices. Thus, the vortices are weakened and become larger further downstream, which in turn reduces the polymer stretch.

Figure 5.12 further shows the evolution of drag reduction and the maximum polymer extension in the y–z plane along the streamwise direction. The maximum of the polymer extension is reached just after the inlet and then decreases gradually along the streamwise direction. On the other hand, the drag reduction increases with the streamwise

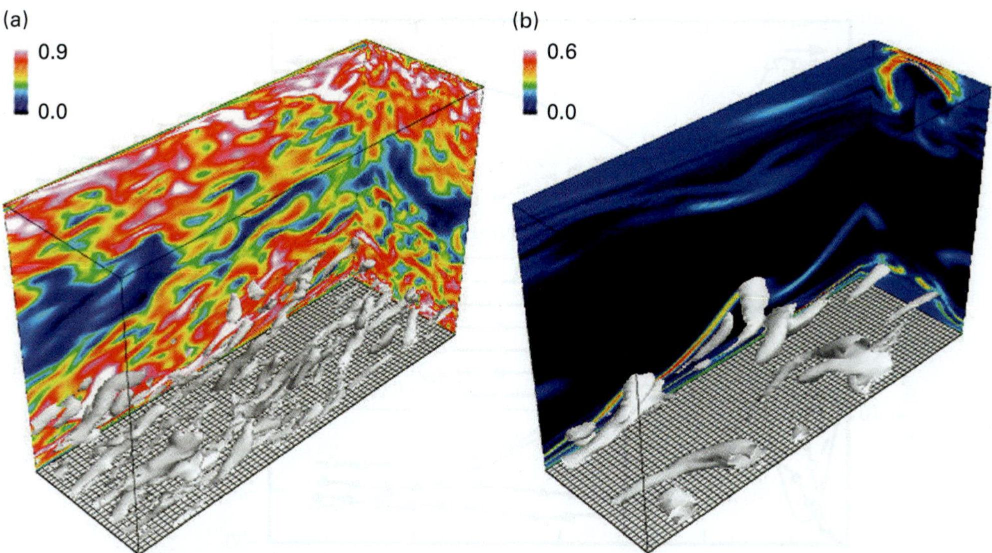

Figure 5.10 Instantaneous visualizations of near-wall vortex structures identified using isosurfaces of the positive second invariant of the velocity gradient tensor, high-speed velocity streaks, and low-speed velocity streaks for the basic Newtonian flow (a) and drag reduction of 60% (b). Flow is from bottom left to top right (White and Mungal 2008). Reproduced with permission of Annual Review of Fluid Mechanics, Volume © by Annual Reviews.

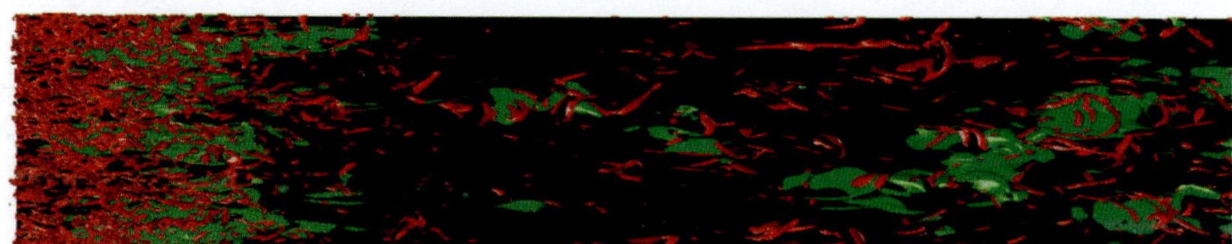

Figure 5.11 Isosurfaces of vortices ($Q = 0.043$, red) and of polymer stretch ($c_{kk} = 15\%L^2$, green) (Dimitropoulos et al. 2005). Reproduced with permission from AIP Publishing.

direction. It is indicated that the high drag reduction is followed by a high polymer extension effect. Thus, the drag reduction is also synchronous with the vortex evolution, both of which are asynchronous with the polymer stretch. In addition, the effect of the Weissenberg number is also reflected. Higher drag reduction and polymer stretch occur at higher Weissenberg numbers.

Similar variations are shown in Figure 5.13 for the evolution of the hairpin vortices with different Weissenberg numbers. The formation and evolution of hairpin vortices are shown in Figure 5.13(a) for the basic Newtonian flow, indicating that the secondary hairpin vortex is generated upstream of the primary hairpin. For low drag reduction at $We_\tau = 25$ (Figure 5.13 (b)), the evolution of hairpin vortices also resembles that of the Newtonian flow. However, the details of vortex formation are slightly different. For high drag reduction at $We_\tau = 100$ (Figure

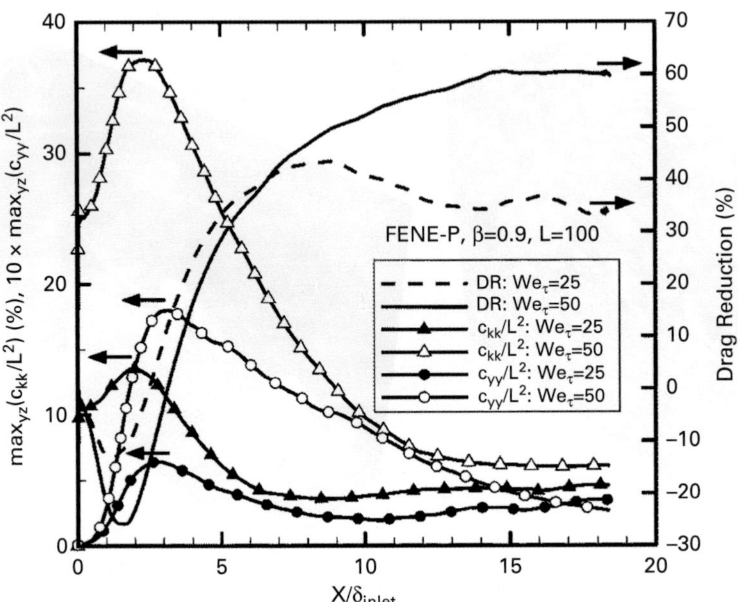

Figure 5.12 Evolution of drag reduction and *y-z* plane maximum polymer extension along the length of the boundary layer (Dimitropoulos et al. 2005). Reproduced with permission from AIP Publishing.

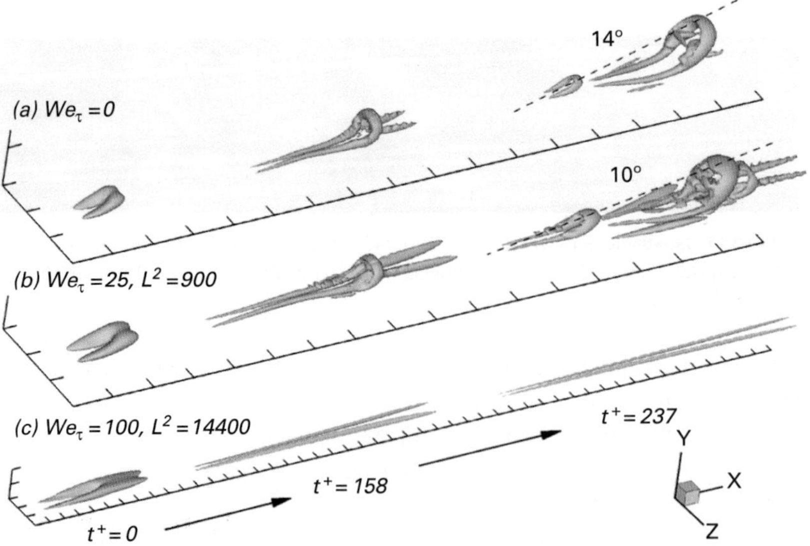

Figure 5.13 Evolution of vortical structure represented by isosurfaces of the vortex swirling strength λ_{ci}. (a) Newtonian flow; (b) $We_\tau = 25$; (c) $We_\tau = 100$ (Kim et al. 2008). Reprinted figure with permission, copyright © 2008 American Physical Society.

5.13(c)), the vortex evolution is completely different than the baseline case. The initial vortex is elongated in the streamwise direction. The swirling strength of vortical structures decreases with time, while there is no hairpin vortex formed in the further downstream region.

5.5   Concluding Remarks

Polymer is a passive but effective control technique for turbulent drag reduction. We mainly introduce the applications of polymer for drag reduction in pipe and channel turbulent flows in this chapter.

It is indicated that the Reynolds number and the polymer concentration are the two important parameters to determine the drag reduction of polymer additives, though there is a maximum drag reduction asymptote. For drag reduction cases, the mean velocity distribution in the viscous sublayer is nearly the same as the baseline Newtonian flow, while that at the logarithmic region is shifted upwards with polymer additives. Meanwhile, the velocity fluctuations, the production and dissipation of turbulent kinetic energy, may also be reduced. Statistical study of the coherent structures indicates that the drag reduction is usually accompanied with a modification of the near-wall structures. In particular, the coherent structures become weakened, which is beneficial for drag reduction in turbulent flow.

References

Berman, N. S. Flow time scales and drag reduction. *Physics of Fluids*, 1977, 20(10): S168–S174

Dimitropoulos, C. D., Dubief, Y., Shaqfeh, E. S. G., Moin, P., and Lele, S. K. Direct numerical simulation of polymer-induced drag reduction in turbulent boundary layer flow. *Physics of Fluids*, 2005, 17(1): 011705

Guan, X. L., Yao, S. Y., and Jiang, N. A study on coherent structures and drag-reduction in the wall turbulence with polymer additives by TRPIV. *Acta Mechanica Sinica*, 2013, 29(4): 485–493

Kim, K., Adrian, R. J., Balachandar, S., and Sureshkumar, R. Dynamics of hairpin vortices and polymer-induced turbulent drag reduction. *Physical Review Letters*, 2008, 100(13): 134504

Min, T., Yoo, J. Y., Choi, H. S., and Joseph, D. D. Drag reduction by polymer additives in a turbulent channel flow. *Journal of Fluid Mechanics*, 2003: 213–238

Procaccia, I., L'vov, V. S., and Benzi, R. Colloquium: Theory of drag reduction by polymers in wall-bounded turbulence. *Reviews of Modern Physics*, 2008, 80(1): 225–247

Ptasinski, P. K., Nieuwstadt, F. T., Den Brule, B. H. A. A., and Hulsen, M. A. Experiments in turbulent pipe flow with polymer additives at maximum drag reduction. *Flow Turbulence and Combustion*, 2001, 66(2): 159–182

Toms, B. A. Some observations on the flow of linear polymer solutions through straight tubes at large Reynolds numbers. *Proceeding of 1st International Congress on Rheology*, North Holland, 1948

Virk, P. S., Merrill, E. W., Mickley, H. S., Smith, K. A., and Mollochristensen, E. The Toms phenomenon: turbulent pipe flow of dilute polymer solutions. *Journal of Fluid Mechanics*, 1967, 30(2): 305–328

White, C. M. and Mungal, M. G. Mechanics and prediction of turbulent drag reduction with polymer additives. *Annual Review of Fluid Mechanics*, 2008, 40: 235–256

Yang, S. and Dou, G. Turbulent drag reduction with polymer additive in rough pipes. *Journal of Fluid Mechanics*, 2010: 279–294

6   Biological Techniques

6.1   Background

As described in the Introduction, there exist different kinds of passive flow control techniques. In general, we can further distinguish them into different groups based on their origin. One group is developed based on our understanding of flow physics, which is called the fundamental technique. For example, we already know that vortical structures may be induced by the instability of the separated shear layer for flow over an object with adverse pressure gradient, thus a small inclined plate placed in the crossflow could induce the streamwise vortex. The control idea of vortex generators is therefore proposed.

The other group may be derived from nature, from animals to plants. Through natural selection living organisms develop their own shapes and organs with specific features, which are expected to be optimal and adaptive for organisms in certain environments. Thus, we can learn from them to copy or mimic these features to improve the performance of man-made systems. Some examples are shown in Figure 6.1. Accordingly, different kinds of biomimetic flow control techniques have been proposed.

We can observe the raising of bird feathers during its landing (Figure 6.1(a)). Researchers have proved the effects of hairy coating on bluff bodies and airfoils, confirming the function of the raising of feathers during bird landing. This function (the raising of feathers) is thought to improve the aerodynamic performance of the bird when landing.

The leading edge of the flippers of humpback whales is not smooth, but has bumpy tubercles (Figure 6.1(b)). Experimental and numerical studies have suggested that the leading-edge tubercles are beneficial for aerodynamic performance at high angles of attack. A recent review on this field has been completed by Aftab et al. (2016).

From a close-view, we can see that sharkskin is comprised of small riblet structures (Figure 6.1(c)), which might be related to reducing the drag of the shark when swimming. Thus, the riblets have been used as a useful drag reduction technique in the past, such as in the so-called sharkskin swimsuit.

Similarly, riblets also occur on one kind of saguaro with typical diameter of the order of 0.5 m and height of 10 m (Figure 6.1(d)). It is suggested that the riblets are related to

Figure 6.1 Some specific features on organisms that might be related to high aerodynamic performance. (a) Self-adaptive hairy flaps of a falcon (Brücker and Weidner 2014), reproduced with permission, copyright © 2014 Elsevier Ltd. All rights reserved. (b) Tubercles on the leading edge of a humpback whale. (c) Riblet on sharkskin (Liu and Li 2012), reproduced with permission from Elsevier. (d) Riblets on a saguaro (Talley et al. 2001), reproduced with permission.

a suppression of vortex-induced-vibration, which is beneficial for the saguaro as it lives up to 150 years of age with high wind velocity in their natural habitat.

In this chapter, we will present some typical biomimetic flow control techniques, including their basic configurations, control effects, and mechanisms.

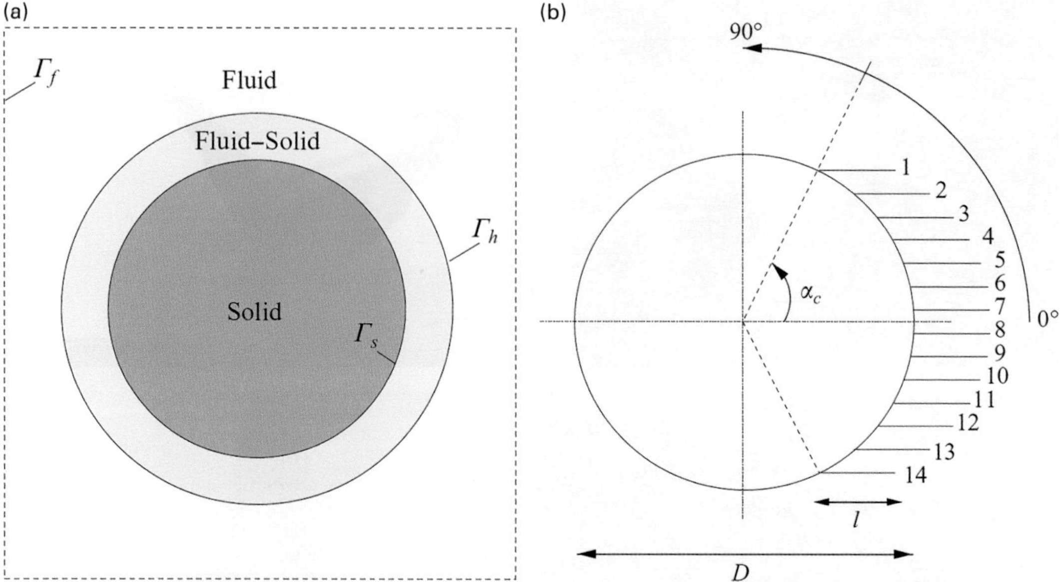

Figure 6.2 (a) The three zones of the computational circular cylinder for hairy coating; and (b) dominant parameters of the hairy coating (Favier et al. 2009). Reproduced with permission, copyright © Cambridge University Press 2009.

6.2   Hairy Coating

6.2.1   Bluff Body

Following the example of the bird feathers, self-adaptive hairy coating was tested to control the flow over a circular cylinder by Favier et al. (2009) at Reynolds number $Re = 200$. They solved the problem by direct solution of the Navier–Stokes equations together with a nonlinear set of equations describing the dynamics of the coating. The hairy coating had similar characteristics to the bird feathers, such as being porous, non-isotropic and compliant. Thus, the flow field was decomposed into three zones, including a solid body, a surrounding fluid and a mixed fluid–solid portion representing the hairy coating, as shown in Figure 6.2(a). The influence parameters can be considered as the inner density of the hairy layer, angular position of the control zone, and the hairy length, as shown in Figure 6.2(b).

The time histories of the drag and lift coefficients for the natural case and one typical control case are presented in Figure 6.3. For the control case, the drag and lift coefficients gradually reach the fully developed regime with time. Then, the mean drag coefficient is reduced by 11.5%, while the amplitude of the drag fluctuation is reduced by approximately 50%. In addition, there is a reduction of approximately 33% on the amplitude of the lift fluctuation. From the two force curves, we can also see the basic trend of the lift and drag coefficients. Since the aerodynamic forces are induced by the asymmetric shedding process of the Kármán vortex, the variation period of the lift

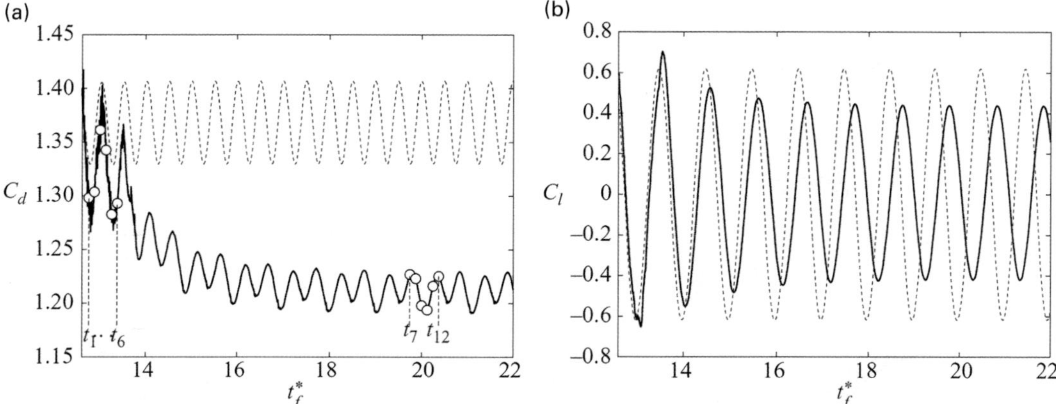

Figure 6.3 Aerodynamical coefficients without control (dashed lines) and with control (bold lines). (a) Drag coefficient; and (b) lift coefficient (Favier et al. 2009). Reproduced with permission, copyright © Cambridge University Press 2009.

coefficient is twice that of the drag coefficient. In general, the maximum drag coefficient occurs at the moment of the maximum or minimum lift coefficient, while the minimum drag coefficient corresponds to the time of the zero lift coefficient.

The control mechanism for the reduction in the drag and lift fluctuations with hairy coating can be analyzed by snapshots of vertical velocity contours and the force of hairy coating on the fluids, as shown in Figure 6.4. For the time corresponding to the maximum drag and lift coefficient, namely t_7, the fluids flow over the upper leeward surface of the cylinder, resulting in a suction there and thus a positive lift coefficient. It is indicated that the hairy coating induces positive forces mainly towards the upwards direction, thus the hairy coating will also be affected by the negative forces mainly towards the downwards direction according to Newton's first law. As a result, the lift coefficient will be decreased due to the opposing force exerted by the surrounding fluids. This phenomenon can also be clearly seen at other snapshots, which accounts for the reduction in both the drag and lift coefficients.

Favier et al. (2009) numerically studied the parameter influence on a two-dimensional circular cylinder, while Niu and Hu (2011) experimentally studied the influence on a disk. The results are summarized in Figure 6.5. In general, the experiment data show a more detailed variation for the influence of the hairy density and coating angle, though there are more numerical data than experimental data for the influence of the hairy length. From the experimental data, a disk with 0.041% density increases drag by approximately 15% in comparison with the natural case, while a disk with hair density larger than 0.081% leads to a drag reduction by approximately 5%–10%. With an increase in the coating angle, the drag coefficient increases firstly by about 95% and then decreases gradually to the value less than the natural case when the coating angle is larger than 19°. The drag coefficient firstly decreases with an increase in the hairy length, and reaches the minimum at about $1D$, and then it increases with the hairy length.

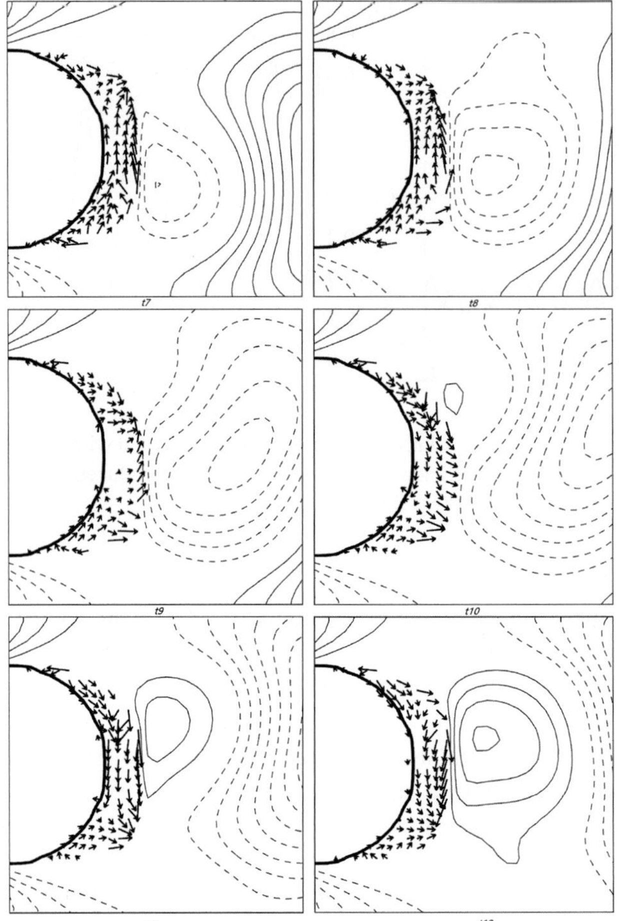

Figure 6.4 Force exerted by the hairy layer on the fluid at the six time instants t_7 to t_{12} shown in Figure 6.3 (Favier et al. 2009). Reproduced with permission, copyright © Cambridge University Press 2009.

6.2.2   Straight Wing

Hairy coating has also been used to control the flow over an airfoil by both experimental and numerical methods. Brücker and Weidner (2014) conducted an experimental study in a water tunnel. A NACA 0020 airfoil was used as the experimental model with chord length of $c = 0.2$ m and span of 0.5 m. For the control case, the flexible hairy flaps with a thickness of 2 mm were covered over the airfoil suction surface, as shown in Figure 6.6(a). The flaps had Young's modulus of 1.7 MPa and a density of 1.2 kg/m^3. The free-stream velocity was $U_\infty = 0.38$ m/s, corresponding to $Re = 7.7 \times 10^4$. The patterns of the hairy coating at a typical angle of attack of $\alpha = 17.5°$ are shown in Figure 6.6(b), where we can see that the hairy coating adapts to the flow.

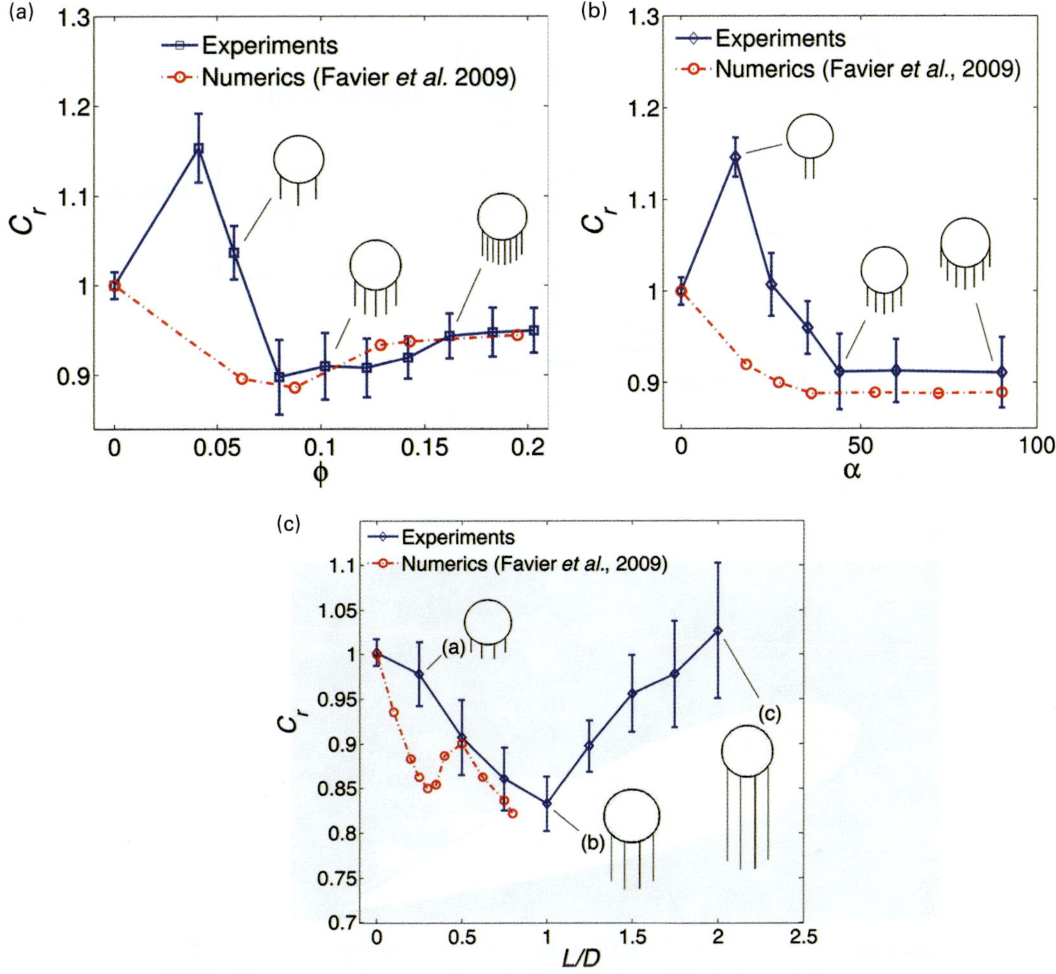

Figure 6.5 Variation of the relative drag coefficient with hairy density (a), coating area (b), and hairy length (c) (Niu and Hu 2011). Reproduced with permission from AIP Publishing.

The time-averaged streamwise velocity and instantaneous vortex development are shown in Figure 6.7. The flow over the airfoil has already separated, forming a large-scale separation region over the airfoil suction surface (Figure 6.7(a)). With hairy coating control, the large-scale recirculation region is eliminated, as shown in Figure 6.7(b), suggesting an increase in the lift coefficient. Such variation can also be found from the instantaneous vortex pattern. For the natural case, the shear layer separates from the leading edge of the airfoil, and develops into small-scale vortical structures shedding downstream, which exhibits irregular vortex distribution in the field of view (Figure 6.7(c)). For the control case, the separated vortex starts to form from approximately $x/c = 0.3$, and increases its strength as it convects downstream (Figure 6.7(d)). The vortex strength becomes larger than the natural case, resulting in a decrease in the separation region.

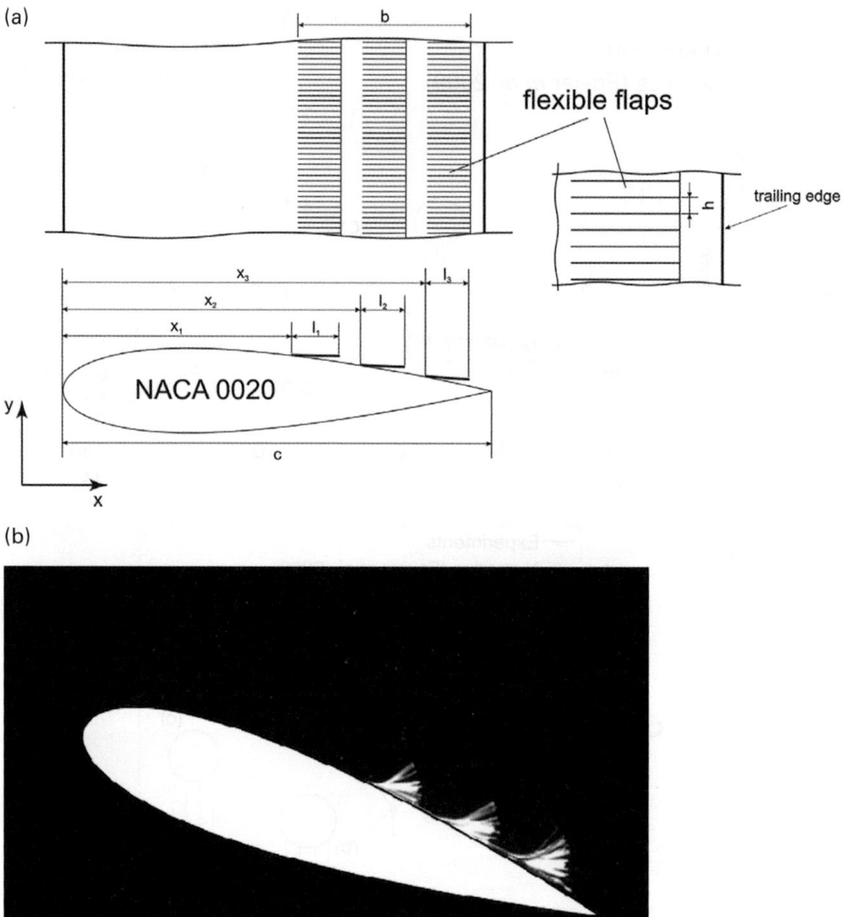

Figure 6.6 (a) Sketch of NACA 0020 airfoil with hairy coating; and (b) view for a hairy flap configuration at $\alpha = 17.5°$ (Brücker and Weidner 2014). Reproduced with permission, copyright © 2008 Elsevier Ltd. All rights reserved.

Hairy coating can also show significant control effects on the airfoil at ultra high angles of attack. One result is presented by Venkataraman and Bottaro (2012), as shown in Figure 6.8. For both natural and control cases, the drag and lift coefficients show periodic variation characteristics. At $\alpha = 22°$, the hairy coating may greatly increase the lift coefficient by shifting it upwards in comparison with the natural case (Figure 6.8(b)), though the drag coefficient is also slightly increased (Figure 6.8(a)). At $\alpha = 45°$ and 70°, the flow control can decrease the drag coefficient while increasing the lift fluctuation, as shown in Figure 6.8(c–f). Venkataraman et al. (2014) further indicated that hairy coating near the leading edge of the airfoil is better than that positioned near the trailing edge, and a larger covering region is also beneficial for a higher lift coefficient.

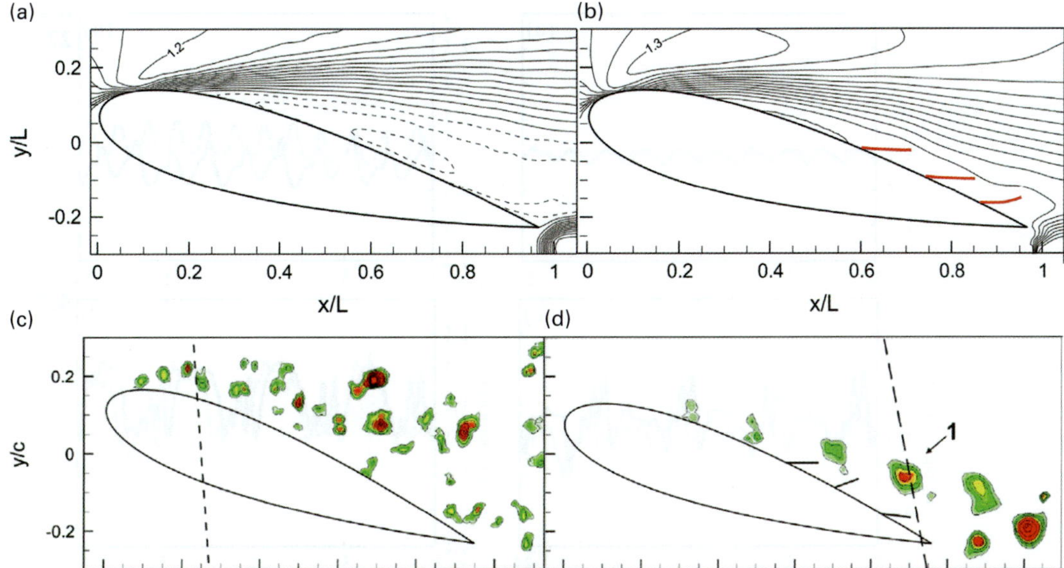

Figure 6.7 Mean streamwise velocity (a, b) and instantaneous vortical structures detected by Q-criterion (c,d) over the NACA 0020 airfoil at $\alpha = 17.5°$ without and with hairy coating (Brücker and Weidner 2014). Reproduced with permission, copyright © 2014 Elsevier Ltd. All rights reserved.

6.3   Leading-Edge Tubercles

6.3.1   Humpback Whale Flipper Model

Bushnell and Moore (1991) mentioned in a review paper that the leading-edge tubercles of the flipper on the humpback whale might be associated with drag-due-to-lift reduction. Fish and Battle (1995) compared the lift coefficient of different manufactured airfoils, and one airfoil was similar to the cross-section of the whale flipper. They suggested that the morphology and placement of leading-edge tubercles functioned as lift-enhancement devices to control the flow and maintain lift at high angles of attack. However, these studies did not provide direct evidence for the improved aerodynamic performance with leading-edge tubercles.

Miklosovic et al. (2004) conducted an innovative work in this field. They constructed two humpback whale flipper models based on the NACA 0020 airfoil section, one with smooth leading edge and the other with sinusoidal leading edge profiles approximating the pattern of the whale, as shown in Figure 6.9. The flipper model had a maximal chord of 16.19 cm and a span of 56.52 cm. Force measurement was conduced in a wind tunnel at around $Re = 5.05 \times 10^5$ to 5.20×10^5 based on the mean chord.

Figure 6.10 compares the aerodynamic forces between the smooth and scalloped models. It is shown from Figure 6.10(a) that the maximum lift coefficient is increased by approximately 6% for the scalloped leading edge, in comparison with the smooth one. The stall angle is also delayed from the original $\alpha = 12°$ to $\alpha = 16.3°$ with an increment of

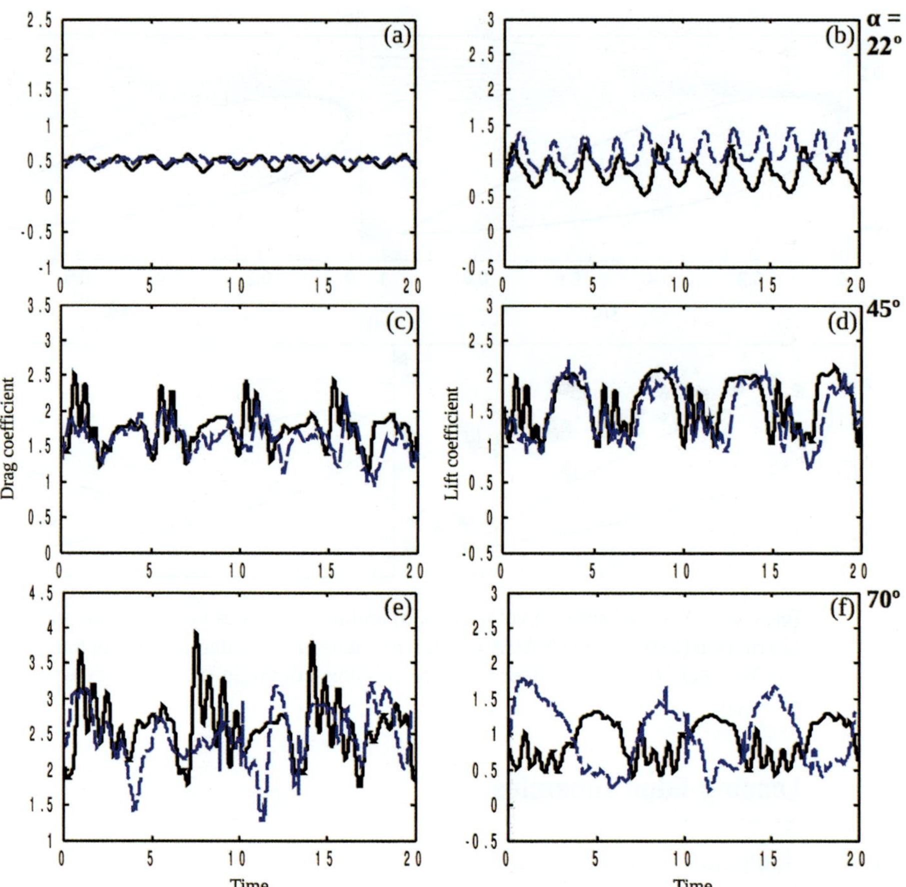

Figure 6.8 Time evolution of the drag (left) and lift (right) coefficients for a NACA 0012 airfoil without (black solid lines) and with (blue dashed lines) hairy coating at $\alpha = 22°$ (top), 45° (middle), and 70° (bottom) (Venkataraman and Bottaro 2012). Reproduced with permission from AIP Publishing.

approximately 36%. Thus the lift coefficient at high angles of attack after stall is greatly increased with leading-edge tubercles, though there is hardly any change at the linear stage. Figure 6.10(b) suggests that the scalloped flipper model produces lower drag than the smooth model, by as much as about 32% at high angles of attack, though there is also hardly any difference at linear stage. The lift-to-drag ratio for the control case at the range of $\alpha = 5°$ to 8.5° and larger than $\alpha = 12°$ may be higher than the smooth case. In particular, the maximum lift-to-drag ratio is increased by approximately 13.1% with scalloped model. Such differences in aerodynamic forces give us direct evidence that the scalloped flipper performs better than the smooth case, in particular at high angles of attack.

Variations in the aerodynamic performance suggest significant changes in the flow field around the flipper model. We can see clearly that streamwise vortex forms from each trough portion (Figure 6.11), which is beneficial for momentum mixing between the outside and inside of the boundary layer, resulting in a delay or an elimination of flow

Figure 6.9 Flipper models with smooth (left) and sinusoidal (right) leading edges (Miklosovic et al. 2004). Reproduced with permission from AIP Publishing.

separation. Thus, the scalloped model could increase the lift coefficient at post stall angles, while the stall angle could also be postponed. However, at small angles of attack, the flow over the flipper model is completely attached; even the streamwise vortex induced by tubercles can hardly change the basic flow condition, thus the leading-edge tubercles have nearly no influence on the lift and drag coefficients.

6.3.2 Straight Wing

The effects of leading-edge tubercles on a straight wing were also studied by many researchers, such as Miklosovic et al. (2007), Johari et al. (2007) and Yoon et al. (2011). Johari et al. (2007) studied the influence of the protuberance's wavelength and amplitude on

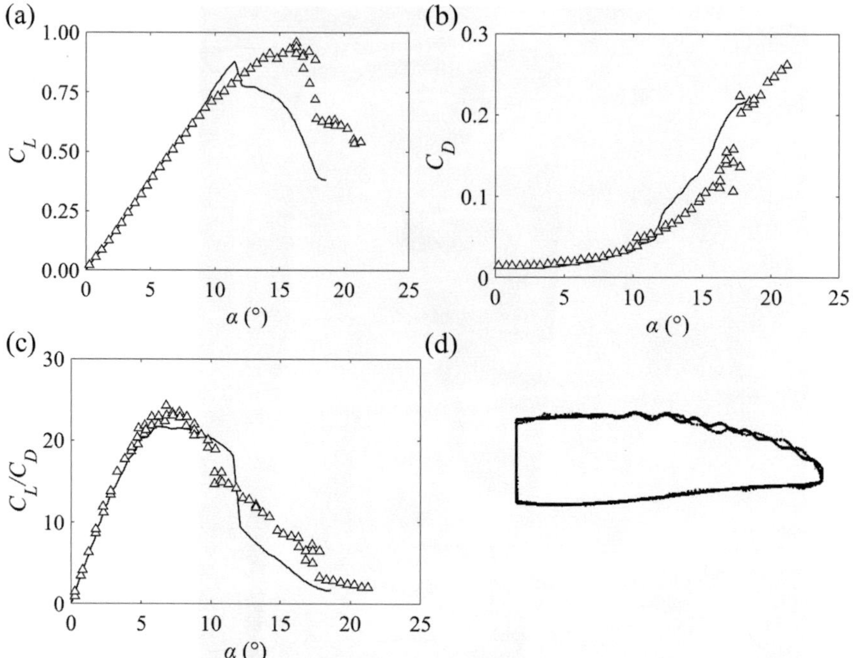

Figure 6.10 Aerodynamic forces for the smooth and scalloped whale flipper models. (a) Lift coefficient; (b) drag coefficient; (c) lift-to-drag ratio; and (d) profile of the idealized scalloped model (solid line) superimposed on the actual whale flipper profile (dotted line) (Miklosovic et al. 2004). Reproduced with permission from AIP Publishing.

the aerodynamic forces. The basic wing was based on the NACA 634–021 airfoil with chord length of $c = 102$ mm and span of $b = 203$ mm. The other six models with sinusoidal leading edge were also made with different wavelengths and amplitudes, as shown in Figure 6.12. Three amplitudes of $0.025c$, $0.05c$, and $0.12c$ were chosen with two wavelengths of $0.25c$ and $0.50c$. The free-stream velocity was 1.8 m/s, corresponding to $Re = 1.83 \times 10^5$.

The influence of the amplitude and wavelength on the lift and drag coefficients of straight wing is shown in Figure 6.13. For the baseline wing, the stall angle is $\alpha = 21°$ with the maximum lift coefficient of 1.13. After the stall angle, the lift coefficient drops sharply, and then the post-stall lift is nearly constant at 0.57 for $22° \leq \alpha \leq 28°$.

The lift coefficient for the straight wing with smallest leading-edge tubercles (8S) follows the baseline case up to $\alpha = 17$ °, and then drops to an intermediate plateau at $C_L = 0.94$ until $\alpha \leq 17°$ (Figure 6.13(a)). Then the lift coefficient decreases gradually with the angle of attack to approach the lift coefficient for the natural case. For the 8L wing, the lift coefficient remains constant at 0.82 for the range of $10° \leq \alpha \leq 26°$, when it breaks from linear growth. The maximum lift coefficient is reduced by about 28% than the natural case, but it was about 40% greater than the natural case at post stall angles. In particular, it is suggested that the 8L wing does not stall in the traditional sense that the lift coefficient decreases sharply from a maximum value. The lift behavior for the 8M

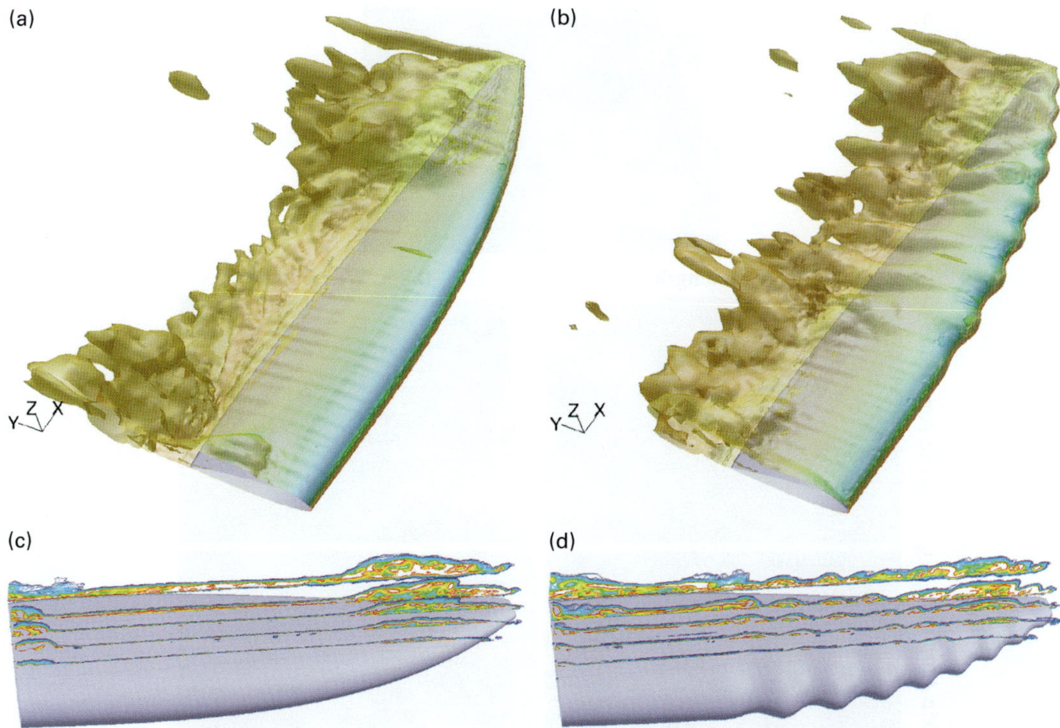

Figure 6.11 (a, b) Instantaneous vorticity magnitude isosurface colored by pressure and (c, d) instantaneous vorticity magnitude slices in the chord-wise direction at $\alpha = 15°$. The left and right columns are for the smooth and scalloped models, respectively (Pedro and Kobayashi 2008). Reproduced with permission, copyright © The American Institute of Aeronautics and Astronautics.

wing falls between that of 8S and 8L, while the post stall lift coefficient is approximately 50% greater than the baseline case. In summary, all the sinusoidal leading-edge cases have a much softer stall characteristic than the baseline case, which is beneficial for the aerodynamic performance at high angles of attack. The drag coefficient for the wing with leading-edge tubercles are greater than the baseline case in the range of $10° < \alpha < 22°$, which further increases with the protuberance amplitude (Figure 6.13(b)). There is no change beyond this range, suggesting that the greater lift-to-drag ratio can be obtained at post stall angles with leading-edge modification.

The influence of the protuberance wavelength on the lift and drag coefficients is shown in Figure 6.13(c) and (d), respectively. The behaviors of the lift and drag with various wavelengths are similar to those with various amplitudes. In general, the wing with smaller wavelength has greater lift coefficient than that with the larger wavelength at post stall angles. However, the latter case has a much smoother stall characteristic. On the other hand, the wing with shorter wavelength tubercles has slightly less drag than the wing with longer wavelength over the majority of angles of attack examined.

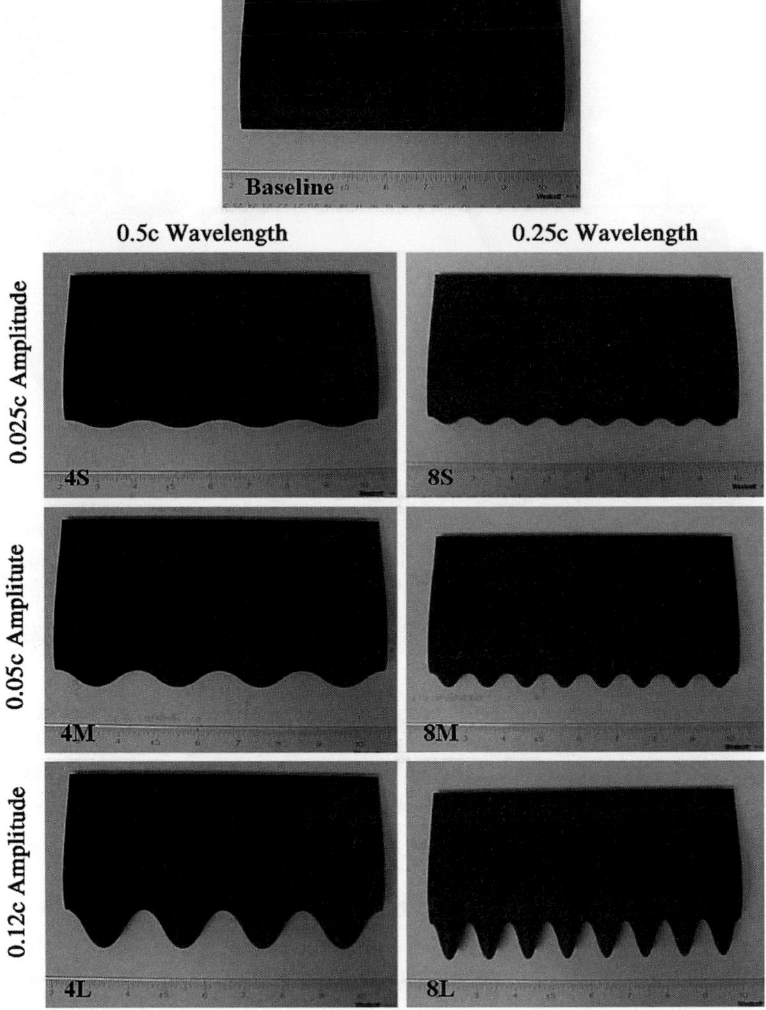

Figure 6.12 Photographs of the baseline and modified airfoils with leading-edge tubercles (Johari et al. 2007). Reproduced with permission, copyright © The American Institute of Aeronautics and Astronautics.

6.3.3   Delta Wing

Chen et al. (2013) and Chen and Wang (2014) conducted a series of investigations on the control of flow around a delta wing with leading-edge modification. The basic delta wing model was made of a 3 mm thickness flat plate with a swept angle of 52° without beveling on either side of the wing (Figure 6.14). The root chord length was 200 mm. Nine models of sinusoidal leading edges with different amplitude and wavelength were tested. The wavelength was selected as $\lambda = 63.5$ mm, 42.3 mm, and 31.7 mm, corresponding to 4, 6, and 8 crests in the whole leading-edge span. The amplitude was

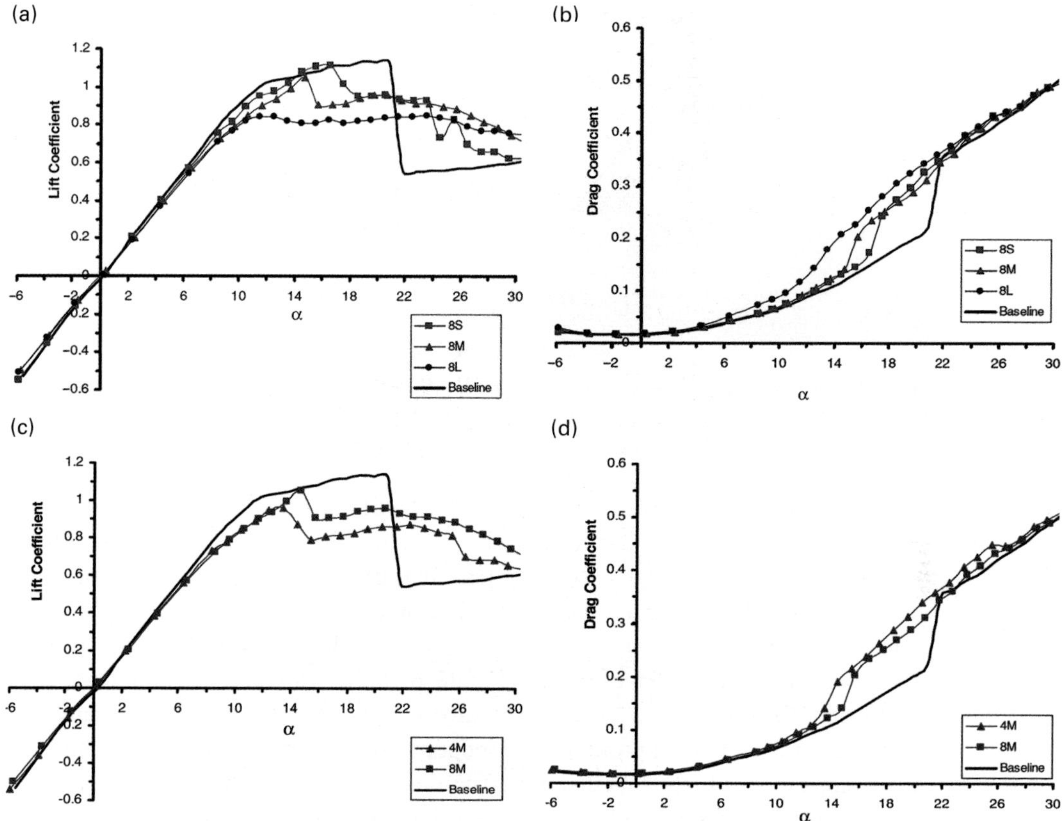

Figure 6.13 Lift (a, c) and drag (b, d) coefficients of the baseline, "8" series (a, b), and "M" series (c, d) straight wings (Johari et al. 2007). Reproduced with permission, copyright © The American Institute of Aeronautics and Astronautics.

selected as A = 2.5%c, 5%c, and 12%c. The free-stream velocity was 20 m/s, corresponding to Re = 2.7×10^5 based on the root chord length.

The influence of the amplitude on the aerodynamic forces on the delta wing is shown in the left column of Figure 6.15. For the baseline case, the lift coefficient increases to the maximum of approximately C_L = 1.09 at the stall angle of α = 26°, as shown in Figure 6.15(a). For all cases with leading-edge modification, the lift slopes for different amplitudes are slightly less than that of the baseline case, indicating that the lift in the linear stage of these models is generally a bit smaller than that of the baseline case. However, the stall angle is postponed from α = 26° to α = 28°, 30°, and 34°for the 4S, 4M, and 4L models, respectively. The maximum lift coefficient is also increased by approximately 1.0% and 1.2% for the 4S and 4M cases, respectively. However, the maximum lift for the 4L model is approximately 4% less than the baseline case. After stall, the lift coefficient for the 4L model declines rapidly by approximately 12.9% and 13.6% with an increment of 1°, which are slightly greater than the value of 11.7% for the baseline case, suggesting a severe stall behavior similar to the baseline case. However,

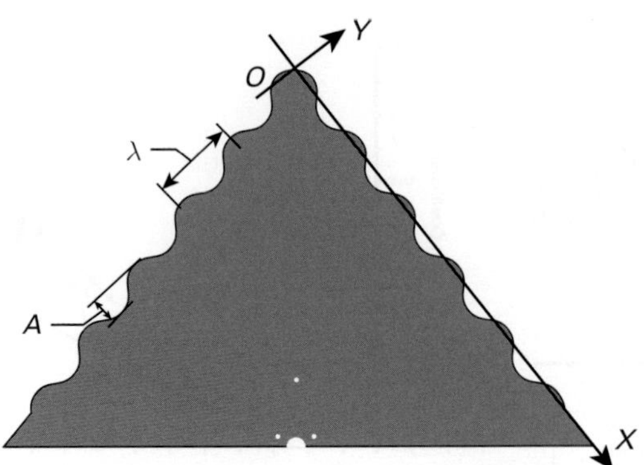

Figure 6.14 Sketch of the delta wing model with sinusoidal leading edge (Chen et al. 2013). Reproduced with permission, copyright © Springer 2013.

a gradual stall can be observed for the 4L case. The lift coefficient decreases only about 8.1% in magnitude from $\alpha = 34°$ to 35°, which is much less than the natural case. For all the leading-edge modification cases, the post stall lift can be enhanced in comparison with the natural case. After $\alpha = 34°$, the lift coefficient increases with the amplitude.

Before the stall angle of $\alpha = 26°$, there is no distinct drag penalty for all the leading-edge modification cases investigated (Figure 6.15(c)). However, a significant drag penalty appears after $\alpha = 26°$. Chen et al. (2013) suggested that this drag penalty was attributed to the additional protuberance-induced lift in the post-stall region. This drag penalty corresponds to the fact that the lift-to-drag ratios for the three control cases are all equal to the baseline case in this region (Figure 6.15(e)). However, the lift-to-drag ratio at small angles of attack is decreased for leading-edge modification in comparison with the baseline case, and it is further decreased with an increase in the amplitude.

The influence of the wavelength on the delta wing forces is shown in the right column of Figure 6.15. The maximum lift coefficient C_L for the 4M, 6M, and 8M models is 1.10, 1.06, and 1.02, respectively, in comparison with the value of 1.09 for the baseline case (Figure 6.15(b)). The lift coefficient for the 8M model is much less than that of 4M and 6M models in the range from $\alpha = 26°$ to $\alpha = 30°$, though it becomes slightly greater than the latter two cases after $\alpha = 30°$. The variation of the drag coefficient follows that of the baseline case before the stall angle (Figure 6.15(d)). Thus, we can see that the lift-to-drag ratio decreases as wavelength becomes smaller (Figure 6.15(f)).

The surface oil-flow pattern for the baseline and 4M control cases at $\alpha = 20°$ and 29° are shown in Figure 6.16. For the baseline case at $\alpha = 20°$ (Figure 6.16(a)), there is a distinct primary attachment line (A1) located along the leading edge, indicating the existence of the primary leading-edge vortex of the delta wing. Outwards from the primary attachment line, there are the secondary separation line (S2) and secondary attachment line (A2), while there are also tertiary separation (S3) and attachment lines (A3) between S2 and A2. These lines indicate the existence of a dual leading-edge vortex system. For the 4M model (Figure 6.16(b)), the primary attachment line A1 can be seen with its length nearly the same as that of the baseline case, while the secondary

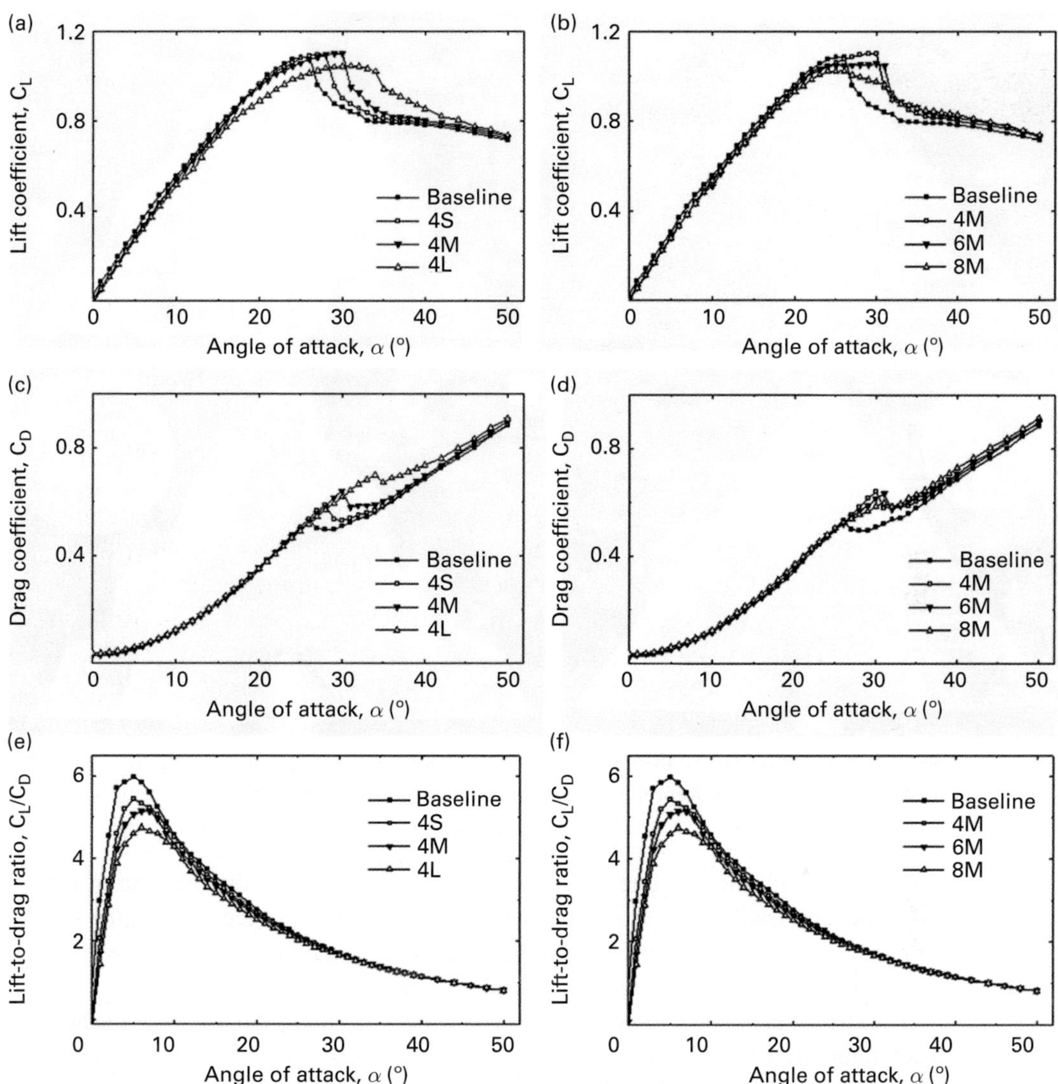

Figure 6.15 Lift coefficient (a,b), drag coefficient (c,d), lift-to-drag ratio (e,f) of the baseline, "4" series (left column), and "M" series (right column) delta wings (Chen et al. 2013). Reproduced with permission, copyright © Springer 2013.

separation line (S2) seems to form from each trough of the sinusoidal leading edge but the main location is also similar to that of the baseline case.

At $\alpha = 29°$, only the secondary separation line can be observed for the baseline case (Figure 6.16(c)), suggesting that the breakdown of the primary leading-edge vortex has moved upstream to the apex of the model. We can see the large-scale separation region between the separation line and the leading edge. However, a pair of short attachment lines can still be observed around the apex of the 4M model (Figure 6.16(d)), indicating that the vortex breakdown is delayed. The separation region is significantly reduced in

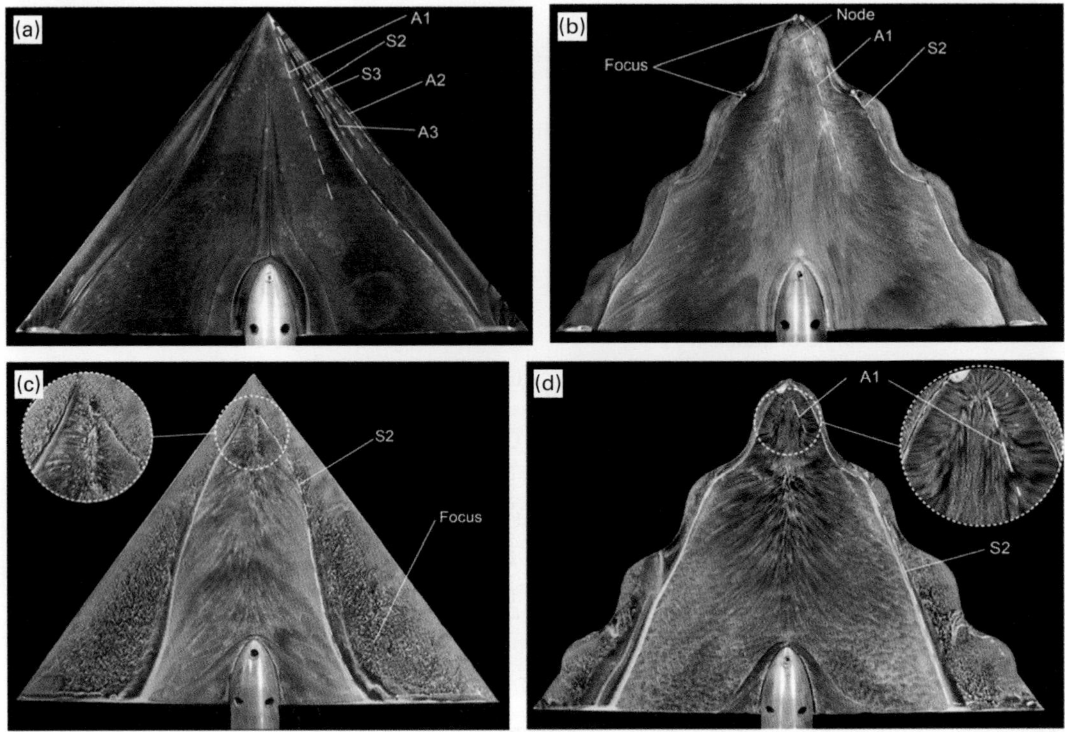

Figure 6.16 Oil-flow pattern for the baseline (a,c) and 4M (b,d) models at $\alpha = 20°$ (a,b) and $\alpha = 29°$ (c,d) (Chen et al. 2013). Reproduced with permission, copyright © Springer 2013.

comparison with the baseline case, which coincides well with the increase of the lift coefficient by leading-edge modification. Chen et al. (2013) suggested that the separation delay was related to the larger effective swept angle around the tip, which was caused by the first crest of sinusoidal leading edges.

The change in the flow topology on a delta wing with sinusoidal leading edge has also been found by PIV measurement of Goruney and Rockwell (2009). It was found that when the amplitude reached 4% of the chord length, it was possible to completely eradicate the negative focus of large-scale separation.

The SPIV measurement was conducted to present the spatial development of the leading-edge vortex for both baseline and 4M control cases. The distributions of the mean vorticity field in ten different streamwise planes are shown in Figure 6.17, with the baseline case (left) and the 4M case (right). For the baseline case, the dual leading-edge vortex can be clearly identified, indicating that the vortex core of the primary leading-edge vortex expands as it develops towards the trailing edge through the $x/c = 85\%$ plane tested in the study. For the 4M model, three leading-edge vortices can be clearly observed. The leading-edge vortex forms from each peak of the sinusoidal wave. The vorticity strength of each leading-edge vortex for the 4M case is comparable with that for the baseline case. The interaction of those leading-edge vortices for the 4M case is comparable

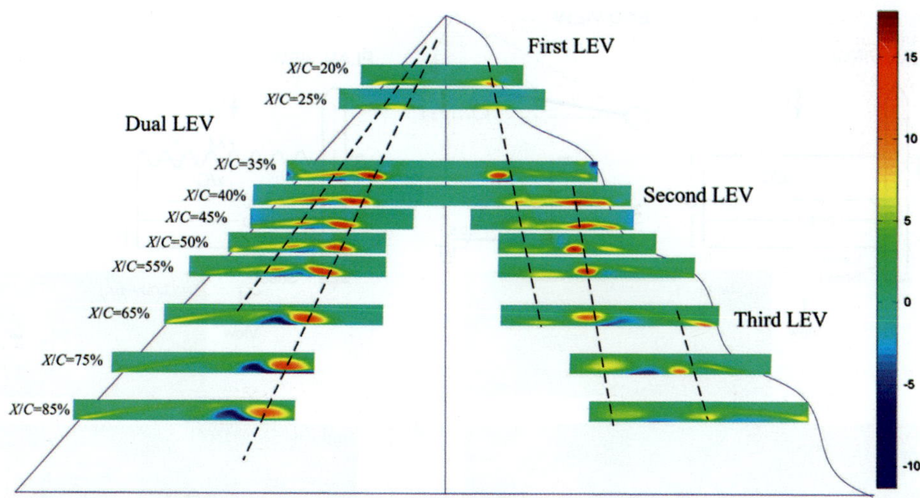

Figure 6.17 Vortex structures of baseline (left part) and 4M (right part) models at $\alpha = 10°$ (Chen and Wang 2014). Reproduced with permission, copyright © Springer 2014.

with that of the dual leading-edge vortices for the natural case. It helps to explain why lift coefficient is nearly the same for the baseline and control cases at this angle of attack.

6.3.4   Flapping Wing

The leading-edge modification can also be used in a flapping wing (Ozen and Rockwell 2010) and a plunging airfoil (Zhang et al. 2013). One study was conducted by Ozen and Rockwell (2010), as shown in Figure 6.18. A rectangular flat plate with chord length of $c = 50.8$ mm, span of $b = 101.6$ mm, and thickness of 1.59 mm was mounted on a revolution body with diameter of $D = 12.7$ mm. The free-stream velocity was $U_\infty = 25.4$ mm/s, corresponding to $Re = 1300$. The flapping motion had a triangular sawtooth form, and the flapping angle varied in a periodic manner between maximum positive and negative values of $\pm30°$. The dimensionless oscillation frequency was $k = \pi fc/U_\infty = 0.3$. Two models were tested, one with a smooth leading edge and the other with a sinusoidal leading edge. The amplitude and the wavelength of the sinusoidal leading edge were $0.098c$ and $0.246c$, respectively.

Figure 6.18 also presents the flow field at a typical angle during the downstroke motion corresponding to the largest-scale vortical structures at inboard locations. The streamwise vorticity at $x/c = 0.1$ is presented in the first row for the above two cases. For the natural case, we can see the positive tip vortex and the positive and negative shear layers over the upper surface of the flat plate. For the sinusoidal leading-edge case, an ordered pattern of small-scale streamwise vorticity can be observed, which is produced from the sinusoidal leading edge. At $x/c = 0.5$, the tip vortex grows in scale, and large-scale concentrations of positive and negative vorticities can be seen at the inboard location for the natural case. With leading-edge modification, the tip vortex is

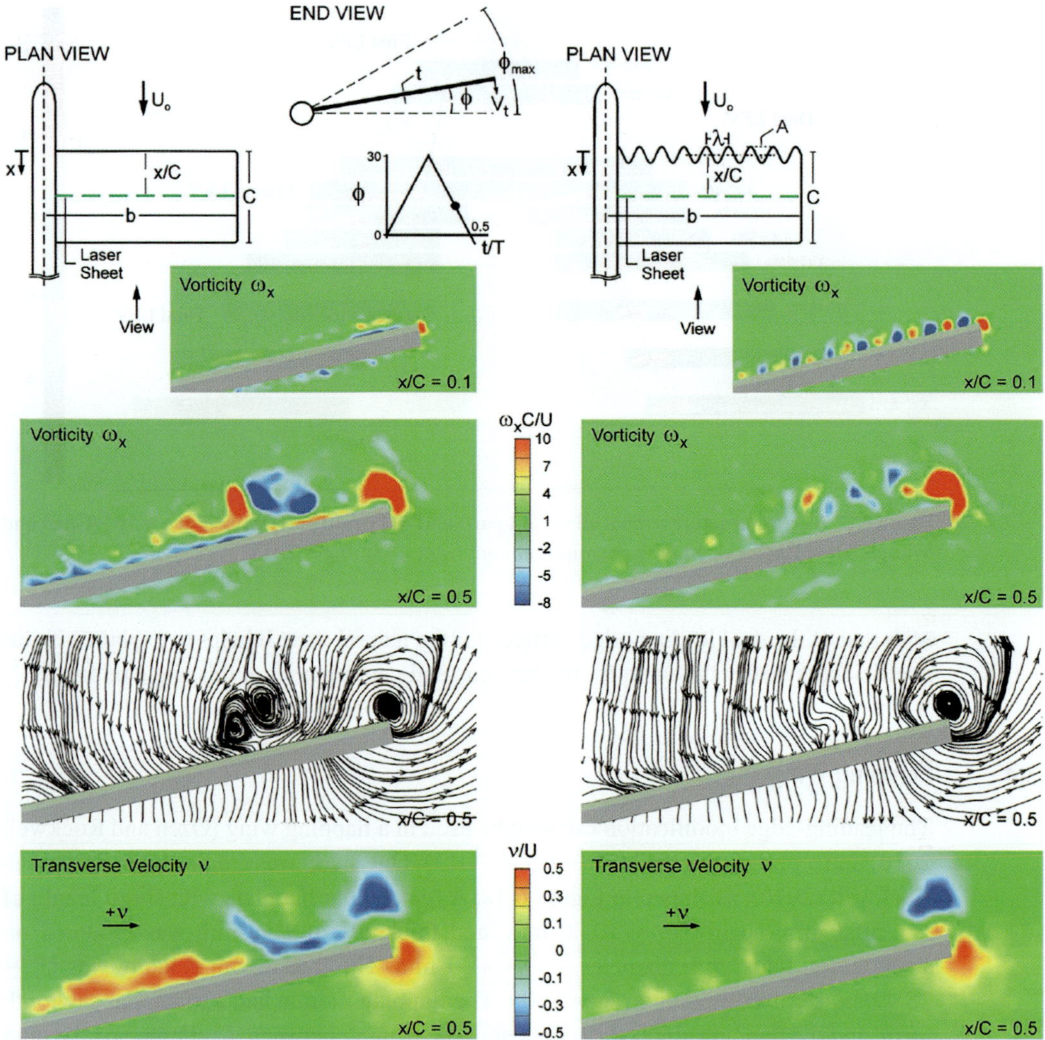

Figure 6.18 Comparison of flow structure on a flapping wing without and with a sinusoidal leading edge at flapping angle of 13.3° and angle-of-attack of 8° (Ozen and Rockwell 2010). Reproduced with permission from AIP Publishing.

similar to the natural case, while the ordered patterns of the streamwise vortices can also be observed. However, the large-scale vorticity concentrations seen from the natural case are eliminated. This can also be validated from the distributions of streamline and transverse velocity. A large-scale recirculation region forms in the inboard position because of flow separation, which is eliminated by the leading-edge modification.

It is suggested that the control mechanism on a dynamic wing is similar to that on a stationary wing. The sinusoidal leading edge can induce a series of streamwise vortices to control the unsteady separated flow. This has also been observed for a plunging airfoil by Zhang et al. (2013).

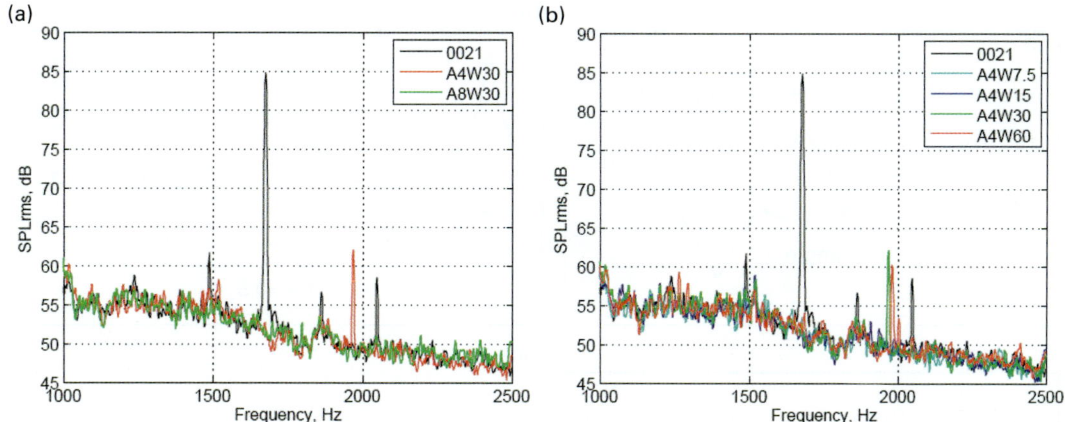

Figure 6.19 SPL versus frequency for variation of amplitude (a) and wavelength (b) of tubercles (Hansen et al. 2010). Reproduced with permission, copyright © The American Institute of Aeronautics and Astronautics.

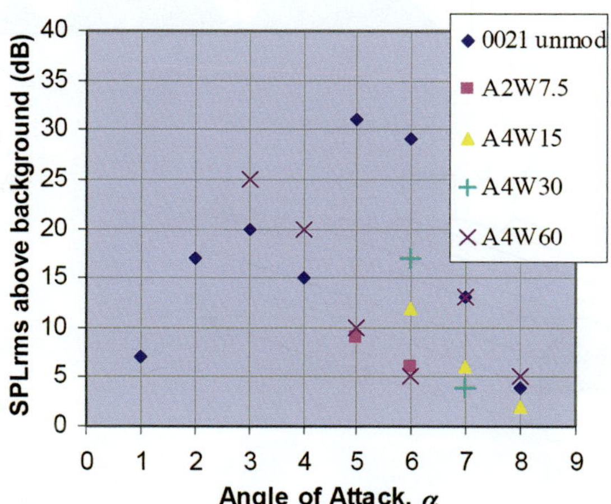

Figure 6.20 SPL versus angle of attack for leading-edge protuberance configurations with tonal noise (Hansen et al. 2010). Reproduced with permission, copyright © The American Institute of Aeronautics and Astronautics.

6.3.5 Noise Reduction

In addition to improving aerodynamic performance, the leading-edge modification was also found to be effective in noise reduction. Hansen et al. (2010) conducted a wind tunnel measurement. The NACA 0021 airfoil was made of aluminum with chord length of $c = 70$ mm and span of $s = 495$ mm. Nine sinusoidal-protuberance configurations were tested with varying amplitude and wavelength, which was labeled as "AW," where the "A" and "W" were followed by the actual scale of the amplitude and wavelength in dimension of millimeters, respectively. Acoustic measurements were carried out using

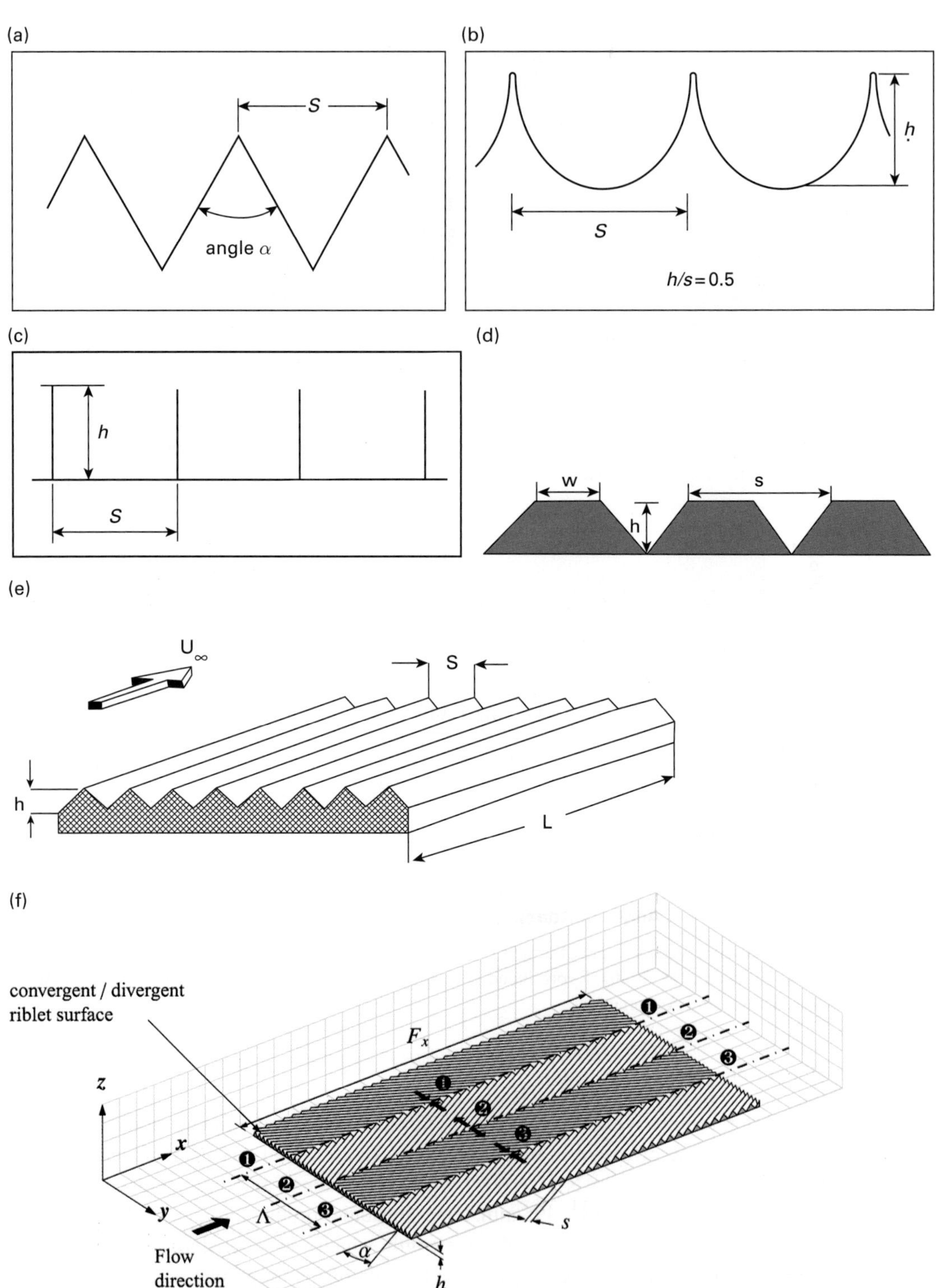

Figure 6.21 Cross-section of various riblets. (a) Sawtooth riblets. (b) Semi-circular scalloped riblets. (c) Blade riblets (Bechert et al. 1997), reproduced with permission, copyright © Cambridge

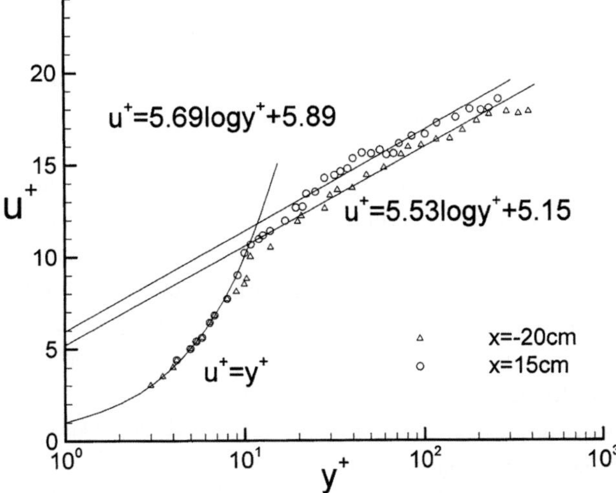

Figure 6.22 Mean velocity profiles over smooth and riblet surfaces (Wang et al. 2000). Reproduced with permission, copyright © 2000 Elsevier Ltd. All rights reserved.

two microphones outside of the wind tunnel test section. The free-stream velocity was $U_\infty = 25$ m/s, corresponding to $Re = 1.2 \times 10^5$.

Figure 6.19 shows the effects of amplitude and wavelength on the acoustic spectra of the airfoil. For all control cases, the SPL amplitude is much lower than that for the baseline airfoil, though the peak frequency might be changed. Besides, the airfoils with the smallest wavelength and the largest amplitude tubercles are the most effective at eliminating noise. Such an effect can also be found at other angles of attack, which is shown in Figure 6.20. Thus, the leading-edge modification shows the great ability to reduce the noise.

6.4 Riblet

6.4.1 Basic characteristics

The idea of riblets originates from sharkskin, and various types have also been studied in the lab. As shown in Figure 6.21(a–d), there are many configurations, such as sawtooth,

---

Caption for Figure 6.21   (cont.)

University Press 1997. (d) Trapezia riblets (Wang et al. 2000), reproduced with permission, copyright © 2000 Elsevier Ltd. All rights reserved. Schematic of the riblets on a flat plate: (e) arrangement in streamwise direction (Viswanath 2002), reproduced with permission, copyright © 2002 Elsevier Ltd. All rights reserved; and (f) convergent and divergent arrangement (Nugroho et al. 2013), reproduced with permission, copyright © 2013 Elsevier Ltd. All rights reserved.

semi-circular scalloped, blade, and trapezia, etc. The shape of the riblets is determined by the height and the spanwise spacing of the cavity. The riblets are usually arranged in the streamwise direction (Figure 6.21(e)), while recent studies have also arranged the riblets in the convergent and divergent form (Figure 6.21(f)).

A typical velocity profile of the riblet surface is shown in Figure 6.22, in comparison with that over a smooth surface. The velocity profile can also be divided into three regions: viscous sublayer, buffer region, and logarithmic region. The velocity is a linear distribution in the viscous sublayer, namely $u^+ = y^+$. The actual thickness of the viscous sublayer can be determined by the distance between the coordinate's origin and the intersection point of the viscous sublayer and the buffer region. It is suggested that the thickness of the viscous sublayer for riblet surface is increased by about 10% in comparison with the smooth surface. Bechert and Hoppe (1985) and Choi (1989) suggested that the riblet surface created a distance between the streamwise vortex and the wall in the sublayer. Some fluid worked as the lubricator in the bottom of the riblet surface, which was equivalent to an increase of the sublayer thickness. Thus, the friction could be reduced. Wang et al. (2000) found that the increase of viscous sublayer thickness made the buffer region and the logarithmic region move upwards. Thus, the velocity in the same position was increased in comparison with the smooth plate case, leading to a drag reduction.

For the logarithmic region, the velocity distribution can be described as $u^+ = A \lg y^+ + C$. The factor of C for the riblet surface is higher than that for the smooth surface, showing that the velocity for the former case is shifted upwards in comparison with the latter. This also indicates that the riblet surface could reduce the friction drag.

The drag reduction for some typical types of riblets is shown in Figure 6.23. For all cases, there is an optimal spanwise spacing to obtain the maximum drag reduction. In general, the drag decreases as the spanwise spacing of the riblets increases, and the maximum drag reduction corresponds to the spanwise spacing from 10 to 20. Then, the drag increases with the spanwise spacing when the spanwise spacing is larger than about 20. It seems that greater drag reduction may be obtained with smaller heights for all riblets. The comparison among these three kinds of riblets suggests that the blade riblets are most effective in drag reduction with a maximum value of about 10%.

6.4.2   Flow Physics

Figure 6.24 shows the flow visualization of the streamwise vortices developed over the riblet surface for drag decreasing and drag increasing cases. For the drag decreasing case ($s^+ = 25.2$), the large-scale streamwise vortex stays above the riblets and the spanwise motion is restricted by the riblets. In general, the streamwise vortices are larger than the riblet spacing. For the drag increasing case ($s^+ = 40.6$), the size of the streamwise vortices is smaller than the riblet spacing; thus they can easily penetrate into the riblet valleys.

The streamwise vortices play an important role in the development of the turbulent boundary layer, while the riblets could modify and reduce the momentum exchange properties of the streamwise vortices effectively. A schematic of the interaction between the

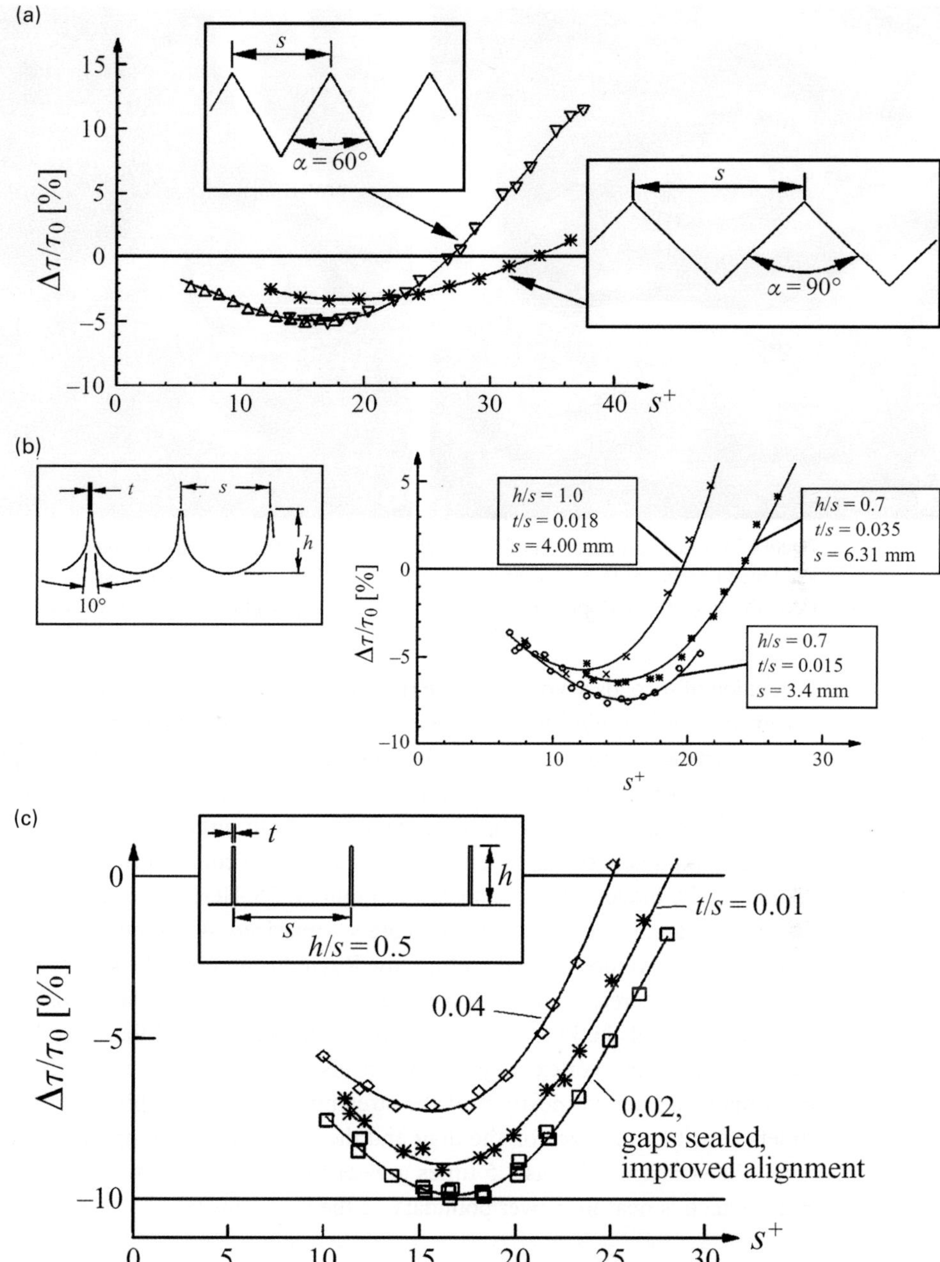

Figure 6.23 Friction drag for various riblets. (a) Sawtooth riblets, (b) semi-circular scalloped riblets, (c) blade riblets (Bechert et al. 1997). Reproduced with permission, copyright © Cambridge University Press 1997.

streamwise vortex and the riblet surface is shown in Figure 6.25. It is shown that the riblet peaks are speculated to accelerate the development of cross-stream secondary vorticity, causing the formation of discrete vortices within the riblet grooves. It is suggested that the

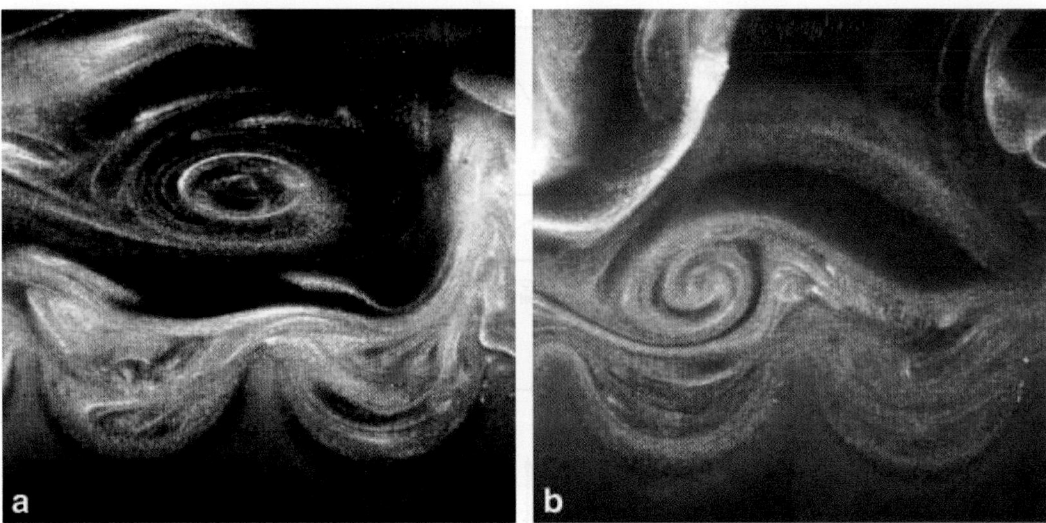

Figure 6.24 Flow visualization of streamwise vortices in the vertical cross-section of semi-circular scalloped riblets. (a) Drag decreasing case ($U_\infty = 3$ m/s); and (b) drag increasing case ($U_\infty = 5$ m/s) (Lee and Lee 2001). Reproduced with permission, copyright © Springer-Verlag 2001.

generation of secondary vortices at the riblet peaks serves two functions to interact with the streamwise vortices. First, they can preserve low-speed fluid within the grooves. Second, they can weaken the streamwise vortices which induce the secondary vortices. This process helps to retard the development of the turbulent boundary layer from the riblet surface and thus reduces the friction drag in comparison with the smooth surface.

Many parameters of the riblets could influence the control effects, and some typical results for the blade riblets are shown in Figure 6.26. Here, the effects of the kinematic viscosity, velocity, spacing, and height are summarized. At a high ratio of the kinematic viscosity to the velocity and therefore low Reynolds number, the size of the streamwise vortex will be decreased, thus the vortex is lifted by the riblets. The opposite trend occurs for a low ratio of the kinematic viscosity to the velocity and therefore high Reynolds number, where the vortex may fall down inside the riblets. The height of the riblets determines the normal position of the streamwise vortices, while the thickness of the riblets has little influence on the drag and the vortex size. The optimal control appears when the vortex width was 1.5 times greater than the spacing with a riblet height $h^+ \approx$ 8–10, which is near the lower boundary of the buffer layer.

6.5   Cactus-Shape Modification

It was found that adult saguaros have one main cylindrical stem ranging from 0.3 m to 0.8 m in diameter and over 8 m to 15 m in height. However, there were always 10 to 30 v-shaped cavities around the circumference, as shown in Figure 6.27(a). One important finding was that the depth of the cavity divided by the diameter of the

Table 6.1 The control effects with different cavity depths for four Reynolds numbers (Talley et al. 2001).

L/D	C_d	C_l	S_t	L/D	C_d	Cl	S_t
0	1.339±0.010	±0.330	0.160	0	1.365±0.037	±0.664	0.175
0.035	1.304±0.011	±0.325	0.161	0.035	1.361±0.045	±0.713	0.172
0.070	1.309±0.011	±0.334	0.162	0.070	1.364±0.057	±0.742	0.172
0.105	1.318±0.012	±0.336	0.161	0.105	1.381±0.049	±0.740	0.170

$Re = 100$ $Re = 200$

L/D	C_d	C_l	S_t	L/D	C_d	Cl	S_t
0	1.683±0.164	±1.923	0.217	0	1.644±0.113	±1.791	0.228
0.035	1.452±0.076	±1.562	0.221	0.035	1.464±0.120	±1.462	0.224
0.070	1.419±0.083	±1.245	0.224	0.070	1.401±0.131	±1.128	0.221
0.105	1.359±0.052	±0.987	0.223	0.105	1.325±0.079	±0.864	0.221

$Re = 20\ 000$ $Re = 100\ 000$

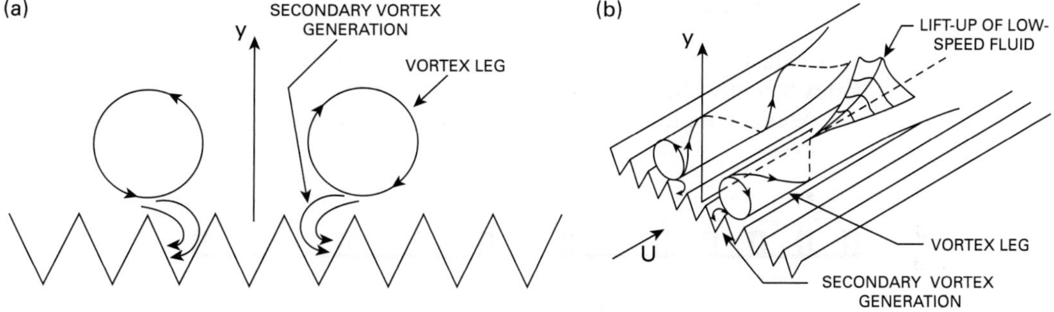

Figure 6.25 Schematic showing the hypothesized interaction of streamwise vortex with riblet surface. (a) cross-section view; and (b) perspective view (Bacher and Smith 1985). Reproduced with permission, copyright © The American Institute of Aeronautics and Astronautics.

cylinder was always kept approximately as $L/D = 0.07 \pm 0.0015$, as shown in Figure 6.27(b). Following this rule, Talley et al. (2001) and Talley and Mungal (2002) conducted a series of experimental and numerical studies for a circular cylinder with v-shaped cavities. The numerical simulation was conducted at four Reynolds numbers at 100, 200, 20 000 and 100 000, while the cavity depth was ranging from $L/D = 0.0$ to $L/D = 0.105$.

The histories of drag and lift coefficients at $Re = 100$ and $Re = 2 \times 10^4$ are shown in Figure 6.28 for the natural case and the control case of $L/D = 0.07$. At both Reynolds numbers, the modified cylinder could reduce the drag coefficient, and the drag reduction becomes more significant at higher Reynolds numbers. The lift coefficient for the control

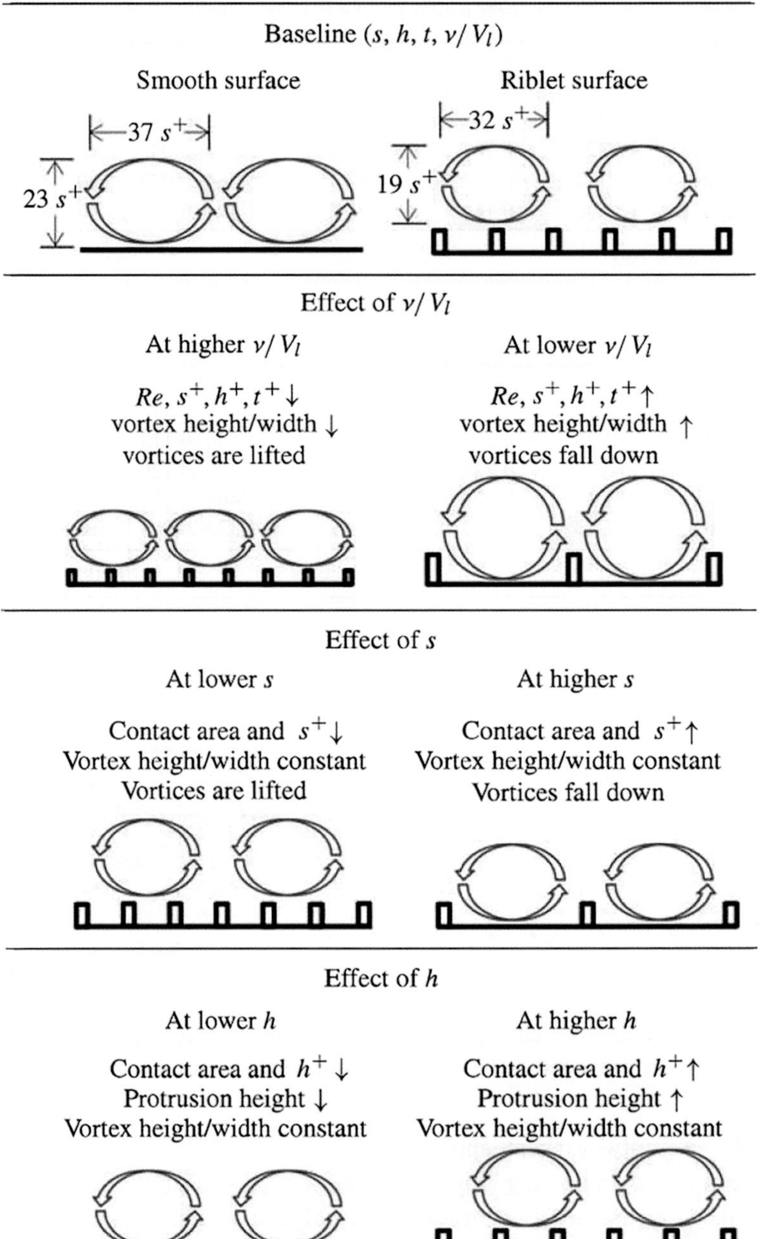

Figure 6.26 Conceptual models of parameter effects on streamwise vortices for the blade riblets (Martin and Bharat 2014). Reproduced with permission, copyright © Cambridge University Press 2014.

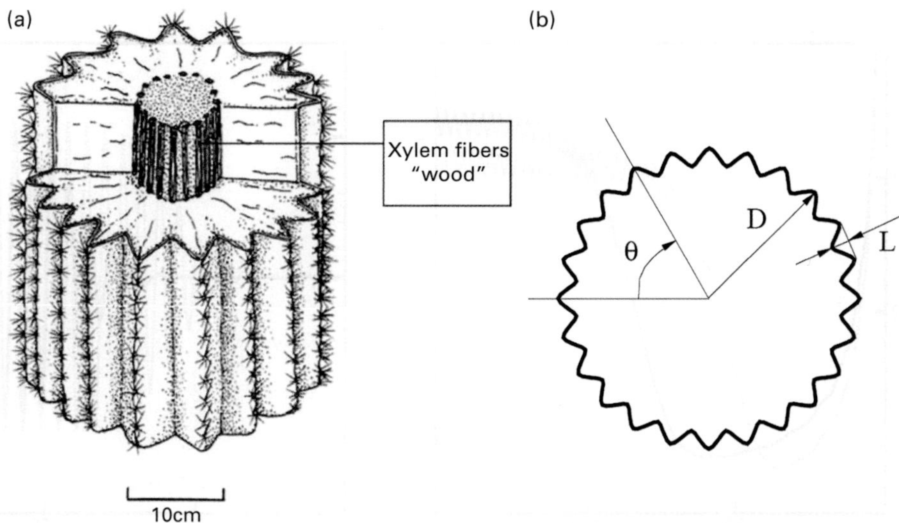

Figure 6.27 (a) Saguaro stem anatomy; (b) sketch of the cross-section (Talley et al. 2001). Reproduced with permission.

case was almost the same as that for the natural case at $Re = 100$, while an obvious decrease in the lift fluctuation can be observed at $Re = 2 \times 10^4$.

The control effects with other cavity depths are summarized in Table 6.1 for four Reynolds numbers. The control cases can decrease the drag at all Reynolds numbers tested, except for the $L/D = 0.105$ case at $Re = 200$, where the drag coefficient is increased. At low Reynolds numbers, the maximum drag reduction occurs for the smallest cavity depth, while it corresponds to the largest cavity depth case for largest Reynolds numbers. The control may even increase the lift fluctuation at $Re = 100$ and 200. However, the lift fluctuation is decreased at $Re = 2 \times 10^4$ and 1×10^5, and the fluctuation reduction increases with an increase of the cavity depth. For all control cases, the Strouhal number changes a little in comparison with the natural case, with the maximum less than 3%.

The detailed flow over the smooth and grooved cylinders with flow visualization is shown in Figure 6.29. We can see that small-scale recirculation regions are formed inside the grooves on the windward side of the cylinder, which has also been found by Babu and Mahesh (2008). Such flow modification is suggested to be beneficial in decreasing the viscous and pressure forces on the grooved cylinder in comparison with the smooth one. Furthermore, the large-scale Kármán vortices for the smooth cylinder are converted into smaller-scale ones for the grooved cylinder, as shown in Figure 6.30. In addition, it seems that the vortices for the grooved cylinder start to form from a further downstream position with a weakened mutual interaction. Thus, it can lead to a decrease in the lift fluctuation, which is advantageous to the saguaros in reducing the damage from the vortex-induced-vibration.

In addition to the V-shaped modifications of the circular cylinder, some researchers (such as Yamagishi and Oki 2007; El-Makdah and Oweis 2013) have also studied

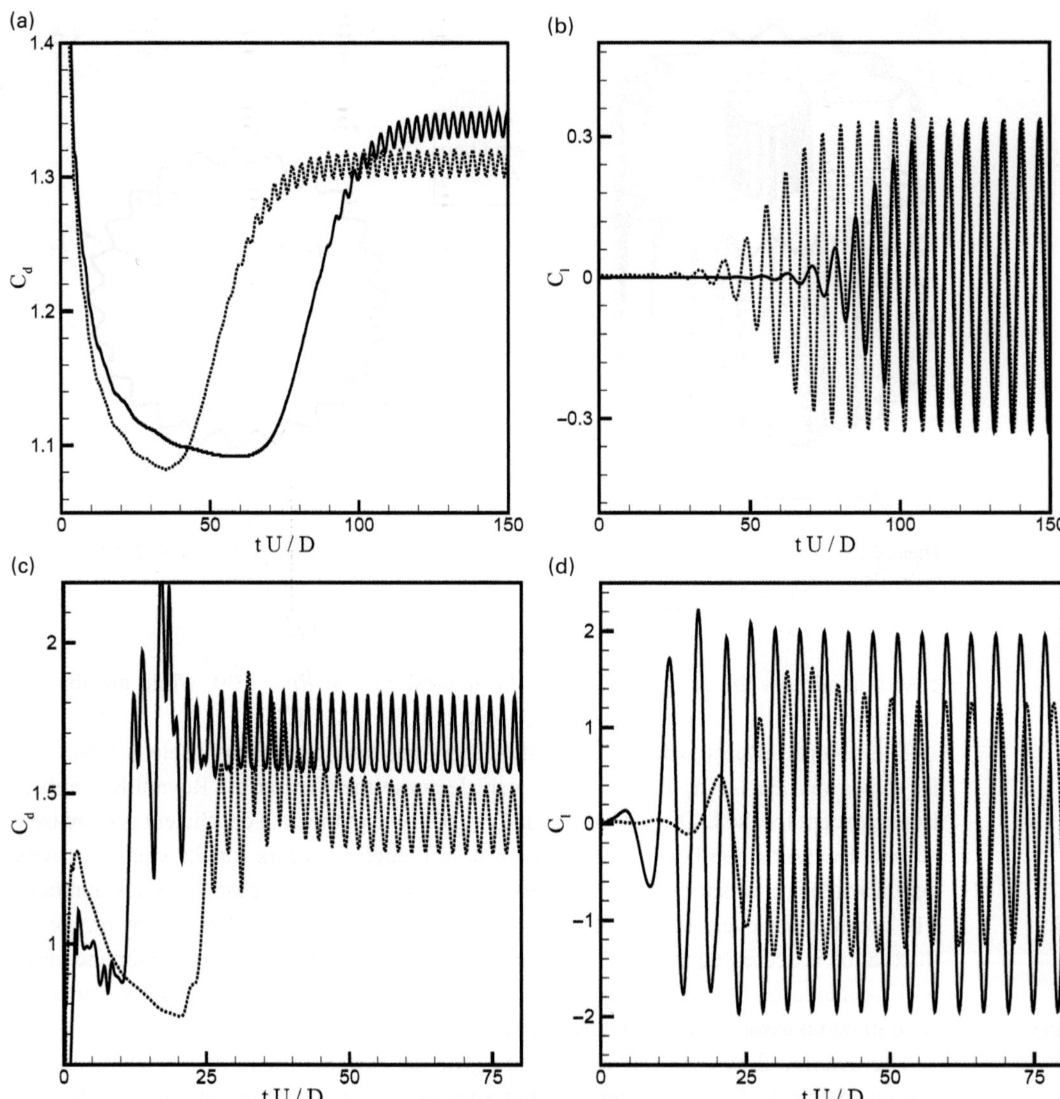

Figure 6.28 History of drag coefficient (a,c) and lift coefficient (b,d) at $Re = 100$ (a,b) and 2×10^4 (c,d) (Talley et al. 2001). Reproduced with permission.

a similar modification, such as the U-shaped cylinder shown in Figure 6.31. The size of the U-groove cylinder in the experiment by El-Makdah and Oweis (2013) is shown in Figure 6.31(b). One test was conducted at $Re = 50\ 000$. It is suggested that the surface modification can reduce the strength of the wake vortices and weaken their mutual interaction. Thus the RMS velocity fluctuation of the U-groove cylinder is greatly reduced in comparison with the natural case, as shown in Figure 6.32.

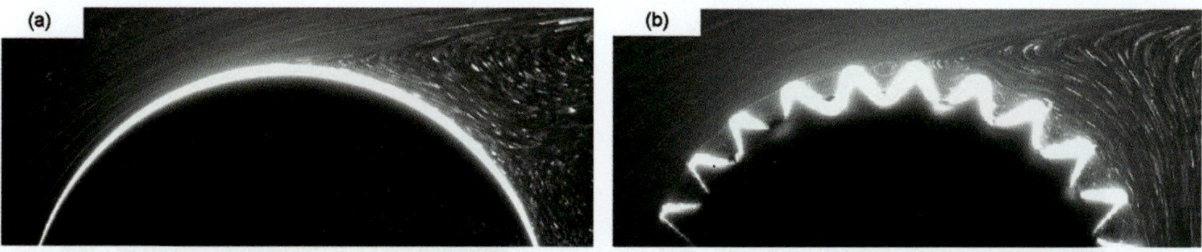

Figure 6.29 Flow visualization of the near-surface flow field around a smooth cylinder (a) and a grooved cylinder (b) (Liu et al. 2011). Reproduced with permission, copyright © 2010 Elsevier Ltd. All rights reserved.

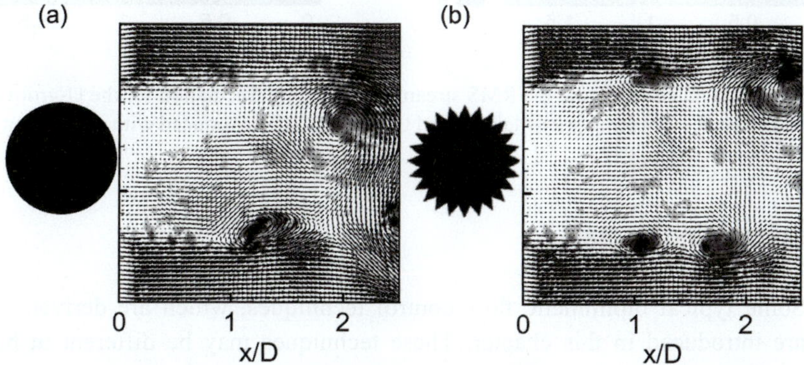

Figure 6.30 Instantaneous vector field and swirling strength field around a smooth cylinder (a) and a grooved cylinder (b) (Liu et al. 2011). Reproduced with permission, copyright © 2010 Elsevier Ltd. All rights reserved.

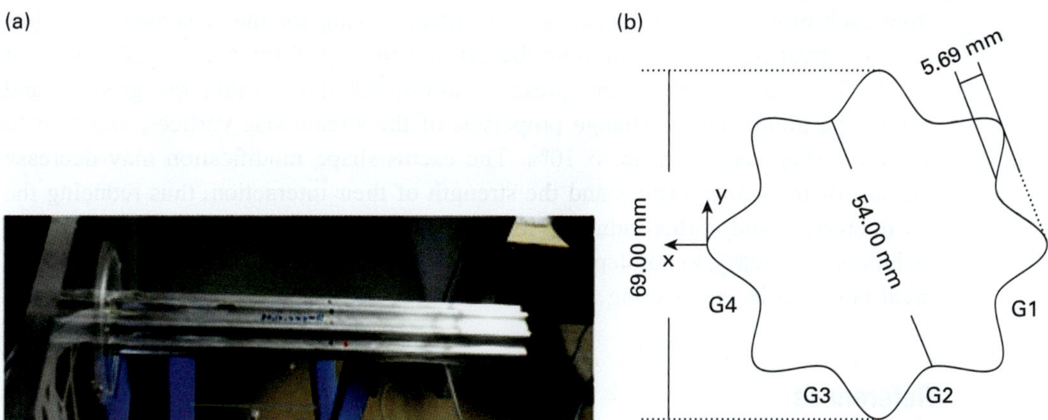

Figure 6.31 (a) A picture of the U-groove cylinder installed in the wind tunnel; and (b) sketch of the cross-section (El-Makdah and Oweis 2013). Reproduced with permission, copyright © Springer-Verlag 2013.

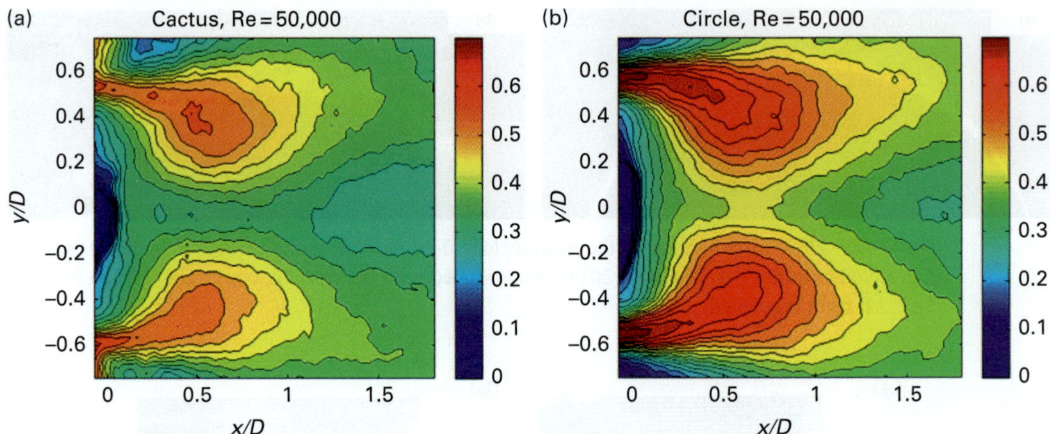

Figure 6.32 Distribution of the RMS streamwise velocity fluctuation for the U-groove cylinder (a) and circular cylinder (b) (El-Makdah and Oweis 2013). Reproduced with permission, copyright © Springer-Verlag 2013.

6.6   Concluding Remarks

Some typical biomimetic flow control techniques, which are derived from nature, are introduced in this chapter. These techniques may be different in basic configurations and the control mechanism, but in general, they can achieve the goal of increasing the lift or decreasing the drag. The hairy coating can adapt to the flow, causing a reverse force to the flow, thus it may reduce the drag coefficient and the lift fluctuations of bluff bodies while increasing the lift coefficient of airfoils. The leading-edge modification with tubercles, can induce a streamwise vortex from each protuberance to enhance momentum mixing for the separated flow, thus it shows great potential to improve the aerodynamic performance at high angles of attack. The surface riblets can preserve low-speed fluid within the grooves and reduce the momentum exchange properties of the streamwise vortices, and lead to a friction drag reduction up to 10%. The cactus-shape modification may decrease the size of the wake vortices and the strength of their interaction, thus reducing the lift fluctuation and vortex-induced vibration. In conclusion, biomimetic flow control techniques are easy to implement with high control effectiveness, and thus show great potential in engineering applications.

References

Aftab, S. M. A., Razak, N. A., Rafie, A. S. M., and Ahmad, K. A. Mimicking the humpback whale: an aerodynamic perspective. *Progress in Aerospace Sciences*, 2016, 84: 48–69

Babu, P. and Mahesh, K. Aerodynamic loads on cactus-shaped cylinders at low Reynolds numbers. *Physics of Fluids*, 2008, 20(3): 035112

Bacher, E. V. and Smith, C. R. A combined visualization-anemometry study of the turbulent drag reducing mechanisms of triangular micro-groove surface modifications. AIAA Paper 1985–0548

Bechert, D. W., Bruse, M., and Hage. W., Van Der Hoeven, J. G. T., and Hoppe. G. Experiments on drag-reducing surfaces and their optimization with an adjustable geometry. *Journal of Fluid Mechanics*, 1997, 338: 59–87

Bechert, D. W., Hoppe, G., and Reif, W. E. On the drag reduction of the shark skin. AIAA Paper 1985–0546

Brücker, C. and Weidner, C. Influence of self-adaptive hairy flaps on the stall delay of an airfoil in ramp-up motion. *Journal of Fluids and Structures*, 2014, 47: 31–40

Bushnell, D. M. and Moore, K. J. Drag reduction in nature. *Annual Review of Fluid Mechanics*, 1991, 23(1): 65–79

Chen, H., Pan, C., and Wang, J. J. Effects of sinusoidal leading edge on delta wing performance and mechanism. *Science China Technological Sciences*, 2013, 56(3): 772–779

Chen, H. and Wang, J. J. Vortex structures for flow over a delta wing with sinusoidal leading edge. *Experiments in Fluids*, 2014, 55(6):1761

Choi, K. S. Near-wall structure of a turbulent boundary layer with riblets. *Journal of Fluid Mechanics*, 1989, 208: 417–458

El-Makdah, A. M. and Oweis, G. F. The flow past a cactus-inspired grooved cylinder. *Experiments in Fluids*, 2013, 54(2):1464

Favier, J., Dauptain, A., Basso, D., and Bottaro, A. Passive separation control using a self-adaptive hairy coating. *Journal of Fluid Mechanics*, 2009, 627: 451–483

Fish, F. E. and Battle, J. M. Hydrodynamic design of the humpback whale flipper. *Journal of Morphology*, 1995, 225(1): 51–60

Goruney, T. and Rockwell, D. Flow past a delta wing with a sinusoidal leading edge: near-surface topology and flow structure. *Experiments in Fluids*, 2009, 47(2): 321–331

Hansen, K. L., Kelso, R. M., and Doolan, C. J. Reduction of flow induced tonal noise through leading edge tubercle modifications. AIAA Paper 2010–3700

Johari, H., Henoch, C. W., Custodio, D., and Levshin, A. Effects of leading-edge protuberances on airfoil performance. AIAA Journal, 2007, 45(11): 2634–2642

Lee, S. J. and Lee, S. H. Flow field analysis of a turbulent boundary layer over a riblet surface. *Experiments in Fluids*, 2001, 30(2): 153–166

Liu, Y. and Li, G. A new method for producing "Lotus Effect" on a biomimetic shark skin. *Journal of Colloid and Interface Science*, 2012, 388(1): 235–242

Liu, Y. Z., Shi, L. L., and Yu, J. TR-PIV measurement of the wake behind a grooved cylinder at low Reynolds number. *Journal of Fluids and Structures*, 2011, 27(3): 394–407

Martin, S. and Bharat, B. Fluid flow analysis of a shark-inspired microstructure. *Journal of Fluid Mechanics*, 2014, 756: 5–29

Miklosovic, D. S., Murray, M. M., Howle, L. E., and Fish, F. E. Leading-edge tubercles delay stall on humpback whale (*Megaptera novaeangliae*) flippers. *Physics of Fluids*, 2004, 16(5): 39–42

Miklosovic, D. S., Murray, M. M., and Howle, L. E. Experimental evaluation of sinusoidal leading edges. *Journal of Aircraft*, 2007, 44(4): 1404–1408

Niu, J. and Hu, D. L. Drag reduction of a hairy disk. *Physics of Fluids*, 2011, 23(10): 101701

Nugroho, B., Hutchins, N., and Monty, J. P. Large-scale spanwise periodicity in a turbulent boundary layer induced by highly ordered and directional surface roughness, *International Journal of Heat and Fluid Flow*, 2013, 41: 90–102

Ozen, C. A. and Rockwell, D. Control of vortical structures on a flapping wing via a sinusoidal leading-edge. *Physics of Fluids*, 2010, 22(2): 021701

Pedro, H. T. C. and Kobayashi, M. H. Numerical study of stall delay on humpback whale flippers. AIAA Paper 2008–0584

Talley, S., Iaccarino, G., Mungal, G., and Mansour, N. An experimental and computational investigation of flow past cacti. Annual Research Briefs. Center for Turbulence Research, NASA Ames/Stanford University, 2001: 51–63

Talley, S. and Mungal, G. Flow around cactus-shaped cylinders. Annual Research Briefs. Center for Turbulence Research, NASA Ames/Stanford University, 2002: 363–376

Venkataraman. D., Bottaro, A., and Govindarajan, R. A minimal model for flow control on an aerofoil using a poro-elastic coating. *Journal of Fluids and Structures*, 2014, 47: 150–164

Venkataraman, D. and Bottaro, A. Numerical modeling of flow control on a symmetric aerofoil via a porous, compliant coating. *Physics of Fluids*, 2012, 24(9): 093601

Viswanath, P. R. Aircraft viscous drag reduction using riblets. *Progress in Aerospace Sciences*, 2002, 38(6–7): 571–600

Wang, J. J., Lan, S. L., and Chen, G. Experimental study on the turbulent boundary layer flow over riblets surface. *Fluid Dynamics Research*, 2000, 27(4): 217–229

Yamagishi, Y. and Oki, M. Numerical simulation of flow around a circular cylinder with curved sectional grooves. *Journal of Visualization*, 2007, 10(2): 179–186

Yoon, H. S., Hung, P. A., Jung, J. H., and Kim, M. C. Effect of the wavy leading edge on hydrodynamic characteristics for flow around low aspect ratio wing. *Computers & Fluids*, 2011, 49(1): 276–289

Zhang, X. W., Zhou, C. Y., Zhang, T., and Ji, W. Y. Numerical study on effect of leading-edge tubercles. *Aircraft Engineering and Aerospace Technology*, 2013, 85(4): 247–257

7 Jet

7.1 Background

The jet is also called the free jet, the steady jet, or the continuous jet, and is one of the earliest techniques used for boundary layer flow control. It usually issues from an orifice and powered by a pump or a high-pressure air source. Thus, it is also known as blowing control. The issuing jet can enhance momentum mixing between inner and outer boundary layer, which is beneficial for separation delay. Similarly, suction control is another approach by drawing the low momentum fluids away from the near-wall region. Therefore, it can also be used for flow control. In particular, both these techniques are sometimes applied simultaneously to the object.

The origin of boundary layer flow control can be traced back to 1904, when boundary layer theory was developed by Prandtl. He was also considered to be the first to use steady suction control to delay flow separation of diffusers and a circular cylinder (Joslin and Miller 2009). Betz (1961) reviewed the early history of boundary layer control (in German). In particular, suction control has been applied in flight test by the Aerodynamic Research Institute of Göettingen, showing that the flow separation over the flap was completely eliminated. From then on, blowing/suction was applied for flow control in various fields. In the early stage, free jets were used as the traditional approach for flow control. Recently, several novel control conceptions based on free jets to modify the effects of other techniques have been proposed, such as the jet vortex generator and jet Gurney flap.

In this chapter, we will first introduce the fundamental characteristics of the free jet and its interaction with the crossflow. Then the jets used in airfoils and wings are presented for steady and unsteady control. Finally, the novel control techniques based on free jets are introduced.

7.2 Fundamental Characteristics

7.2.1 Free Jets

The jet is usually issued from a nozzle, and typical flow patterns are shown in Figure 7.1. Figure 7.1(a) shows the typical jet development during the transitional stage at moderate

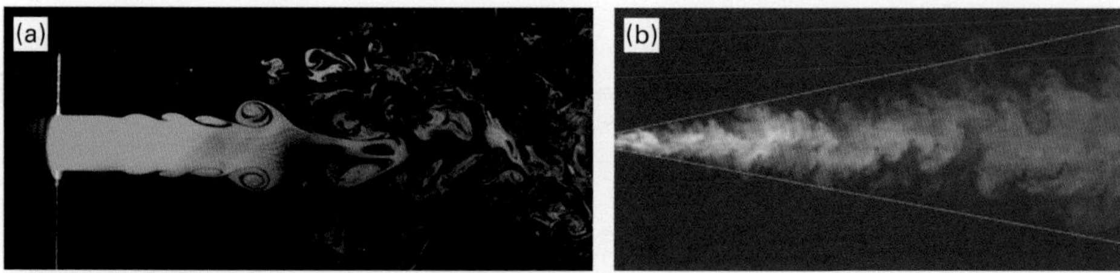

Figure 7.1 Flow pattern of free jets at different Reynolds numbers. (a) $Re_j = 1620$ (Todde et al. 2009), reproduced with permission, copyright © Springer-Verlag 2009; and (b) $Re_j = 10^4$ (Cater and Soria 2002), reproduced with permission, copyright © Cambridge University Press 2002.

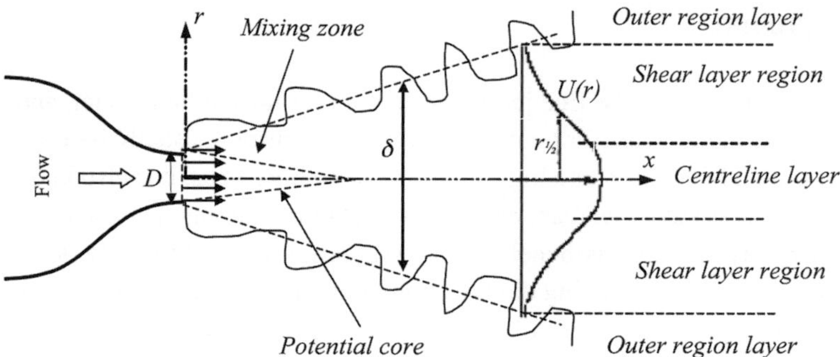

Figure 7.2 Schematic of a free turbulent jet issuing from a nozzle (Ball et al. 2012). Reproduced with permission, copyright © 2012 Elsevier Ltd. All rights reserved.

Reynolds number. A laminar jet forms from the orifice, which undergoes flow transition resulting in turbulent flow. However, the jet becomes fully turbulent at high Reynolds numbers, as shown in Figure 7.1(b).

For a turbulent free jet, the typical flow development is shown in Figure 7.2. There are three different regions: the near field ($0 \leq x/D \leq 7$, where D is the diameter of the nozzle), the intermediate field ($7 \leq x/D \leq 30$), and the far field ($x/D \geq 30$). The near field is featured by the potential core for flow establishment, and followed by the mixing zone, which finally develops into turbulence.

At the same time as the jet development, vortical structures are also induced. The flow separates from the orifice, forming the shear layer, as also shown in Figure 7.2. The generation mechanism of the vortices depends on the state of the boundary layer at the jet exit (Toyoda and Hiramoto 2009). When the flow is laminar, the shear layer rolls up into the vortical structures due to the Kelvin–Helmholtz instability. The vortices merge repeatedly and evolve into large-scale vortices near the end of the potential core zone. The formation and evolution of the large-scale vortices show periodic characteristics, where a dominant frequency f can be detected. Despite the difference in the diameter of the orifice D and the jet velocity U_j, the Strouhal number $St = fD/U_j$ derives a nearly constant value of 0.24–0.51 at around $x/D = 4$. However, when the flow is turbulent, the vorticity in

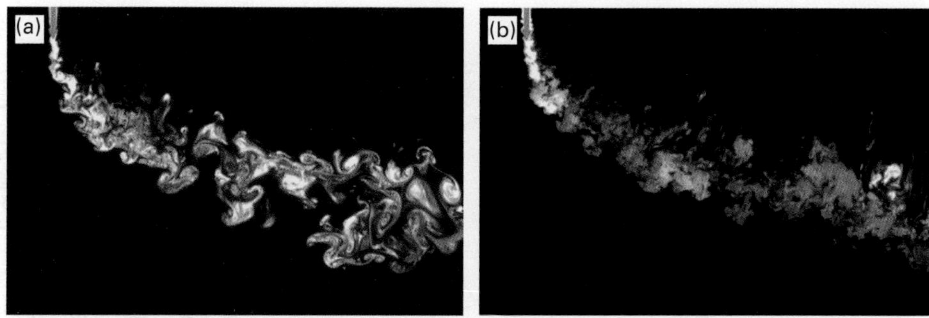

Figure 7.3 Instantaneous flow pattern of a free jet in crossflow at $R = 10$ and (a) $Re_j = 10^3$ and (b) $Re_j = 10^4$ (Shan and Dimotakis 2006). Reproduced with permission, copyright © Cambridge University Press 2002.

the shear layer is concentrated into clusters, forming large-scale vortices. The Strouhal number for the jet developing from fully developed pipe flow is about 0.38.

7.2.2   Interaction of Jets with Crossflow

As a basis for flow control with free jets, it is important to understand the interaction of the jet with crossflow (Mahesh 2013). There are several important dimensionless parameters that can determine the flow pattern. One is the ratio of the jet velocity to the free-stream velocity, $R = U_j/U_\infty$. However, the strength of the jet can also be scaled by the jet Reynolds number based on the jet velocity and the orifice diameter, $Re_j = U_j D/v$. There is also another parameter to scale: the boundary-layer thickness to the jet diameter δ/D. The free-stream Reynolds number can be obtained based on the jet orifice diameter, $Re = U_\infty D/v$.

A typical flow pattern is shown in Figure 7.3 for two Reynolds numbers at a fixed jet-to-free-stream velocity ratio of 10. The interaction leads to a complex three-dimensional and highly unsteady flow. The free jet is bent with an incline angle to the free stream. At low Reynolds numbers, the jet core remains laminar for several diameters after exiting the orifice, as shown in Figure 7.3(a). However, at high Reynolds numbers, the jet's potential core transits and mixes within approximately one diameter from the orifice, as shown in Figure 7.3(b). The full development of the jet in a free stream is featured with small-scale structures, which are much smaller at higher Reynolds numbers.

Figure 7.4 presents sketches of flow topology for a round jet in crossflow at high-velocity-ratio, with the time-averaged case (Figure 7.4(a)) and the instantaneous case (Figure 7.4(b). The interaction between the jet and crossflow leads to the formation of different wake patterns, including the counter-rotating vortex pair, the wake vortex, the horseshoe vortex, the hovering vortex or inner vortex, the leading-edge and trailing-edge vortex. They are introduced as follows:

(1) The counter-rotating vortex pair is usually considered to be the main vortical feature, however, only for the time-averaged flow topology. It originates from

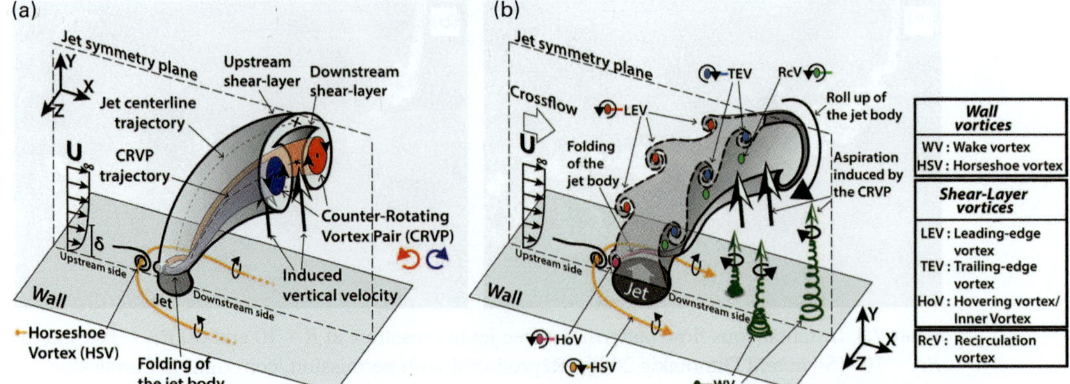

Figure 7.4 (a) Time-averaged and (b) instantaneous flow topology of the high-velocity-ratio round jet in a crossflow (Cambonie and Aider 2014). Reproduced with permission from AIP Publishing.

a folding of the jet body and shear layers near the base of the jet. It also induces vertical velocity towards the upstream shear layer as well as the wake vortices shown in the instantaneous topology. Such a flow feature contributes to the flow entrainment and momentum mixing of the surrounding flow field. Thus, the counter-rotating vortex pair is usually given particular attention by many researchers.

(2) As indicated above, the wake vortex is also called the upright vortex. It exists in a limited velocity ratio range of 2 to 10, when the trajectory of a counter-rotating vortex pair is high enough to lift them up.

(3) The horseshoe vortex, which is formed by the roll-up of the boundary layer, wraps around the jet. It is similar to that formed in the junction flow upstream of the object.

(4) The hovering vortex is located between the horseshoe vortex and the jet base, which is also called the inner vortex.

(5) The leading-edge vortices are induced from the upstream shear layer while the trailing-edge vortices develop from the downstream shear layer. They are both induced because of the Kelvin–Helmholtz instability.

(6) The recirculation vortex develops in the shear layer on the downstream side of the low velocity area located just behind the jet.

Figure 7.5 shows the summary of flow pattern at different Reynolds numbers and velocity ratios. The wake vortex or upright vortex does not occur at smaller Reynolds numbers or at smaller velocity ratios, while it can be observed to shed when the Reynolds number and velocity ratio reach the threshold. However, the hovering vortex can occur while the Kelvin–Helmholtz shear layer rolls-up at small Reynolds numbers, while the large-scale shear layer rolls-up at high Reynolds numbers.

As we can see from the flow visualization, the jet is bent by the free stream. Thus the jet trajectories can be calculated, as shown in Figure 7.6. The curves are scaled by the velocity ratio and the orifice diameter. Differences can be seen from the curves with

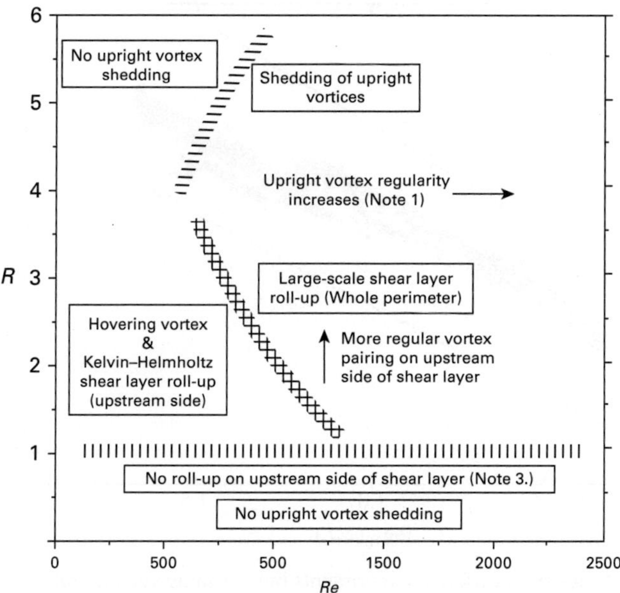

Figure 7.5 Summary of flow phenomenon of a free jet in crossflow as a function of the velocity ratio and the Reynolds number (Kelso et al. 1996). Reproduced with permission, copyright © Cambridge University Press 1996.

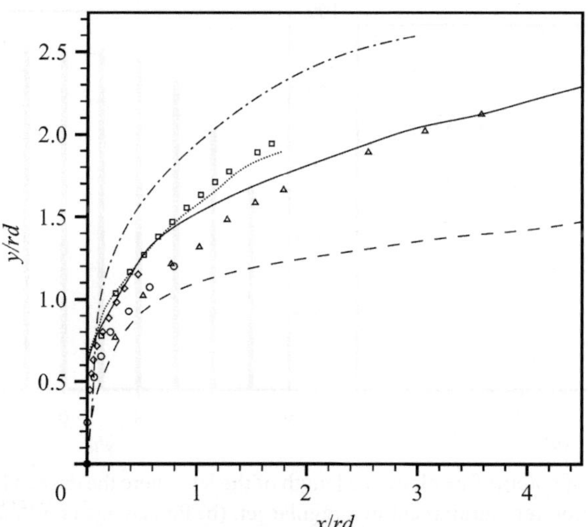

Figure 7.6 Jet trajectories from different experiments scaled by the velocity ratio and the orifice diameter. Data from different researchers are summarized by Muppidi and Mahesh (2005). Reproduced with permission, copyright © Cambridge University Press 2005.

different velocity ratios. However, one can see the linear relationship of the x and y axes in the logarithmic law at the same velocity ratio, while the Reynolds number has less influence, as shown in Figure 7.7.

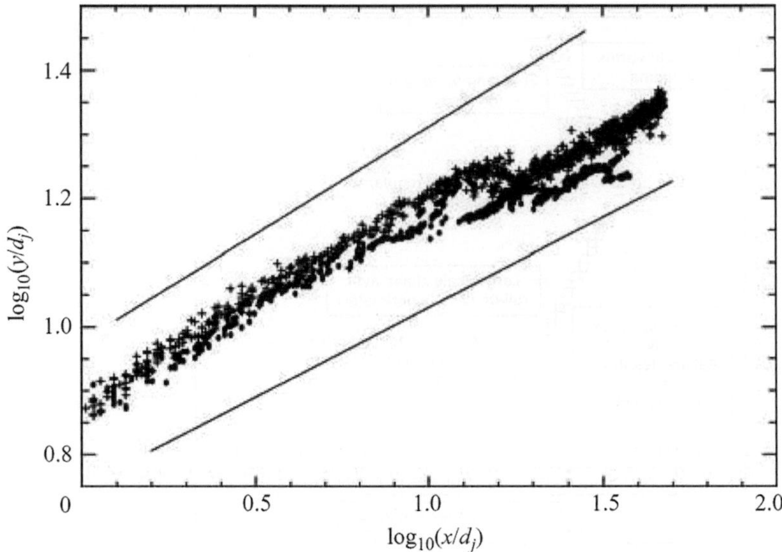

Figure 7.7 Jet trajectories from different experiments but the same velocity ratio of 10 in the logarithmic law. The lower and upper straight lines show the relationship of $y/D \propto (x/D)^{0.28}$ and $y/D \propto (x/D)^{1/3}$, respectively (Shan and Dimotakis 2006). Reproduced with permission, copyright © Cambridge University Press 2006.

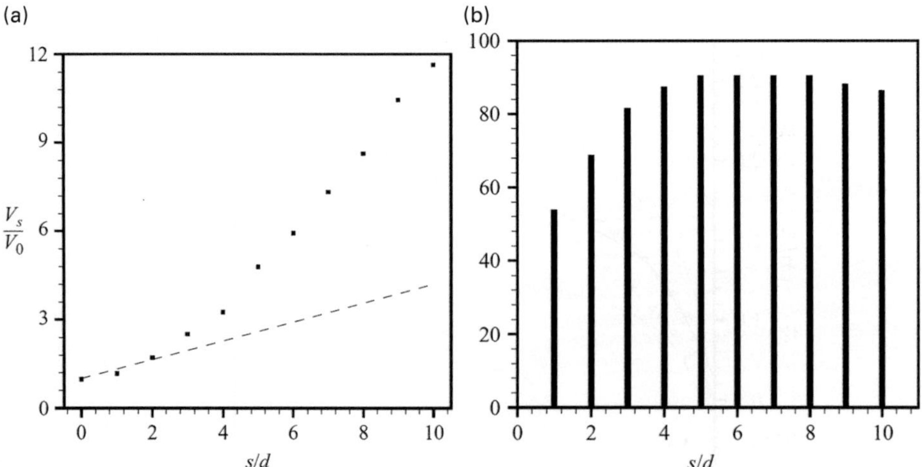

Figure 7.8 (a) Variations of volume flux along the length of the jet, where the dashed line shows the Ricou–Spalding correlation for entrainment in a regular jet. (b) Percentage contribution of the downstream side of the jet to the total entrainment (Muppidi and Mahesh 2008). Reproduced with permission, copyright © Cambridge University Press 2008.

The jet in a free stream is thought to be more efficient at entrainment and mixing with surrounding fluids than a regular free jet. The decay in the centerline velocity and jet fluid concentration is also more rapid than a free jet. This can be validated from Figure 7.8(a) for the variation of the volume flux with distance from the jet exit.

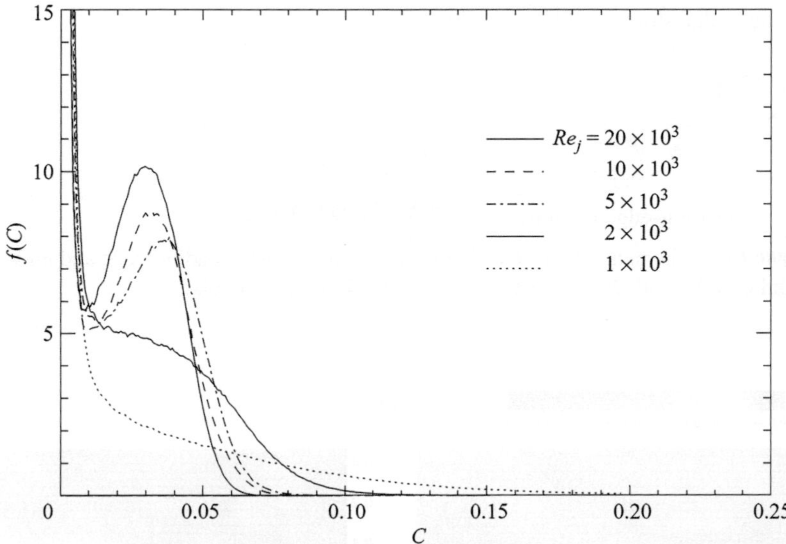

Figure 7.9 Distribution of jet-fluid-concentration for different jet Reynolds numbers at velocity ratio of 10 (Shan and Dimotakis 2006). Reproduced with permission, copyright © Cambridge University Press 2006.

The jet in the crossflow shows a higher entrainment rate past $2D$, and it becomes more than twice that for a regular free jet from $6D$.

Figure 7.8(b) shows the percentage contribution of the downstream side of the jet to the total entrainment. It is found that both the upstream and downstream sides of the jet contribute similar amounts to entrainment close to the jet exit, where the volume flux is similar to that of a free jet. However, the contribution of total entrainment from the downstream side increases further downstream, which reaches about 90% of the total entrainment from about $5D$ from the orifice. It can maintain the high contribution up to $10D$ from the orifice.

The flow mixing ability can be measured by the probability density function (PDF) of the concentration. Figure 7.9 shows scalar PDFs for five different jet Reynolds numbers at the same velocity ratio of $R = 10$. It is clearly shown that the peak value increases with the jet Reynolds number, indicating that the mixed-fluid concentration becomes more homogeneous at higher jet Reynolds numbers.

7.3　Jets for Flow Control

7.3.1　Airfoil

Steady Flow Control

The free jet has been widely used in various fields as a blowing tool. Müller-Vahl and Greenblatt (2015) conducted an experimental study on the control of the flow around a NACA 0018 airfoil in a wind tunnel. The airfoil model had a chord length

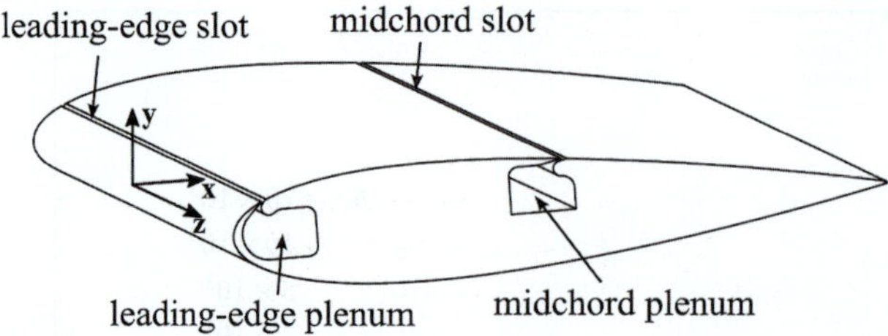

Figure 7.10 Schematic of the NACA 0018 airfoil model with leading-edge and midchord blowing (Müller-Vahl et al. 2015). Data kindly supplied by D. Greenblatt.

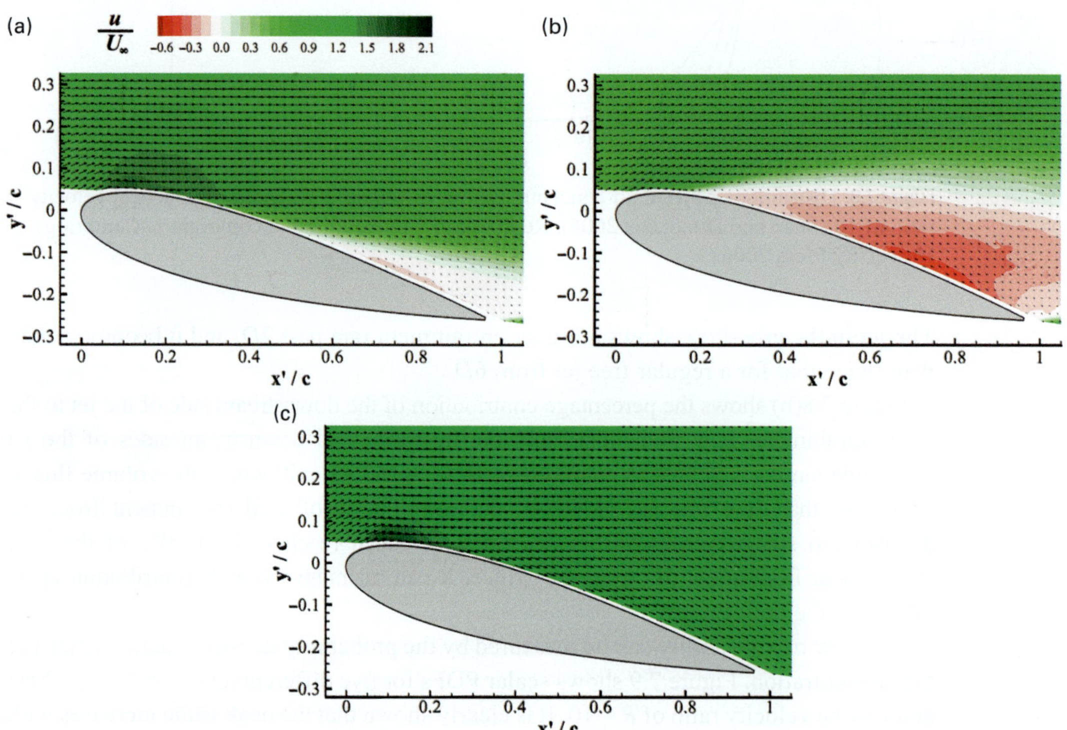

Figure 7.11 Mean streamwise velocity fields for the natural and leading-edge blowing cases at $\alpha = 15°$, $Re = 2.5 \times 10^5$. (a) Natural case; (b) $C_\mu = 0.6\%$; and (c) $C_\mu = 5\%$ (Müller-Vahl et al. 2015). Data kindly supplied by D. Greenblatt.

of $c = 347$ mm and a span of $s = 610$ mm. Two narrow slots with 1.2 mm width were positioned at the 5%c and 50%c suction surface from the leading edge, as shown in Figure 7.10. The free-stream velocities were selected at $U_\infty = 5.6$ m/s, 11.1 m/s, and 16.7 m/s, corresponding to $Re = 1.25 \times 10^5$, 2.5×10^5, and 3.75×10^5, respectively. The momentum coefficient, $C_\mu = wU_j^2/(0.5cU_\infty^2)$, was varied from 0.3% to 5%.

The time-averaged velocity fields over the airfoil at $\alpha = 15°$ without and with leading-edge jet control are shown in Figure 7.11. The flow over the airfoil is nearly attached

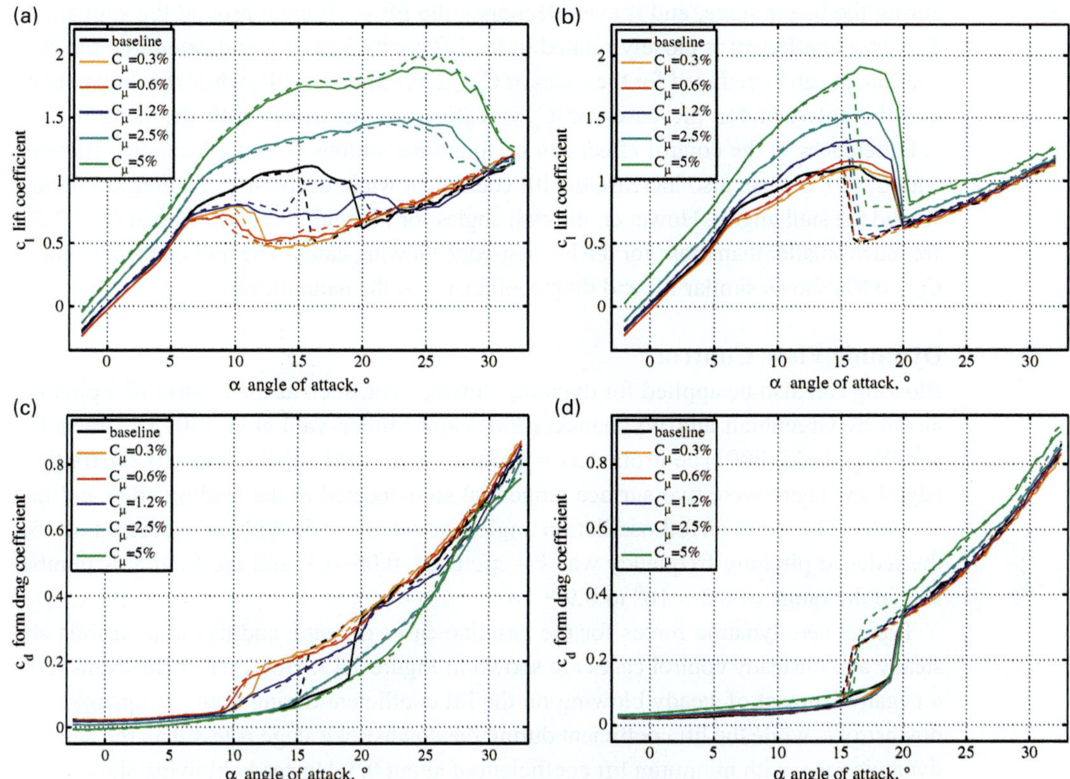

Figure 7.12 Aerodynamic forces on a symmetric airfoil with 5%c leading-edge (a, c) and 50%c midchord (b, d) blowing at different momentum coefficients. (a, b) Lift coefficient: (c, d) drag coefficient. The pitch-up motion is indicated with solid lines, and the pitch-down motion is indicated with dashed lines (Müller-Vahl et al. 2015). Data kindly supplied by D. Greenblatt.

over most of the suction surface, except for a portion of reverse flow near the trailing edge. With steady jet control, opposite effects may occur, depending on the momentum coefficient. The control case with $C_\mu = 0.6\%$ may enlarge the separation region, while the case with $C_\mu = 5\%$ can completely eliminate it, leading to a fully attached flow.

The effects of the momentum coefficient on the aerodynamic forces are shown in Figure 7.12. The left column is for the leading-edge blowing case, while the right is for the mid-chord blowing. When the free jet is issuing from the leading-edge slot, the lift coefficient can be increased at high momentum coefficient ($C_\mu \geq 2.5\%$) for all angles of attack, from the linear stage to the post stall angles. The lift enhancement becomes more significant around the stall angles. In particular, the maximum lift coefficient of the pitch-up motion can be increased by 19% and 51% for $C_\mu = 2.5\%$ and 5% control cases, respectively, while both the stall angles are posted to 24° from the original 17°. For the pitch-down motion, the maximum lift coefficient is increased by 24% and 61% for $C_\mu = 2.5\%$ and 5% control cases, respectively, while the stall angle is posted to 22° and 26° from the original 16°. The blowing for $C_\mu \leq 1.2\%$ may not influence the lift coefficient during the linear stage, and it even decreases the lift coefficient around the stall angle. The drag coefficient is closely related to the lift coefficient. It is indicated that the drag coefficient can be reduced for the cases of $C_\mu \geq 2.5\%$ after $\alpha \geq 19°$, while the drag penalty may be enlarged for the cases of $C_\mu \leq 1.2\%$ in comparison with the natural case.

Differences in the control effect can be found for various blowing locations. Blowing with $C_\mu \geq 1.2\%$ can also increase the lift coefficient while decreasing the drag coefficient around the stall angles. However, the stall angles for the mid-chord blowing at $C_\mu \geq 2.5\%$ are much smaller than those for the leading-edge blowing cases. The mid-chord blowing at $C_\mu \leq 0.6\%$ shows similar lift and drag coefficients to the natural case.

Dynamic Flow Control

Blowing can also be applied for dynamic flow control, such as the control of a pitching airfoil by Greenblatt and Wygnanski (2001) and Müller-Vahl et al. (2015). Greenblatt and Wygnanski (2001) controlled a $c = 365$ mm NACA0015 airfoil with a $0.25c$ trailing-edge flap. There were two surface tangential slots located at the leading edge and flap shoulder. The mean airfoil oscillation angle was set at $\alpha = 5°$ with an excursion of $\pm5°$, the reduced pitching frequency was $k = \pi fc/U_\infty = 0.05$–$0.3$, and the Reynolds number was in the range of 0.3×10^6 to 0.9×10^6.

Typical aerodynamic forces for the baseline cases of static and dynamic airfoils and steady and unsteady control cases are shown in Figure 7.13. However, there seems to be a negative impact of steady blowing on the lift coefficient during both the upstroke and downstroke, while the lift coefficient during the downstroke stage is less than the baseline dynamic case, with minimum lift coefficient of about 0.3. Unsteady blowing shows great ability to improve the aerodynamic forces of the pitching airfoil. It is found that the lift coefficients during both upstroke and downstroke are nearly similar, indicating that it effectively eliminates lift hysteresis. For the drag coefficient, unsteady blowing may slightly increase it in comparison with the baseline dynamic case, while steady blowing greatly increases it. Similar effects also occur for the momentum coefficient.

The above comparison between steady blowing and unsteady blowing indicates that unsteady blowing may be more efficient in some cases. There might be two important reasons for this. First, unsteady control can make use of the excitation frequency. When it is coupled with the natural frequency, the instability of the local boundary layer may be triggered to drive the flow into the ideal state. Second, unsteady blowing may induce more vortical structures, which are also beneficial for flow control. Thus, as a typical unsteady control case, the synthetic jet, which is characterized by periodic blowing and suction, is found to be more efficient than the conventional free jet. It will be introduced in detail in the subsequent chapter.

However, it is not always the case that steady blowing may damage the aerodynamic performance of a pitching airfoil. One of the crucial parameters that can influence the effects is the momentum coefficient. It was shown by Müller-Vahl et al. (2015) that the blowing case with $C_\mu = 0.6\%$ might decrease the lift coefficient at some angles of attack, while the steady blowing with $C_\mu \geq 2.5\%$ could increase the lift coefficient during most stages of the upstroke and downstroke motions. This can be clearly seen from Figure 7.14.

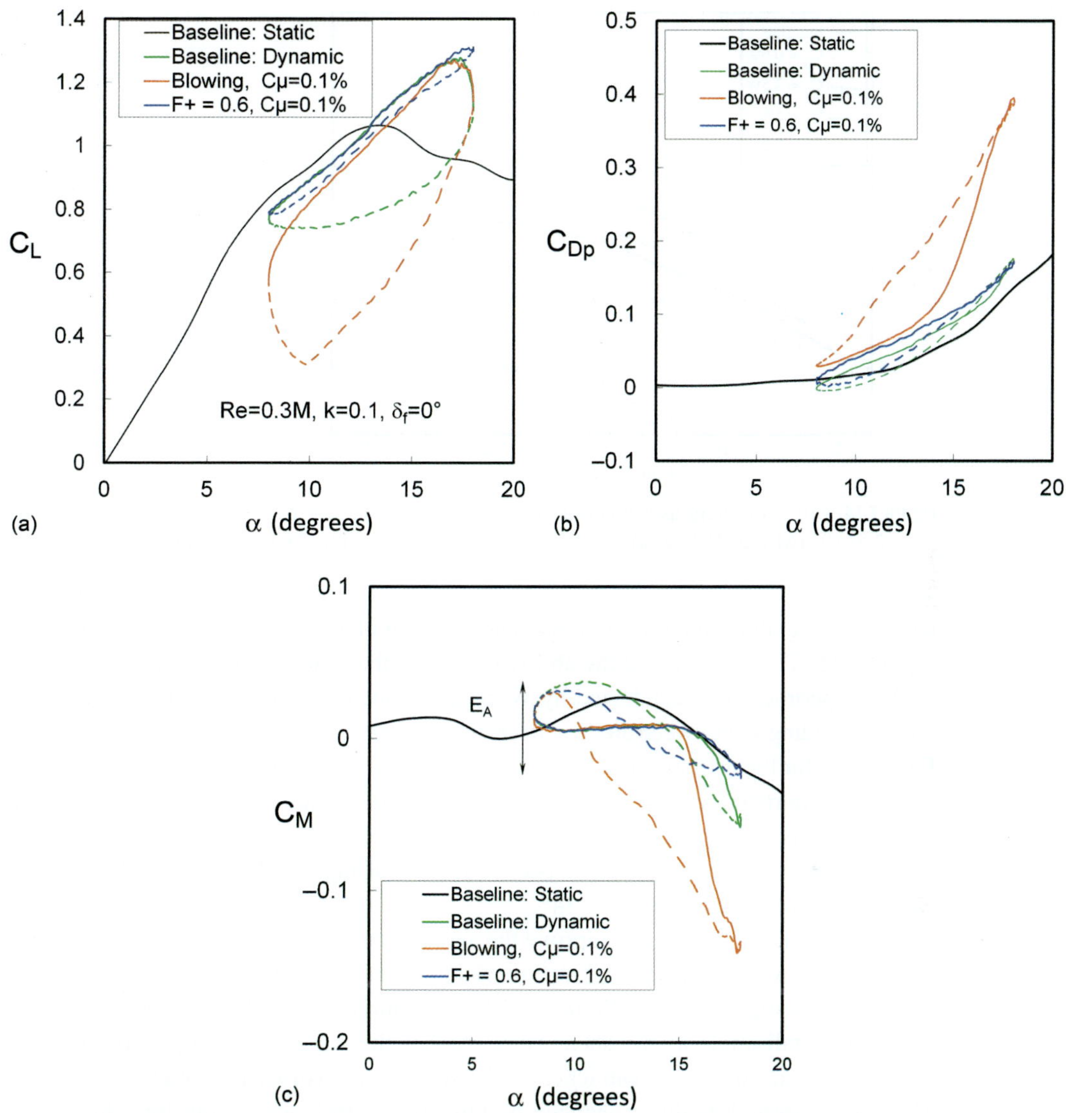

Figure 7.13 Aerodynamic forces on the NACA 0015 airfoil for the natural and control cases. (a) Lift coefficient; (b) form drag coefficient; and (c) pitching moment coefficient (Greenblatt and Wygnanski 2001). Data kindly supplied by D. Greenblatt.

7.3.2   Delta Wing

The leading-edge vortex can provide the lift coefficient for the delta wing. A leading vortex with high strength may provide high lift coefficient, while its breakdown may lead to a sudden drop in lift. Thus, it is essential for delta wing flow control to strengthen the leading-edge vortex (Gursul et al. 2007). There are different approaches to free jet control based on the direction of blowing, such as spanwise blowing along the leading

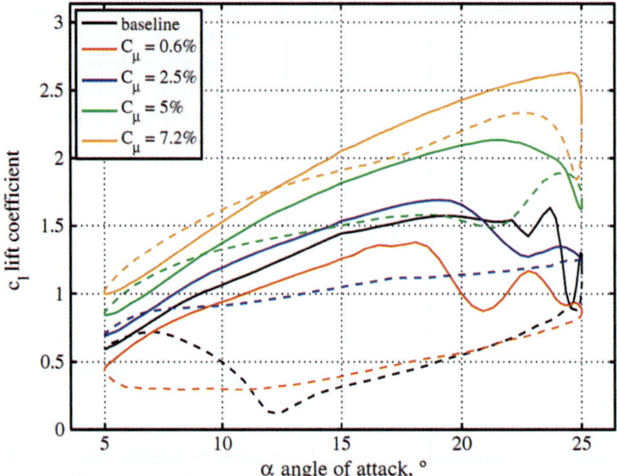

Figure 7.14 Lift coefficient as a function of the angle of attack $\alpha = 15° + 10°\sin(2\pi ft)$, $k = 0.074$, $Re = 2.5\times10^5$ (Müller-Vahl et al. 2015). Data kindly supplied by D. Greenblatt.

edge, blowing along the trailing edge, and blowing along the vortex axis, as shown in Figure 7.15. They all showed the ability to control the flow around a delta wing.

One experiment was conducted by Wood and Roberts (1988) to control a delta wing with 60° leading-edge sweep angle with spanwise blowing. The lift coefficient is shown in Figure 7.16 for the natural and different control cases. It is clearly shown that blowing can increase the normal force after stall, though there is nearly no difference during the linear stage. The maximum lift coefficient is increased by about 9%, 15% and 21% for the $C_\mu =$ 0.02, 0.04 and 0.06 control cases, respectively, the stall angle is also postponed to $\alpha = 39°$, 40° and 45° from the original $\alpha = 34°$.

Unsteady blowing can also improve the aerodynamic performance of the delta wing, in a similar way to steady control. A typical control effect on the leading-edge vortex by spanwise blowing is shown in Figure 7.17 by Williams et al. (2008). The angle of $\alpha = 25°$ is near the stall angle as the shear layer reaches the wing centerline. With unsteady blowing control, the flow reattaches to the wing and a vortex flow pattern develops. The vortex strength is thus enhanced by blowing control, accounting for a higher lift coefficient.

Wang et al. (2003) used trailing-edge blowing to control a delta wing beveled at an angle of 60°, as shown in the sketch of the control configuration in Figure 7.18. The jet was vectoring towards the left side. The flow pattern at $\alpha = 25°$ for the natural case is shown in Figure 7.18(a). The leading-edge vortices can be clearly seen over the suction surface. However, they break down in an asymmetric manner. Blowing control results in a delay of the left vortex breakdown, with an earlier vortex breakdown on the right side. In comparison with the $\beta = 35°$ vectoring-angle control case (Figure 7.18(b)), the $\beta = 60°$ blowing case can lead to an even more significant delay of the left leading-edge vortex (Figure 7.18(c)).

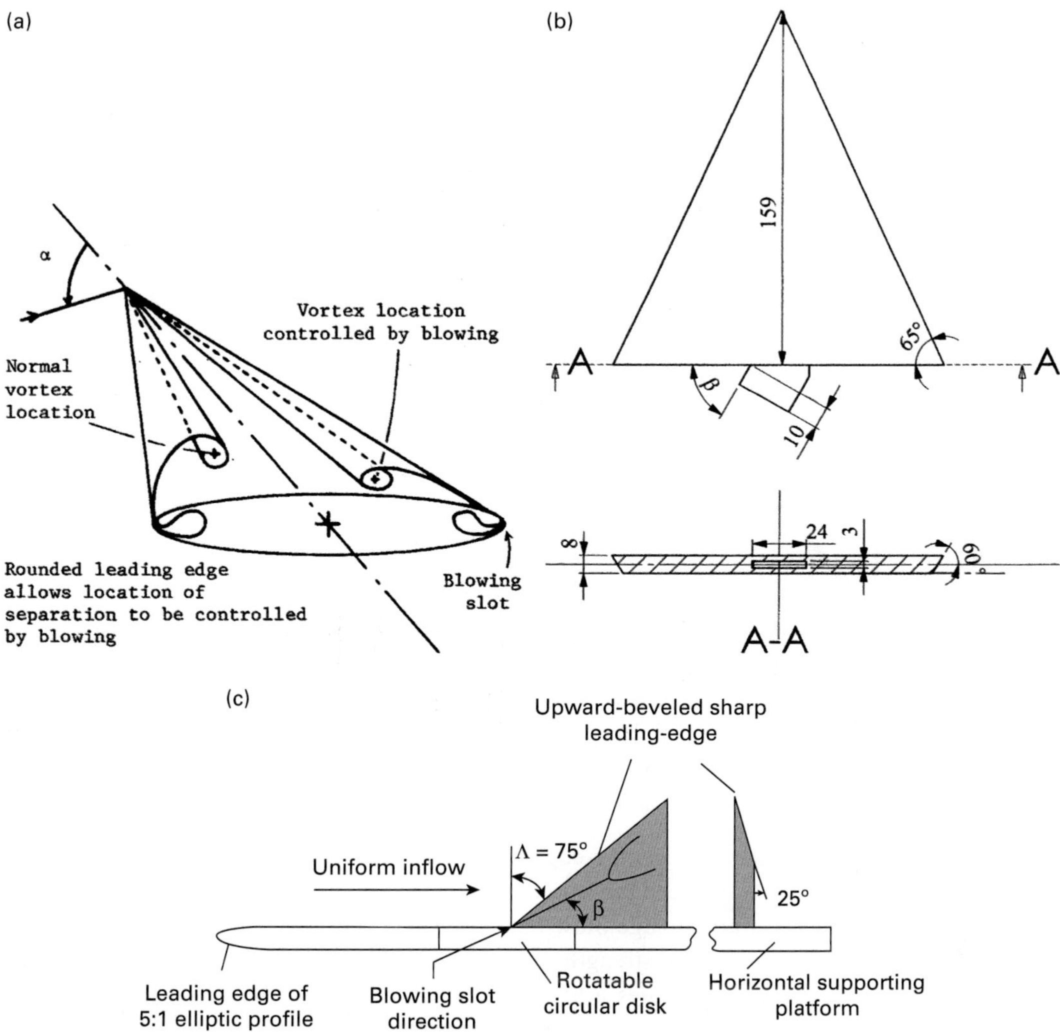

Figure 7.15 Examples to show the different jet-based control approaches for delta wing or similar configuration. (a) Spanwise blowing along the leading edge (Wood and Roberts 1988), reproduced with permission, copyright © The American Institute of Aeronautics and Astronautics. (b) Blowing along the trailing edge (Wang et al. 2003), reproduced with permission, copyright © Springer-Verlag 2003. (c) Blowing along the vortex axis (Kuo and Lu 1998), reproduced with permission, copyright © The American Institute of Aeronautics and Astronautics.

A summary of the variations of the vortex breakdown location with the vectoring angle and velocity ratio is shown in Figure 7.19. For the left leading-edge vortex, the $\beta = 10°$ blowing has nearly no influence, while the control cases of $\beta = 35°$ and $60°$ can greatly delay vortex breakdown. It is suggested that the control effects on the left vortex become more significant with an increase in both the vectoring angle and the velocity

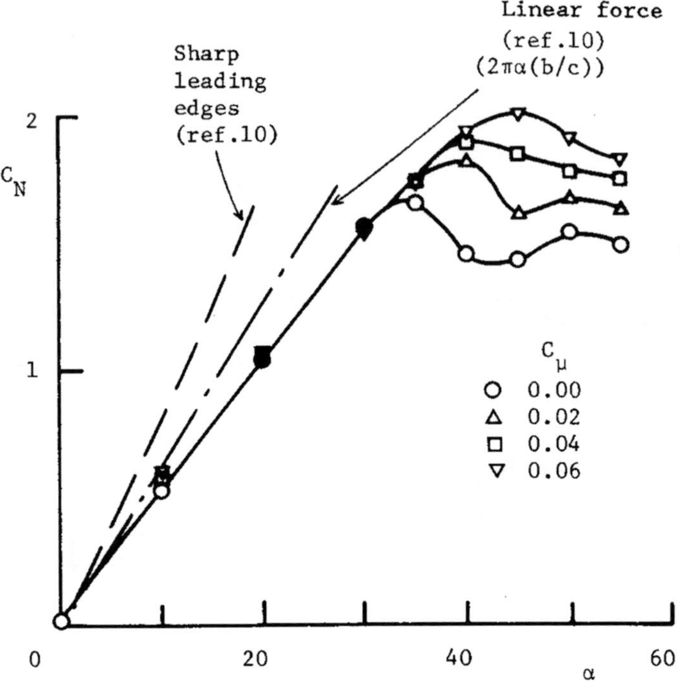

Figure 7.16 Normal force on a delta wing for the natural and leading-edge blowing cases (Wood and Roberts 1988). Reproduced with permission, copyright © The American Institute of Aeronautics and Astronautics.

ratio. However, the right leading-edge vortex breakdown becomes earlier, resulting in an increased distance between the left and right vortices.

The effect of trailing-edge blowing on one side was also validated by Wang et al. (2007). Since the vortex strength over the left and right leading-edge vortices may not be the same, it provides a strategy for roll control of the delta wing. One can also provide enough rolling momentum if increasing the vortex strength of one side only. However, when the vectoring jets are issued from both sides in a symmetric manner, the flow over both sides can be improved, as shown by Yavuz and Rockwell (2006).

7.3.3   Aircraft

Free jets have also been validated for their effects on aircraft configurations, in the forms of spanwise blowing, trailing-edge blowing, and blowing along the vortex axis. Cui et al. (2007, 2008) conducted a series of investigations, and one example is shown in Figure 7.20. A delta wing-body configuration model with a 60° sweep angle was used for flow control. The Reynolds number based on root chord length and the free-stream velocity was about $Re = 8.5 \times 10^4$.

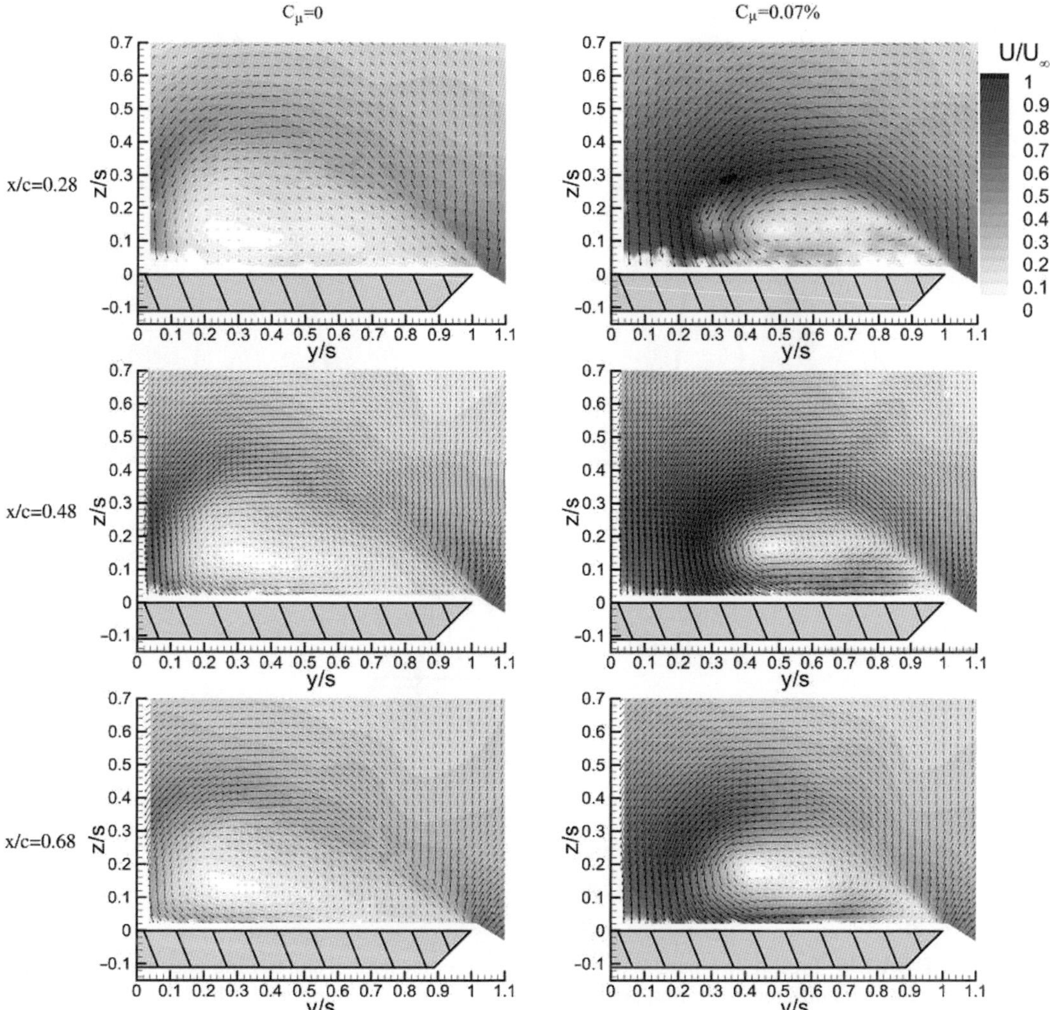

Figure 7.17 Time-averaged crossflow velocity at $\alpha = 25°$ for the unsteady blowing at $C_\mu = 0.07\%$, $k = 1.3$. The left column is for the natural case and the right for the control case (Williams et al. 2008). Reproduced with permission, copyright © 2008 by Ismet Gursul.

Typical flow visualization of the aircraft model to show the vortex breakdown for different momentum coefficients is shown in Figure 7.21. The leading-edge vortex breaks down very early close to the apex point. Clearly, spanwise blowing can delay the vortex breakdown position, which becomes more significant at higher momentum coefficients. Accordingly, the lift coefficient increases with the momentum coefficient, as validated in Figure 7.22. With an increase in the angle of attack, larger blowing momentum is necessary to postpone the vortex breakdown location, and thus to achieve a considerable lift enhancement.

Rahman and Whidborne (2010) also validated that trailing-edge blowing could improve the aerodynamic performance of a blended-wing-body aircraft, which is

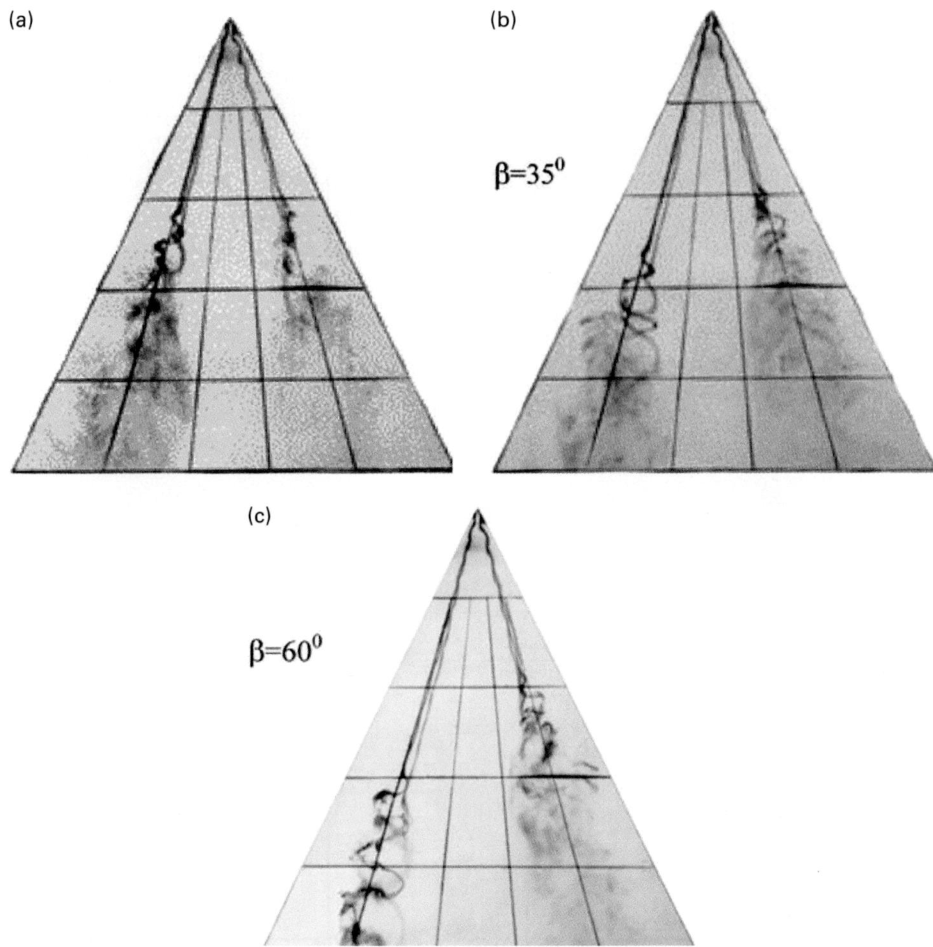

Figure 7.18 Typical flow pattern of the delta wing for the natural and trailing-edge blowing control cases with different vectoring angles at $\alpha = 25°$, $R = 8$. (a) Natural case; (b) $\beta = 35°$; (c) $\beta = 60°$ (Wang et al. 2003). Reproduced with permission, copyright © Springer-Verlag 2003.

shown in Figure 7.23. The lift coefficient, drag coefficient, and nose-down pitching moment are increased with the blowing jet for all the angles of attack tested, and the effect becomes more significant with increasing the momentum coefficient. In addition, the effects of the flap deflection of 20° are also presented, indicating the great potential of replacing the mechanical flap with trailing-edge blowing.

7.4   Novel Conceptions Based on Jets

Blowing based on the free jet is a conventional control approach. The main mechanism is that the jet may enhance the flow entrainment and momentum mixing, thus delaying

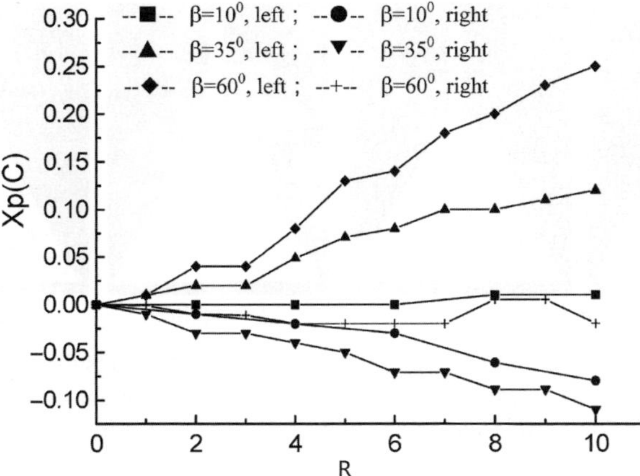

Figure 7.19 The variation of the relative location of the leading-edge vortex breakdown with velocity ratio and vectoring angle (Wang et al. 2003). Reproduced with permission, copyright © Springer-Verlag 2003.

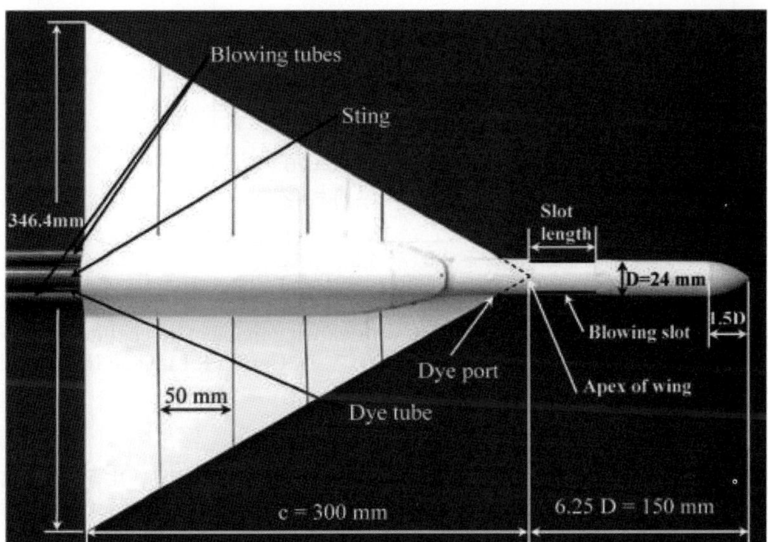

Figure 7.20 Generic delta wing-body model with a 60° sweep angle under spanwise blowing (Cui et al. 2008). Figure reproduced with permission, courtesy of Y. D. Cui, T. T. Lim and H. M. Tsai, National University of Singapore.

flow separation or even reattaching the flow. The aerodynamic performance can be thus improved. The free jet can also be applied for flow control based on other mechanisms, such as increasing the circulation over the airfoils, wings, aircrafts and so on. Besides, the free jet could also simulate the function of some of the conventional passive flow control techniques, such as vortex generators and Gurney flaps. As we will see in the following chapters, jet-based techniques can obtain similar effects as conventional techniques. However, they have additional advantages, such as active control ability.

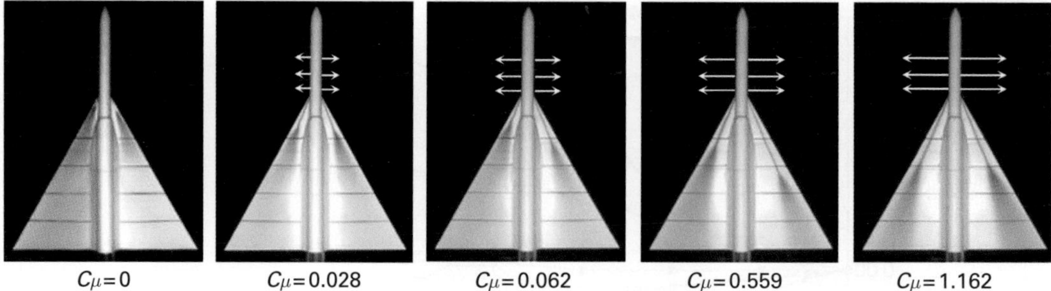

Figure 7.21 Time-averaged flow visualization of vortex breakdown structures for different blowing momentum coefficient at $\alpha = 28°$. Arrows indicate the direction of spanwise blowing (Cui et al. 2008). Figure reproduced with permission, courtesy of Y. D. Cui, T. T. Lim and H. M. Tsai, National University of Singapore.

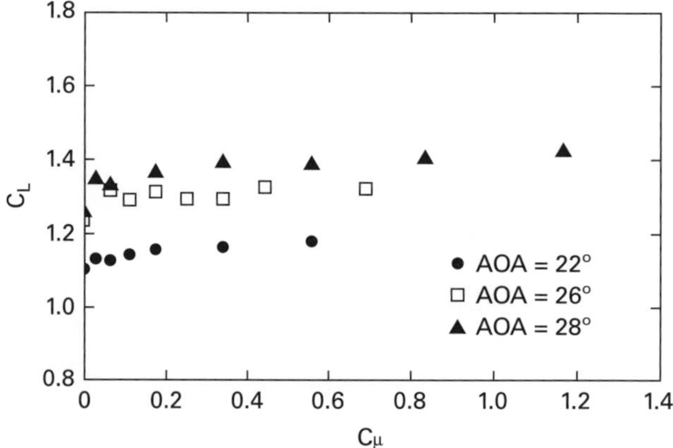

Figure 7.22 Variation of the lift coefficient with blowing momentum coefficient (Cui et al. 2008). Figure reproduced with permission, courtesy of Y. D. Cui, T. T. Lim and H. M. Tsai, National University of Singapore.

In this section, we will introduce the basic characteristics of some representative techniques, such as circulation control, jet vortex generator, and jet Gurney flap.

7.4.1   Circulation Control

The conception of circulation control is based on the Coanda effect observed by Romanian inventor Henri Coanda in the early 1930s. When this effect is used for an airfoil, it requires a rounded trailing edge instead of the traditional sharp one. A slot embedded on the round trailing-edge surface was used as the orifice for the free jet, as shown in Figure 7.24(a). When fluids flow over the curved surface, it attaches to the surface for a long distance, as shown in Figure 7.24(b).

The flow pattern for the airfoil without and with circulation control is compared in Figure 7.25. It is indicated that the tangential jet over the round trailing edge turns the near wake downwards as a result of the increased circulation over the airfoil. Thus, the

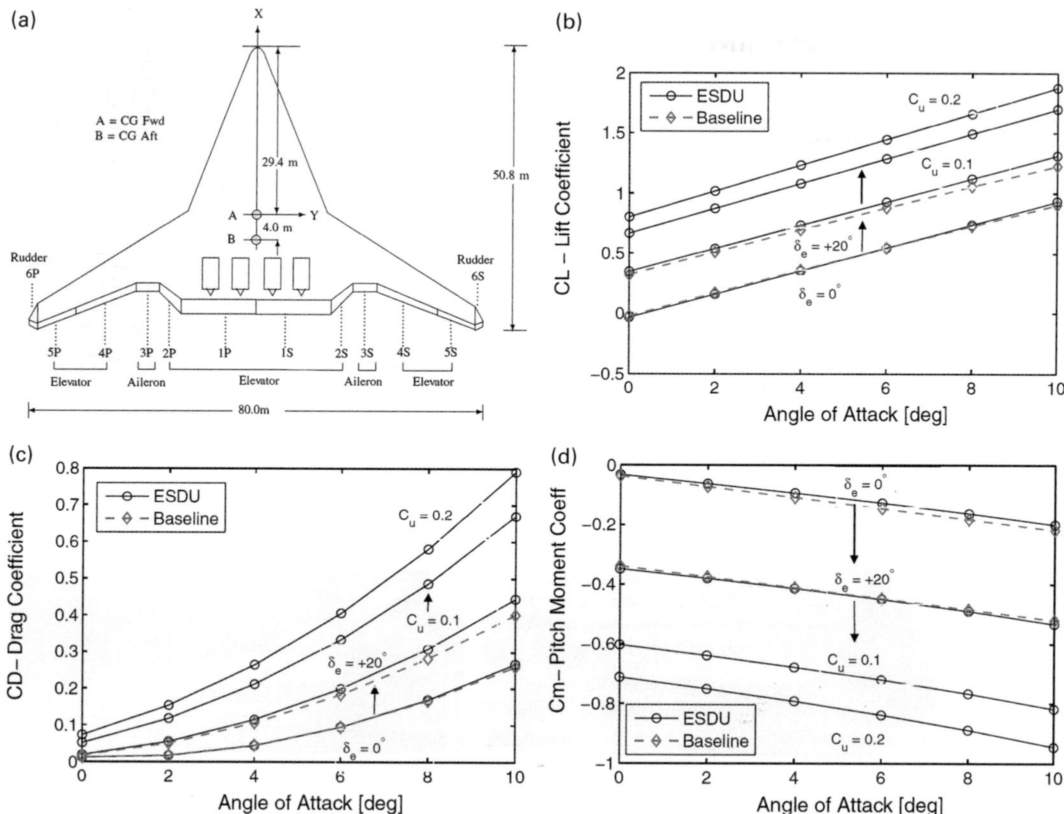

Figure 7.23 (a) Sketch of the trailing-edge blowing used on a blended-wing-body aircraft. Lift coefficient (b), drag coefficient (c), and nose-down pitching moment (d) of the aircraft without and with trailing-edge blowing (Rahman and Whidborne 2010). Reproduced with permission, copyright © The American Institute of Aeronautics and Astronautics.

circulation control can significantly increase the lift coefficient of the airfoil over all angles of attack by shifting the lift curve upwards, as shown in Figure 7.26. Additionally, increasing the momentum coefficient results in a more obvious upwards shifting. As a result, the zero-lift angle becomes more negative at a larger momentum coefficient.

The variation of the lift coefficient with the momentum coefficient at the zero angle of attack is shown in Figure 7.27(a). It is indicated that the lift coefficient increases with the momentum coefficient, resulting in a lift augmentation efficiency of about $\Delta C_L/\Delta C_\mu = 50$. There exist some other researches to show that the maximum coefficient could be up to $\Delta C_L/\Delta C_\mu = 80$. However, the drag may also increase with the momentum coefficient, thus the lift-to-drag ratio reaches a maximum at around $C_L = 1.2$, namely $C_\mu = 0.009$ (Figure 7.27(b)). Then, it decreases with an increase in the lift coefficient as well as the momentum coefficient.

In general, circulation control shows great ability to improve the aerodynamic performance. Thus, many researchers have paid particular attention to this technique. However, Loth (2004) indicated that only two aircraft with circulation control were

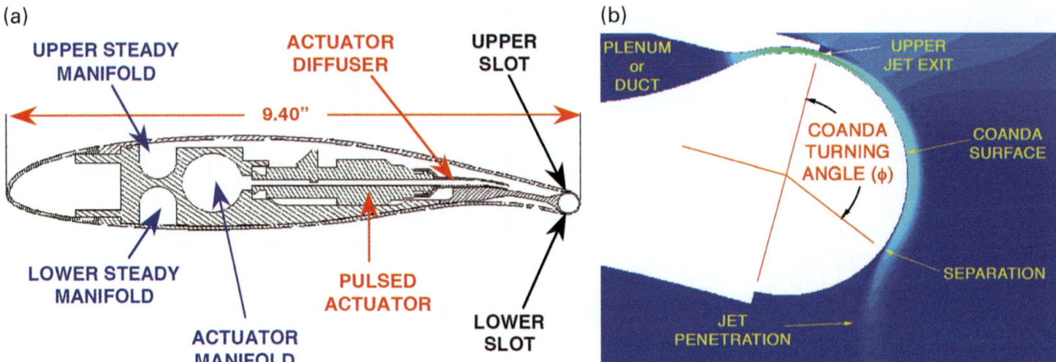

Figure 7.24 (a) Sketch of the rounded trailing-edge airfoil with circulation control; (b) detailed flow pattern near the trailing-edge to show the Coanda effect (Jones et al. 2002). Reproduced with permission, copyright © The American Institute of Aeronautics and Astronautics.

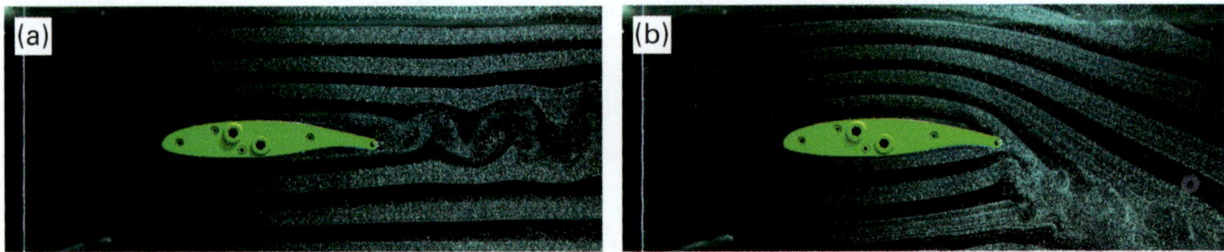

Figure 7.25 Flow visualization over an airfoil without (a) and with (b) circulation control (Jones et al. 2002). Reproduced with permission, copyright © The American Institute of Aeronautics and Astronautics.

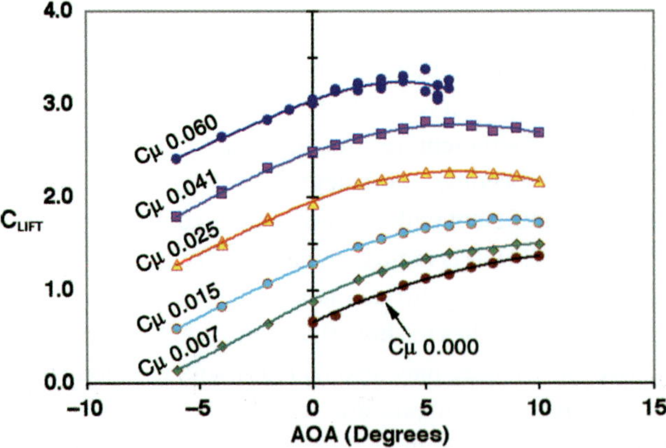

Figure 7.26 Lift coefficient of the 2-D GACC airfoil without and with circulation control (Jones et al. 2002). Reproduced with permission, copyright © The American Institute of Aeronautics and Astronautics.

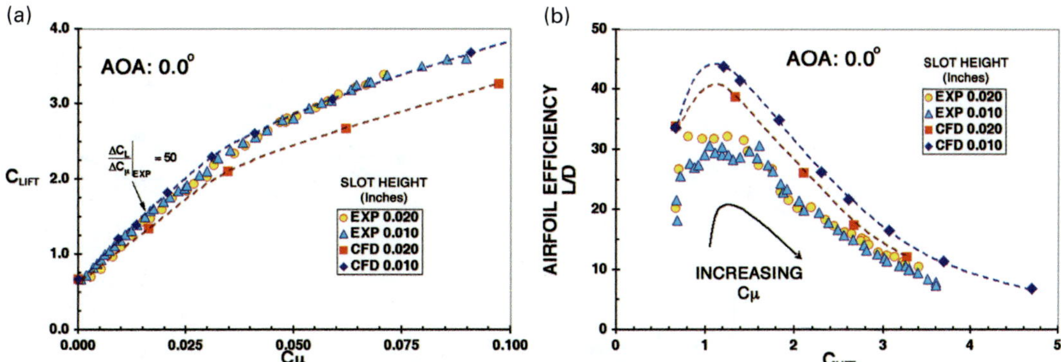

Figure 7.27 (a) Variation of the lift coefficient with the momentum coefficient and (b) variation of the lift-to-drag ratio with the lift coefficient at the zero angle of attack (Jones et al. 2002). Reproduced with permission, copyright © The American Institute of Aeronautics and Astronautics.

flight tested in the period 1974 to 2004, which is mainly attributed to the difficulties of implementing circulation control. First, circulation control needs a complex air source system. In real applications, the blowing jet can be issued from the engine, however, this may reduce the engine efficiency. Second, the circulation control system requires that the airfoil/wing should have a rounded trailing edge with Coanda effect, which may need advanced technology for aircraft manufacture and thus increase the cost.

Although some of the technical difficulties need to be solved, circulation control still has potential applications in aircraft. Thus, it has attracted great interest from NASA (Rich et al. 2005). In its vehicle systems program, there were six vehicle sectors leading the aerospace industry into an exciting future. Circulation control concepts may play a role in several of these projects, such as extreme short takeoff and landing vehicles, personal air vehicles, and supersonic aircraft. Moreover, circulation control also has potential applications in other vehicles, such as noise reduction for subsonic transports and rotor control for rotorcraft.

7.4.2 Jet Vortex Generator

Mechanical vortex generators show great ability to delay flow separation by inducing the streamwise vortices. As a passive flow control technique, however, traditional vortex generators cannot be implemented in real-time control, and it might not be effective in a state that it was not designed for. It has been found that the interaction of the jet with the crossflow may also result in the generation of streamwise vortices. Thus the concept of the jet vortex generator is proposed. The jet can be issued from narrow slots (Figure 7.28(a)) or round orifices (Figure 7.28(b,c)). Similar to mechanical vortex generators, jet vortex generators can also be arranged in counter-rotating (Figure 7.28(a,b)) or co-rotating (Figure 7.28(c)) manner.

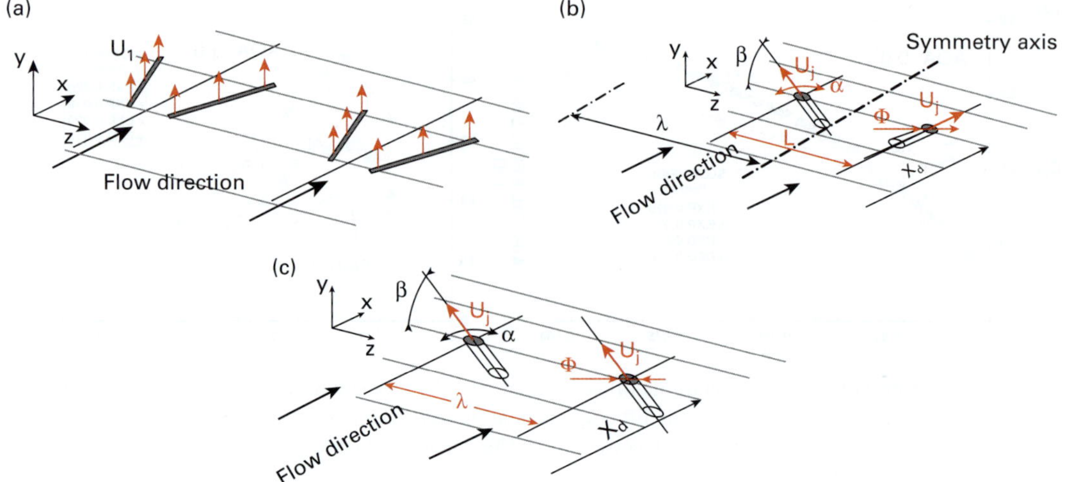

Figure 7.28 Schematic of slotted jets (a) and round jets (b, c) vortex generators in counter-rotating (a, b) and co-rotating (c) arrangements (Godard et al. 2006; Godard and Stanislas 2006). Reproduced with permission, copyright © 2006 Elsevier SAS. All rights reserved.

There are many different parameters to describe jet vortex generators, including the size of the narrow slot or the round orifice, the incidence angles of the jet to the free stream and the wall surface, and the distances between the two neighboring orifices. In addition, the ratio of the jet speed to the free-stream velocity needs also to be considered. As well as steady blowing, the free jets can also be issued in an unsteady manner. Some studies have been conducted by Godard et al. (2006) and Godard and Stanislas (2006).

Meunier and Brunet (2008) controlled a multi-element airfoil by placing a row of round jets parallel to the flap leading edge at $x/c_{flap} = 25\%$, which was slightly upstream of the separation point of about $x/c_{flap} = 30$–35%. The free-stream velocity was $M_\infty = 0.22$, corresponding to $Re = 3.13 \times 10^6$. The momentum coefficient for the free jet was $C_\mu = 0.22$. They could observe the formation of streamwise vortices, which were similar to those induced by mechanical vortex generators. It was also found that the suction pressure upstream of the orifice was increased, including the portion of the main wing near the trailing edge, indicating the delay of flow separation. Thus, the lift coefficient was increased by about 3.8% and the lift-to-drag ratio was increased by about 6.7% in comparison with the natural case.

The effects of the jet vortex generators have also been validated in a turbine blade, as shown in Figure 7.29. The jets were positioned at the separation point of the baseline solution, $x/c = 0.37$ from the leading edge. The distributions of the streamwise velocity and spanwise vorticity indicate that both steady and unsteady jet vortex generators can reduce the separation region over the suction surface. The separation point is delayed to about $x/c = 0.56$ for the steady case, and $x/c = 0.58$ for the unsteady case.

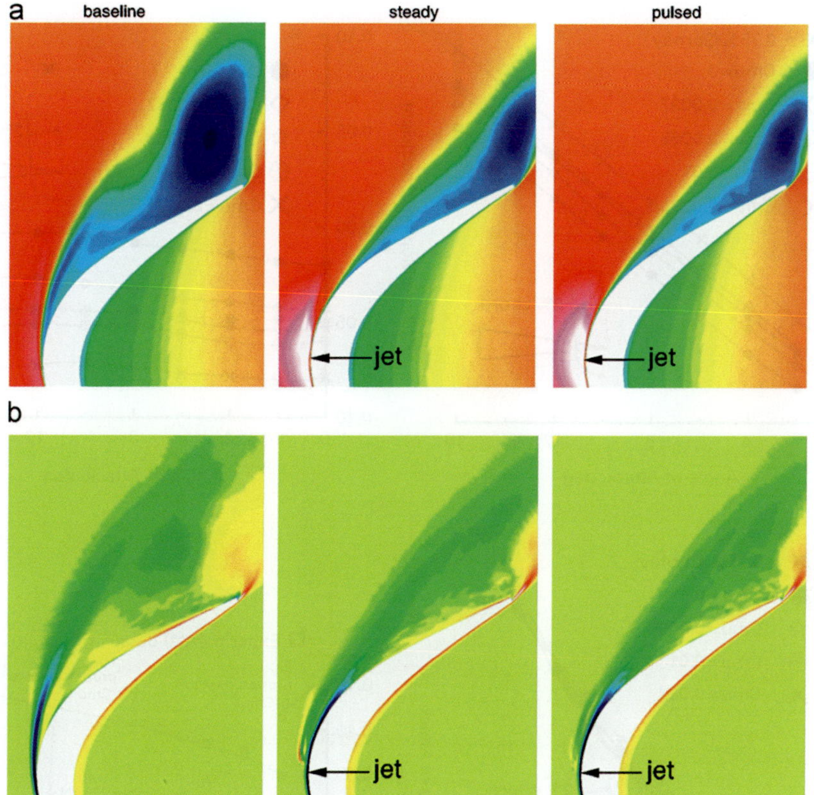

Figure 7.29 Instantaneous flow contours over a turbine blade without and with jet vortex generator control. (a) Streamwise velocity, (b) spanwise vorticity (Rizzetta et al. 2008). Reproduced with permission, copyright © 2008 Elsevier Ltd. All rights reserved.

7.4.3 Jet Flap

The Gurney flap shows great ability to increase the lift coefficient of airfoils, wings, and aircraft. However, accompanied with the lift enhancement, there is also an inevitable drag penalty. Therefore, the Gurney flap could be made more useful if it can be stored during cruise. A good alternative is the jet Gurney flap. The jet can be issued from a slot over the pressure surface near the trailing edge. The interaction of the jet with the free stream can lead to a virtual aeroshaping effect to simulate the mechanical Gurney flap.

Traub et al. (2004) carried out a wind tunnel investigation on a NACA 0015 airfoil with a jet slot located at 2% chord upstream of the trailing edge, as shown in Figure 7.30. The free-stream velocity was 15 m/s, corresponding to $Re = 0.7 \times 10^6$. It is indicated that the jet Gurney flap can shift the lift coefficient over all angles of attack, which is similar to that induced by a mechanical Gurney flap. However, the drag penalty induced by the jet Gurney flap is not obvious. The enhancements of lift and momentum caused by the jet Gurney flap with blowing momentum coefficient of 0.68% are equivalent to those

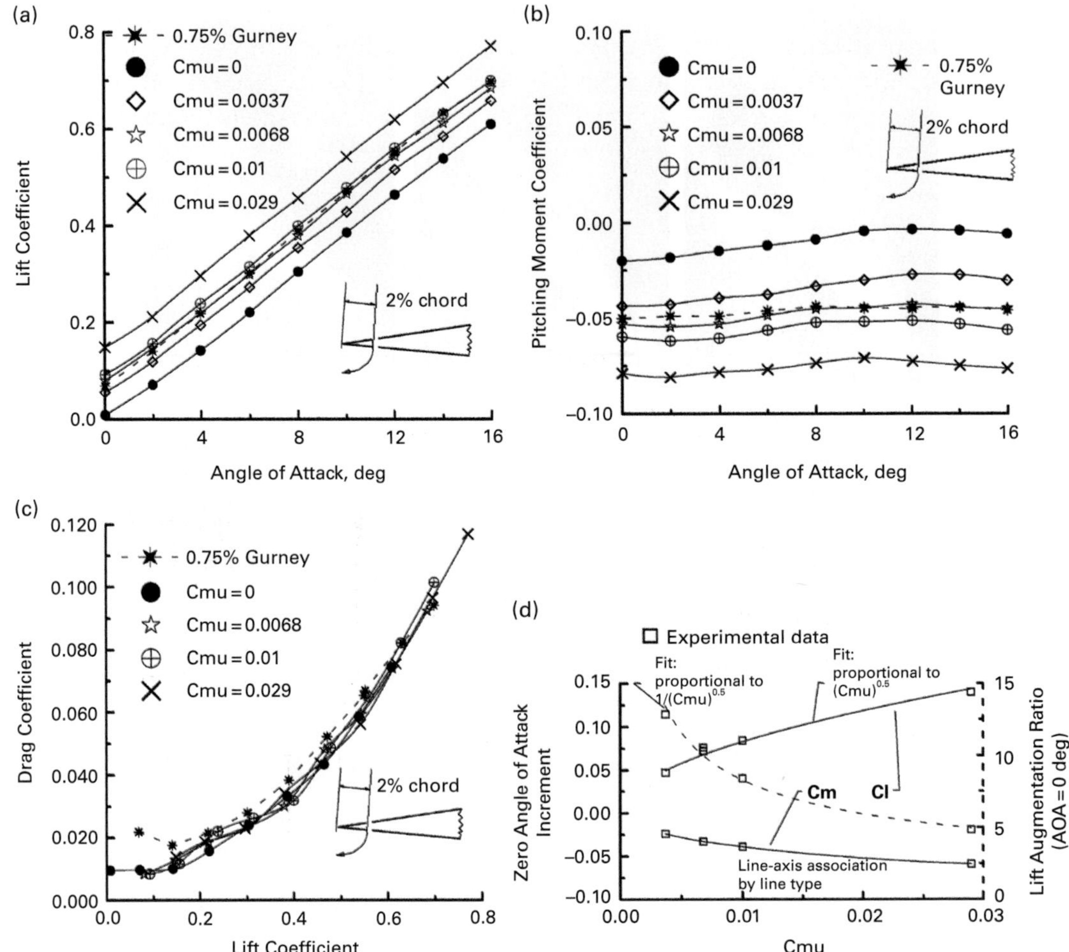

Figure 7.30 Aerodynamic forces on the NACA 0015 airfoil for different momentum coefficients of the jet Gurney flap. (a) Lift coefficient; (b) pitching moment; (c) drag coefficient; and (d) increment of lift coefficient and pitching moment with the momentum coefficient at $\alpha = 0°$ (Traub et al. 2004). Reproduced with permission.

induced by a 0.75% chord mechanical Gurney flap. The nose-down pitching moment is also increased with the jet momentum, though the drag coefficient shows little change in comparison with the natural case.

We can also see the variations of the lift coefficient and pitching momentum compared with the natural case by increasing the momentum coefficient (Figure 7.30(d)). A dependency proportional of $\sqrt{C_\mu}$ is suggested for the lift augmentation. However, the lift augmentation ratio, which is defined as $(C_{LC\mu\neq0} - C_{LC\mu=0})/C_\mu$, decreases with the momentum coefficient, showing the relationship $1/\sqrt{C_\mu}$.

Traub and Agarwal (2008) further undertook an investigation to establish the effect of a mechanical Gurney flap in conjunction with a jet flap at low Reynolds numbers. It was found that the jet forcing would further increase the lift coefficient of the airfoil installed with Gurney flap, and the control effect maintained a theoretically determined dependence on the jet momentum coefficient.

7.5   Concluding Remarks

The free jet is a conventional flow control technique, which was first developed in the early twentieth century. The fundamental control mechanism is that the issuing jet can enhance momentum mixing between inner and outer boundary layer, which is beneficial for separation delay. Thus, the jet has been widely tested in airfoils, wings, and aircraft for flow control. In general, the unsteady jet can provide more efficient control than the steady jet. The free jet can be used as an approach for circulation control, which can increase the lift coefficient significantly. Besides, the interaction of the free jet with free stream can also simulate the function of some conventional passive techniques, such as the vortex generator and Gurney flap. However, in comparison with passive techniques, the control techniques based on the free jet can be conducted in real-time and unsteady control, which is more robust. One problem for the free jet for engineering applications is the source of the air supply. Nonetheless, the jet has great potential applications in flow control once the problem of air supply is solved.

References

Ball, C. G., Fellouah, H., and Pollard, A. The flow field in turbulent round free jets. *Progress in Aerospace Sciences*, 2012, 50: 1–26

Betz, A. *History of Boundary Layer Control in Germany.* Aerodynamische Versuchsanstalt, 1961

Cambonie, T. and Aider, J. L. Transition scenario of the round jet in crossflow topology at low velocity ratios. *Physics of Fluids*, 2014, 26(8): 084101

Cater, J. E. and Soria, J. The evolution of round zero-net-mass-flux jets. *Journal of Fluid Mechanics*, 2002, 472: 167–200

Cui, Y. D., Lim, T. T., and Tsai, H. M. Control of vortex breakdown over a delta wing using forebody spanwise slot blowing. *AIAA Journal*, 2007, 45(1): 110–117

Cui, Y. D., Lim, T. T., and Tsai, H. M. Forebody slot blowing on vortex breakdown and load over a delta wing. *AIAA Journal*, 2008, 46(3): 744–751

Godard, G., Foucaut, J. M., and Stanislas, M. Control of a decelerating boundary layer. Part 2: Optimization of slotted jets vortex generators. *Aerospace Science and Technology*, 2006, 10(5): 394–400

Godard, G. and Stanislas, M. Control of a decelerating boundary layer. Part 3: Optimization of round jets vortex generators. *Aerospace Science and Technology*, 2006, 10(6): 455–464

Greenblatt, D. and Wygnanski, I. Dynamic stall control by periodic excitation. Part 1: NACA 0015 parametric study. *Journal of Aircraft*, 2001, 38(3): 430–438

Gursul, I., Wang, Z., and Vardaki, E. Review of flow control mechanisms of leading-edge vortices. *Progress in Aerospace Sciences*, 2007, 43(7): 246–270

Jones, G. S., Viken, S. A., Washburn, A. E., Jenkins, L. N., and Cagle, C. M. An active flow circulation controlled flap concept for general aviation aircraft applications. AIAA Paper 2002–3157

Joslin, R. D. and Miller, D. N. *Fundamentals and Applications of Modern Flow Control*. Published by the American Institute of Aeronautics and Astronautics, 2009

Kelso, R. M., Lim, T. T., and Perry, A. E. An experimental study of round jets in cross-flow. *Journal of Fluid Mechanics*, 1996, 306: 111–144

Kuo, C. H. and Lu, N. Y. Unsteady vortex structure over delta-wing subject to transient along-core blowing. *AIAA Journal*, 1998, 36(9): 1658–1664

Loth, J. L. Why have only two circulation-controlled STOL aircraft been built and flown in years 1974–2004. *Proceedings of the 2004 NASA/ONR Circulation Control Workshop, Hampton. 2005*

Mahesh, K. The interaction of jets with crossflow. *Annual Review of Fluid Mechanics*, 2013, 45: 379–407

Meunier, M. and Brunet, V. High-lift devices performance enhancement using mechanical and air-jet vortex generators. *Journal of Aircraft*, 2008, 45(6): 2049–2061

Müller-Vahl, H. F., Strangfeld, C., Nayeri, C. N., Paschereit C. O., and Greenblatt, D. Control of thick airfoil, deep dynamic stall using steady blowing. *AIAA Journal*, 2015, 53(2): 277–295

Muppidi, S. and Mahesh, K. Direct numerical simulation of passive scalar transport in transverse jets. *Journal of Fluid Mechanics*, 2008, 598: 335–360

Muppidi, S. and Mahesh, K. Study of trajectories of jets in crossflow using direct numerical simulations. *Journal of Fluid Mechanics*, 2005, 530: 81–100

Rahman, N. U. and Whidborne, J. F. Propulsion and flight controls integration for a blended-wing-body transport aircraft. *Journal of Aircraft*, 2010, 47(3): 895–903

Rich, P., McKinley, B., and Jones, G. S. Circulation control in NASA's vehicle systems program. *Proceedings of the 2004 NASA/ONR Circulation Control Workshop*, NASA/CP-2005-213509, 2005, Part. 1: 1–36

Rizzetta, D. P., Visbal, M. R., and Morgan, P. E. A high-order compact finite-difference scheme for large-eddy simulation of active flow control. *Progress in Aerospace Sciences*, 2008, 44(6): 397–426

Shan, J. W. and Dimotakis, P. E. Reynolds-number effects and anisotropy in transverse-jet mixing. *Journal of Fluid Mechanics*, 2006, 566: 47–96

Todde, V., Spazzini, P. G., and Sandberg, M. Experimental analysis of low-Reynolds number free jets. *Experiments in Fluids*, 2009, 47(2): 279–294

Toyoda, K. and Hiramoto, R. Manipulation of vortex rings for flow control. *Fluid Dynamics Research*, 2009, 41(5): 051402

Traub, L. W. and Agarwal, G. Aerodynamic characteristics of a Gurney/jet flap at low Reynolds numbers. *Journal of Aircraft*, 2008, 45(2): 424–429

Traub, L. W., Miller, A. C., and Rediniotis, O. Comparisons of a Gurney and jet-flap for hinge-less control. *Journal of Aircraft*, 2004, 41(2): 420–423

Wang, J. J., Li, Q. S., and Liu, J. Y. Effects of a vectored trailing edge jet on delta wing vortex breakdown. *Experiments in Fluids*, 2003, 34(5): 651–654

Wang, Z. J., Jiang, P., and Gursul, I. Effect of thrust-vectoring jets on delta wing aerodynamics. *Journal of Aircraft*, 2007, 44(6): 1877–1888

Williams, N. M., Wang, Z., and Gursul, I. Active flow control on a nonslender delta wing. *Journal of Aircraft*, 2008, 45(6): 2100–2110

Wood, N. J. and Roberts, L. Control of vortical lift on delta wings by tangential leading-edge blowing. *Journal of Aircraft*, 1988, 25(3): 236–243

Yavuz, M. M. and Rockwell, D. Control of flow structure on delta wing with steady trailing-edge blow. *AIAA Journal*, 2006, 44(3): 493–501

8 Synthetic Jet

8.1 Principle

The idea of flow control using a synthetic jet originated from acoustic streaming. One example was performed by Ingard and Labate (1950). They used the acoustic signal to drive the air inside a hollow cylinder, resulting in changes of pressure. Thus, fluids were ejected and sucked from the orifice and a series of vortex rings was observed. In subsequent studies on acoustic streaming by Mednikov and Novitskii (1975) and Lebedeva (1980), the maximum blowing velocity of the acoustic-driven jet could reach 17 m/s and 10 m/s, respectively.

However, the modern design of the synthetic jet actuator was proposed in the 1990s. One study was conducted by James et al. (1996). They produced a round turbulent jet normal to the surface of a submerged oscillating diaphragm, which was driven at its resonance frequency by a piezoceramic element. The jet was formed without net mass injection across the actuator surface when the excitation level of the actuator exceeded a given threshold. They also observed the periodic formation of a small cluster of cavitation bubbles near the orifice. It was deduced that the jet was synthesized by a train of vortex rings that were formed during each cycle of the surface oscillation owing to secondary flow induced by the cavitation bubbles. Thus, they called it synthetic jet. This device is considered the basic design of the piezoceramic synthetic jet actuator, which includes piezoceramic diaphragm, cavity, and orifice, as shown in Figure 8.1.

In a subsequent study, Smith and Glezer (1998) confirmed the periodic formation of the vortex ring and also presented its formation mechanism. Based on the basic working principle of the synthetic jet, other types of synthetic jet actuator have also been developed, such as that driven by a piston, a louder-speaker, a ferromagnetic shape memory alloy, and a piezoceramic disk (Luo and Xia 2005). The basic characteristic is that fluids can be periodically ejected and sucked from the orifice by the periodic motion of the device inside the cavity. With the advancement of related technique, synthetic jet actuators have been well developed over the years. The velocity of the jet can reach up to 250 m/s for a piezoceramic synthetic jet actuator (Shaw et al. 2006) and it can reach 124 m/s for a piston-driven one (Gilarranz et al. 2005), which can normally fit the needs of engineering applications.

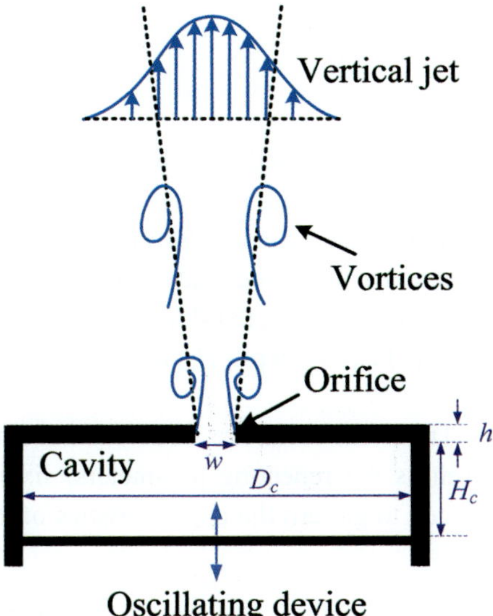

Figure 8.1 Schematic diagram of a synthetic jet actuator.

Despite the difference in the constitution of the actuator, the formation process and formation mechanism of the synthetic jet are the same. (1) During the forward motion of the diaphragm or piston (the blowing cycle), the fluids are ejected from the orifice, forming the shear layer around the orifice due to flow separation. The shear layer grows and develops into vortical structures. The generation mechanism for the vortical structures depends on the condition of the boundary layer at the jet exit (Toyoda and Hiramoto 2009). When the boundary layer is laminar, the shear layer separating from the orifice edge is rolled up owing to the growth of the Kelvin–Helmholtz instability, and vortical structures are formed; when the boundary layer is turbulent, the vorticity in the shear layer is concentrated into clusters, forming large-scale vortices. The vortical structures are convecting downstream due to the self-induced velocity. (2) During the backward motion of the diaphragm or piston (the suction cycle), the fluids outside the orifice are sucked into the cavity; on the other hand, the vortex structures formed during the previous blowing cycles have moved further downstream and are little influenced by the suction process. Thus, a series of vortical structures are formed periodically during the blowing and suction process. For a narrow slot orifice, a vortex pair is generated, while a vortex ring forms from a circular orifice. The net flow flux of the synthetic jet during one period is zero. Thus, it is also called the zero-net-mass-flux jet. Accordingly, there is no need for an external air supply for the application of the synthetic jet. However, the net momentum flux during one period is not zero.

There are many studies on the synthetic jet and its applications. Some developments have been reviewed by Glezer and Amitay (2002), Luo and Xia (2005), and Zhang et al. (2008). More recent and detailed knowledge about synthetic jets will be presented here.

8.2   Influence of Parameters

8.2.1   Dimensional Analysis

The flow characteristics of the synthetic jet is governed by different control parameters, that can be classified into three groups: the operating parameters, the geometric parameters and the fluid parameters. According to Zhong et al. (2007), for a typical synthetic jet actuator with narrow-slot orifice, there are nine independent parameters, including: (1) excitation amplitude A, (2) excitation frequency f_e, (3) slot width w, (4) slot length l, (5) slot depth h, (6) cavity diameter D_c, (7) cavity depth H_c, (8) kinematic viscosity v, and (9) fluid density ρ.

According to the Buckingham-Pi Theorem, there are three fundamental physical units. Thus, if f_e, w, and, ρ are chosen as the repeating parameters, six independent dimensionless parameters can be deduced to govern the characteristics of the synthetic jet flow field,

$$\pi_1 = \frac{f_e w^2}{v}, \quad \pi_2 = \frac{A}{w}, \quad \pi_3 = \frac{l}{w}, \quad \pi_4 = \frac{h}{w}, \quad \pi_5 = \frac{D_c}{w}, \quad \pi_6 = \frac{H_c}{w}. \tag{8.1}$$

The dimensionless parameters π_1 and π_2 are related to the actuator operating conditions, whereas π_3, π_4, π_5 and π_6 are related to the actuator geometry.

The two operating parameters π_1 and π_2 are related to two important dimensionless parameters: stroke length $L = L_0/D_0$ and Reynolds number ($Re_{I_0} = \dfrac{I_0}{v\rho D_0}$ based on the momentum per unit width or $Re_{U_0} = \dfrac{U_0 D_0}{v}$ based on the time-averaged blowing velocity). The stroke length L_0 stands for the distance of the fluid motion during the blowing cycle, namely $L_0 = U_0 T$. Here, T is the excitation period, and U_0 is the time-averaged blowing velocity during the whole period,

$$U_0 = \frac{1}{T} \int_0^{T/2} u_0(t) dt, \tag{8.2}$$

where $u_0(t)$ is the instantaneous velocity at the orifice. Thus, the momentum per unit width I_0 can be obtained by

$$I_0 = \rho D_0 \int_0^{T/2} u_0^2(t) dt. \tag{8.3}$$

According to the slug model proposed by Smith and Glezer (1998), the synthetic jet stroke length and Reynolds number are the two most important normalized parameters to characterize the synthetic jet flow.

We can also deduce that the Strouhal number has a close relationship with the dimensionless stroke length,

$$St = \frac{f_e D_0}{U_0} = \frac{D_0}{U_0 T} = \frac{D_0}{L_0} = \frac{1}{L}. \tag{8.4}$$

Following the above analysis, Zhong et al. (2007) indicated that the dimensionless parameter π_1 can also be written in the form of the Stokes number, namely

$$S = \sqrt{\frac{2\pi f D_0^2}{\nu}}. \tag{8.5}$$

From this definition, the three operating parameters, namely the Stokes number S, the dimensionless Stroke length L and the synthetic jet Reynolds number Re_{U0} have the following relationship,

$$S = \sqrt{\frac{2\pi Re_{U0}}{L}}. \tag{8.6}$$

Thus, only two parameters are independent.

When the synthetic jet is used for flow control, two more parameters should be considered: the length scale of the model L and the free-stream velocity U_∞. Thus, in addition to Equation (8.1), there should be two more dimensionless parameters:

$$\pi_7 = \frac{f_e L}{U_\infty}, \quad \pi_8 = \frac{U_\infty L}{\nu}. \tag{8.7}$$

The dimensionless parameters π_7 and π_8 can be recognized as two transitional operating parameters, namely the reduced excitation frequency and the Reynolds number of the model. The synthetic jet Reynolds number reveals the strength of the exit velocity, while the Reynolds number of the model reveals the strength of the free-stream velocity. In order to compare the strength of the synthetic jet and the free-stream velocity directly, usually another dimensionless parameter is used, namely the equivalent momentum coefficient $C_\mu = 2\overline{U_0}^2 w / U_\infty^2 D$ or the velocity ratio $V_R = U_0/U_\infty$.

The influence of these dominant parameters on the flow field of the synthetic jet is introduced as follows.

8.2.2 Stokes Number

Based on the results of a fully developed oscillating laminar pipe flow, the Stokes number plays an important role in determining the thickness of the Stokes layer inside the orifice and hence the shape of the velocity profile (Zhou et al. 2009). It also determines the strength of vortex roll-up of the synthetic jet issuing from the orifice. Thus, Zhong et al. (2007) found that strong roll-up occurs at the orifice at a large Stokes number of 22, however, no vortex roll-up was observed at $S = 7$ since the Stokes layers merging along the orifice inhibited the curling up of vortex sheets that formed the vortex rings. Thus, a minimum Stokes number of around 10 was required for the exit velocity profile to depart from the orifice.

One example of the Stokes number determining the exit velocity profile is shown in Figure 8.2 by exhibiting the exit velocity profiles at the moment of the middle of the blowing cycle. It is found that the exit velocity profiles with different stroke length or Reynolds number have similar shapes when the Stokes number is the same. Thus, the stroke length and the Reynolds number do not affect the velocity profiles, suggesting that the Stokes number is a dominant parameter that determines the exit velocity profile. As the Stokes number increases, the velocity profile changes from a parabolic shape to a top-hat shape.

8.2.3   Stroke Length

The stroke length represents the distance of the fluid motion during the blowing cycle, thus it determines the distance of the vortex rings/pairs to the orifice (Glezer 1988). At the same synthetic jet Reynolds number, the distance of the vortex rings/pairs to the orifice increases with the stroke length, and the distance between the two vortex rings/pairs also increases accordingly, as shown in Figure 8.3. It is also found that due to the increased distance between vortex rings/pairs, their interaction becomes weak, and thus the vortex breakdown distance to the orifice is also increased. Actually, if the dimensionless stroke length is too small, the synthetic jet vortex is too close to the orifice at the end of the blowing cycle. Thus, the vortex formed during the blowing cycle may be sucked into the cavity during the suction cycle, with the result that the synthetic jet vortex cannot be formed. It is suggested that there might be a critical stroke length to determine the formation of the synthetic jet. One of the necessary conditions for synthetic jet vortex formation is that the dimensionless stroke length should be larger than 0.5 (Holman et al. 2005).

The stroke length also determines the vortex circulation. It was found that the total circulation of the synthetic jet increased with the stroke length. However, when it was

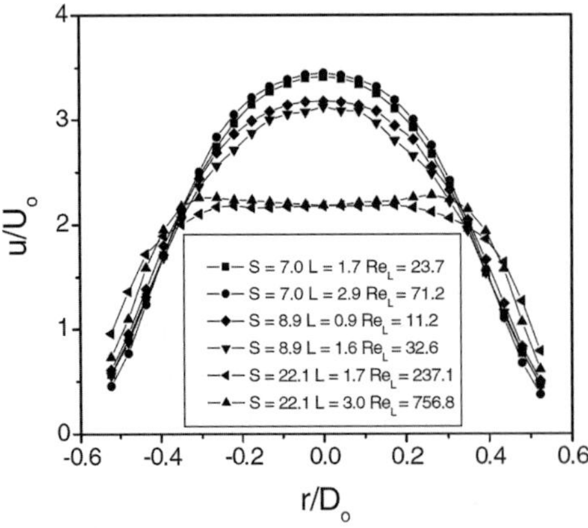

Figure 8.2 Exit velocity profiles at the moment of the middle of the blowing cycle (Zhong et al. 2007). Reproduced with permission, copyright © Springer 2007.

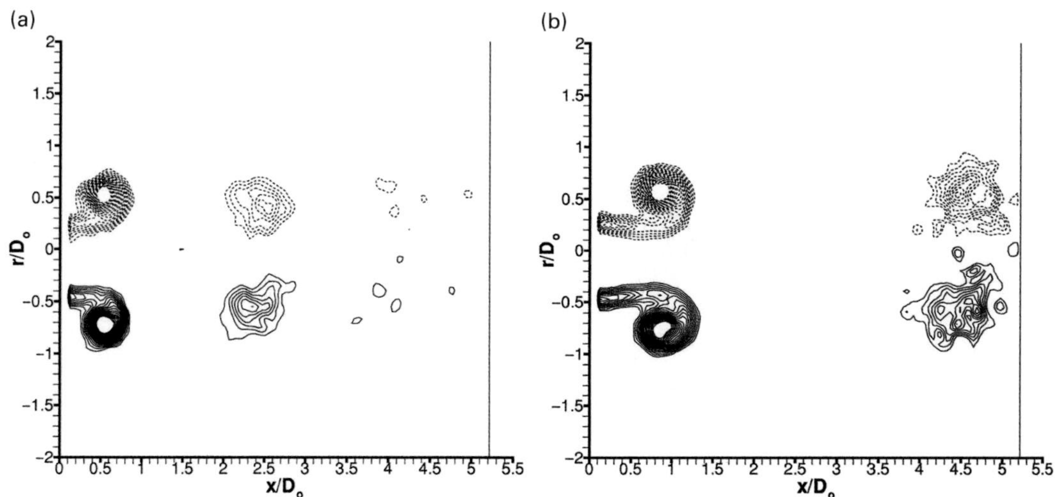

Figure 8.3 Phase-averaged vorticity contours at $\Phi = 162°$, $Re_{U0} = 2500$ for (a) $L = 2$ and (b) $L = 3$ (Shuster and Smith 2007). Reproduced with permission from AIP Publishing.

larger than a threshold, a trailing vortex formed and thus the circulation in the primary vortex remained constant with a larger stroke length. Nevertheless, the circulation of the entire structure continued to increase, as shown in Figure 8.4. Gharib et al. (1998) obtained a threshold value of 4 for a single vortex ring, while Zhong et al. (2007) and Shuster and Smith (2007) indicated a value of 4 and 4 to 5 for the synthetic jet, respectively.

Furthermore, if one plots the curves of time evolution of the vortex ring trajectory ($x/L_0 - t/T$) and vortex ring diameter ($D_r/D_0 - t/T$), the cross-stream profiles of the streamwise velocity ($U/U_0 - r/D_0$), the streamwise variation of the mean jet centerline velocity ($U_0/U_{cl} - x/L_0$), the streamwise variation of the mean jet width ($b/D_0 - x/L_0$), the streamwise variation of the mean jet momentum ($M_j/M_0 - x/D_0$), it would be found that at the same Reynolds number, the curves with different stroke lengths show different variations; in comparison, the curves with different Reynolds numbers but the same stroke length coincide very well (Shuster and Smith 2007). These variations indicate that those flow characteristics are mainly influenced by the stroke length but not affected by the Reynolds number.

8.2.4   Reynolds Number

The Reynolds number determines the strength of the synthetic jet vortex. However, it does not influence the vortex trajectory. One example is shown in Figure 8.5. At the same stroke length, the vortex ring formation position and the mutual distance remain the same even with increasing the Reynolds number, however, the vortex strength increases with the Reynolds number.

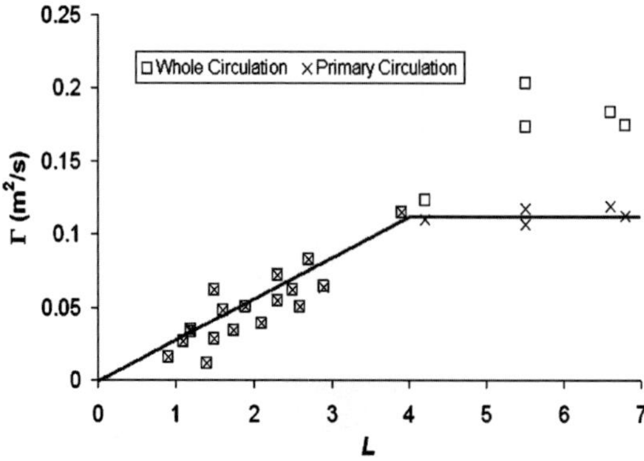

Figure 8.4 Variation of total circulation and primary vortex circulation with stroke length (Zhong et al. 2007). Reproduced with permission, copyright © Springer 2007.

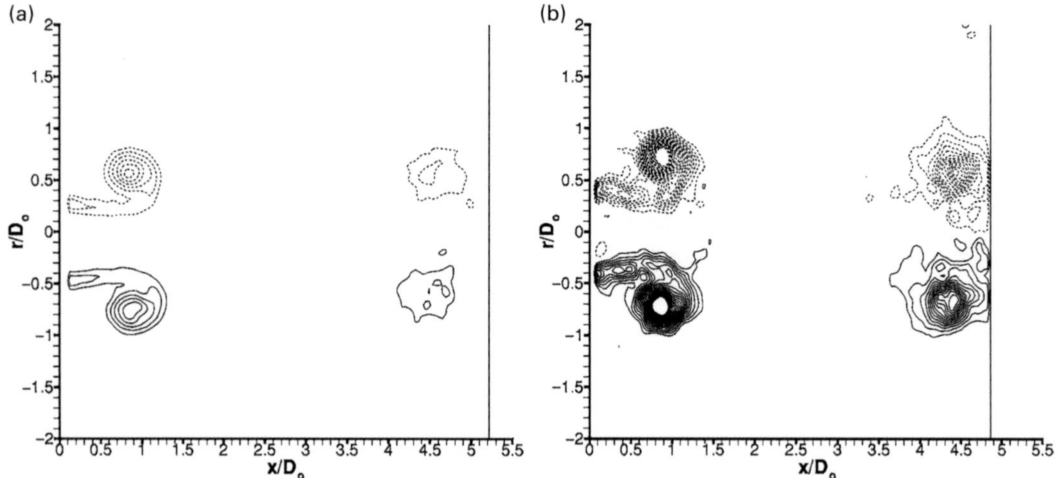

Figure 8.5 Phase-averaged vorticity contours at $\Phi = 162°$, $L = 3$ for (a) $Re_{U0} = 2500$ and (b) $Re_{U0} = 10\,000$ (Shuster and Smith 2007). Reproduced with permission from AIP Publishing.

The Reynolds number together with the stroke length determine the state of the synthetic jet. At the same Strouhal number (stroke length), the laminar jet breaks into individual rings that become turbulent and eventually coalesce to produce a turbulent jet flow as the Reynolds number increases, as shown in Figure 8.6(a) by Cater and Soria (2002). In particular, at high Strouhal numbers, namely low stroke length, a jet-like pattern always forms and there is no obviously vortex ring in the downstream position; at low Strouhal numbers, namely high stroke length, secondary vortex rings may form behind the leading vortex ring. Similarly, Shuster and Smith (2007) found that at the

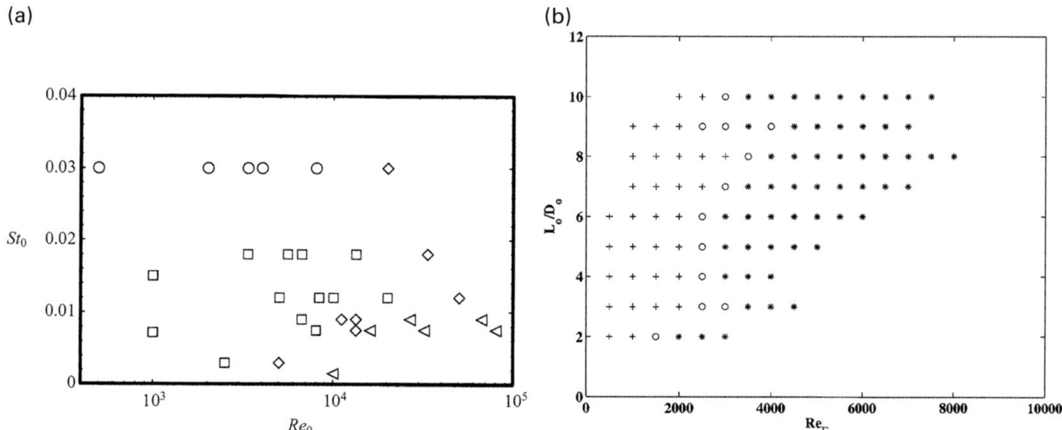

Figure 8.6 Flow state of the synthetic jet vortex ring at different parameters. Figure (a) is from Cater and Soria (2002): ○ laminar jets, □ laminar rings, ◊ transitional jets, ◁ turbulent jets. Reproduced with permission, copyright © Cambridge University Press 2002. Figure (b) is from Shuster and Smith (2007): + laminar, ○ transitional, * turbulent. Reproduced with permission from AIP Publishing.

same stroke length, the synthetic jet undergoes the laminar, transitional, turbulent states when increasing the Reynolds number, as shown in Figure 8.6(b).

The Reynolds number is also an important parameter to determine the formation of the vortex ring. Crook and Wood (2001) found that a vortex ring could not be formed at $Re_{U0} = 330$. When increasing the Reynolds number, the influence of the suction process decreases, and thus the vortex ring might start to form. The diameter of the vortex ring increases with the Reynolds number until a critical value. After that, the diameter of the vortex ring nearly maintains the same and the secondary vortex forms, which is separated from the primary vortex. The critical Reynolds number to form the secondary vortex may differ due to differences in other parameters. For example, Crook and Wood (2001) found it was $Re_{U0} = 2300$, while Cicca and Iuso (2007) found it was between $Re_{U0} = 330$ and $Re_{U0} = 1290$.

8.2.5 Formation Condition

From the above analysis, the development of the synthetic jet is influenced by different control parameters. Thus, they also determine the formation of the synthetic jet. Utturkar et al. (2003) and Holman et al. (2005) discussed the criterion condition to form a synthetic jet based on the slug model. They suggested that the formation of synthetic jet was governed by the self-induced velocity V_I of the vortex formed during the blowing cycle and the mean suction velocity V_S during the suction cycle. By magnitude analysis, they deduced the theoretical jet formation criterion as

$$\frac{V_I}{V_s} \sim \frac{1}{St} = \frac{Re_{U_0}}{S^2} > K, \tag{8.8}$$

where the constant K is approximately 2 and 0.16 for two-dimensional and axisymmetric synthetic jets, respectively. The condition for synthetic jet formation is that the ratio of the self-induced velocity of the vortex to the averaged suction velocity should be larger than the constant K.

8.3   Characteristics of Velocity Field

During the periodic formation and convection of the synthetic jet vortex ring/pair, the related velocity field changes periodically at different positions, as shown in Figure 8.7. Near the orifice, the velocity signal shows a sinusoidal waveform, reflecting the influence of both the blowing and suction process. The magnitude of the induced velocity peak on the centerline decreases as the vortex is convecting downstream. In the further downstream positions, the negative velocity disappears gradually due to the reduced influence of the suction process.

Figure 8.8 shows the time-averaged velocity profile at several streamwise positions for three Reynolds numbers. Here, the mean streamwise velocity is normalized by the mean blowing velocity at the orifice over an entire cycle, and the radial location is scaled by the orifice diameter. It shows a single-peak waveform that is symmetric about the centerline. The velocity reaches the maximum in the near field region and then starts to decrease. The difference in Reynolds number only changes the value of the velocity.

If the mean velocity is normalized by the jet centerline velocity, and the radial distance from the centerline is normalized by the jet half-width, the velocity profiles at different positions collapse together, as shown in Figure 8.9. In particular, they match well with

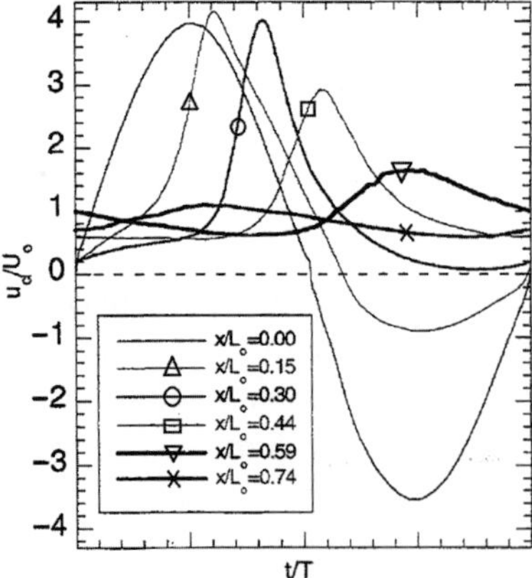

Figure 8.7 The variation of the centerline velocity u_{cl}/U_0 at different streamwise locations during one period for $L = 13.5$, $Re_{U0} = 695$ (Smith and Swift 2001). Reproduced with permission.

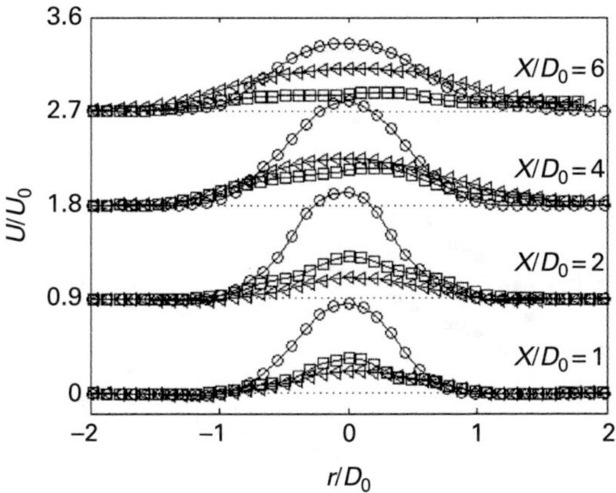

Figure 8.8 Mean streamwise velocity profiles at different streamwise locations for $Re_\Gamma = 1234.5$ (○), 2469 (□), 4938 (◁) (Shan and Wang 2010). Reproduced with permission, copyright © 2010 Elsevier.

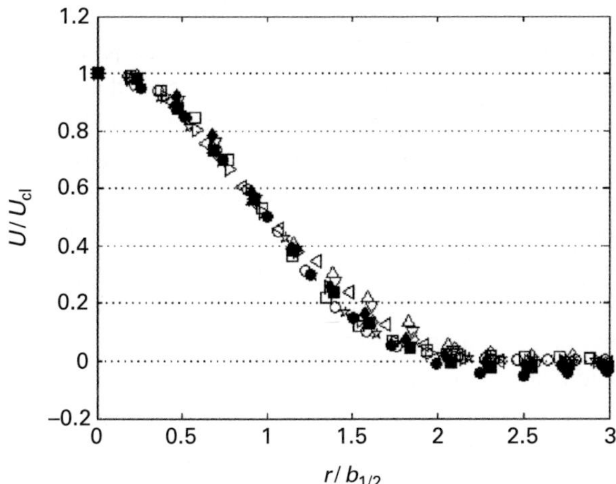

Figure 8.9 Normalized mean streamwise velocity profiles for $Re_{U0} = 315.5$ (Shan and Wang 2010). Reproduced with permission, copyright © 2010 Elsevier.

the hyperbolic cosine function, namely $U = U_{cl}\cosh^{-2}(\eta y)$. Such a characteristic is similar to that of a continuous jet.

Despite of the similarity in some of the velocity properties, the vortical structure of the synthetic jet can enhance the momentum mixing of the surrounding fluids, and thus it is more able to enhance entrainment and mixing. One example is shown in Figure 8.10, where the width and the streamwise volume flux per unit depth for the synthetic jet are much higher than a continuous jet. In particular, the apparent spreading rate for the synthetic jet was found to be 0.13 at $Re_{U0} = 10\,000$, $St = 0.0015$ by Smith and Swift (2001), compared to 0.1 for the equivalent continuous jet by Cater and Soria (2002).

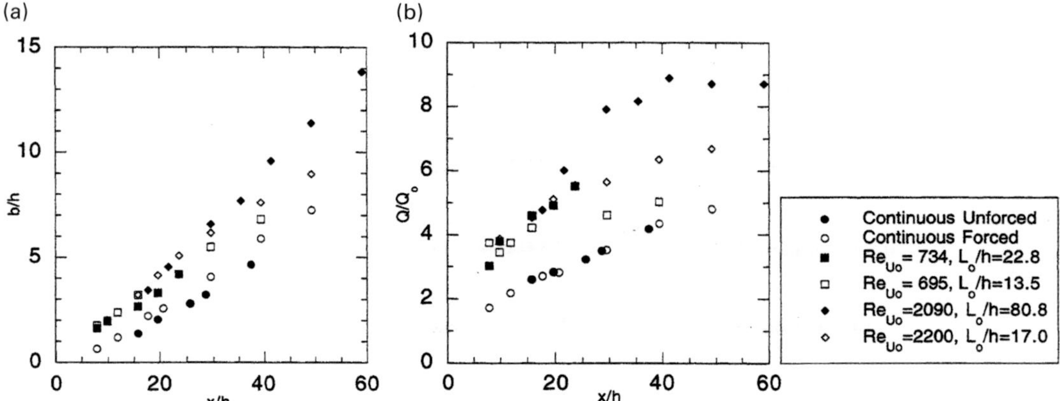

Figure 8.10 Comparison between the synthetic jet and continuous jet for (a) width and (b) streamwise volume flux per unit depth (Smith and Swift 2001). Reproduced with permission.

8.4   Novel Synthetic Jet

In the application of the synthetic jet, the efficiency of synthetic jet control needs to be improved. It has been revealed that the formation mechanism of the synthetic jet is that the blowing velocity should be much higher than the suction velocity. Therefore, if the timescale of the suction cycle increases and the blowing cycle decreases, while the mass flux stays the same in the two cycles, the blowing velocity will be enhanced while the suction velocity will be reduced. Thus, the vortex formed during the blowing cycle has a larger convection speed, while the entrainment of the orifice during the suction cycle on the vortex is reduced. Based on this consideration, Zhang and Wang (2007) proposed a novel actuation signal to generate a more efficient synthetic jet, which was modified from a standard sinusoidal signal, as shown in Figure 8.11. For the novel signal, T_1 and T_2 represent the time duration of the blowing cycle and suction cycle in one period, respectively. The ratio of these two parameters is defined as the suction duty cycle factor $k = T_2/T_1$. According to the definition, $k > 1$ means that the suction cycle is longer than the blowing cycle, and thus the mean blowing velocity will be much higher than the suction velocity. On the other hand, $k < 1$ causes an opposite result.

Zhang and Wang (2007) proved numerically that the entrainment effect of the actuator decreases with an increase in the suction duty cycle factor, so that the vortex pair could propagate farther downstream and coalesce to synthesize a stronger vortex pair, as shown in Figure 8.12. Thus, the entrainment effect of the synthetic jet can be enhanced with the novel synthetic jet by increasing the suction duty cycle factor. It has also been experimentally validated by Wang et al. (2010) and Shan and Wang (2010) for two-dimensional and axisymmetric synthetic jets, respectively.

Luo et al. (2006) proposed a dual synthetic jet actuator, as shown in Figure 8.13(b–d), which differs from the conventional one (Figure 8.13(a)).

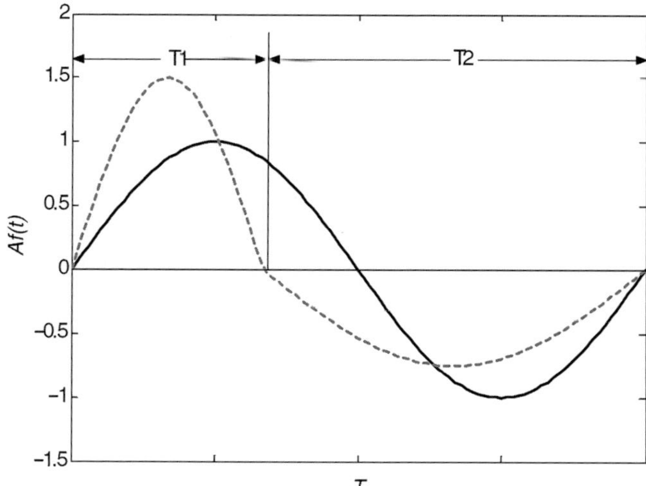

Figure 8.11 Novel actuation signal (Zhang and Wang 2007). Reproduced with permission, copyright © The American Institute of Aeronautics and Astronautics.

It consisted of two cavities, an oscillating device, two exit slots, and a slide block. Two adjacent jets were established under the two exit slots and were driven by the motion of the same oscillating device.

A snapshot of the flow field induced by the dual synthetic jet actuator is shown in Figure 8.14(b–d), while that induced by a conventional synthetic jet actuator is shown in Figure 8.14(a). The slide block regulated the two jets, and then two adjacent jets merged into a single, larger synthetic jet (Figure 8.14(b–d)). In particular, the symmetry of the synthetic jet could be regulated by an extended step, as shown in Figure 8.13(d) and Figure 8.14(d). Such characteristics have also been validated in subsequent studies by Luo et al. (2007) and Luo and Xia (2008), and the dual synthetic jet actuator could be used for vectoring control (Deng et al. 2015) and heat transfer (Luo et al. 2016).

8.5   Numerical Model

We also need to study the synthetic jet and its control accurately by numerical methods. NASA Langley Research Center organized a workshop in 2004 on "CFD Validation of Synthetic Jets and Turbulent Separation Control" (see the link: http://cfdval2004.larc .nasa.gov). They proposed three models for numerical test, namely synthetic jet in a quiescent air (such as Yao et al. 2004), in a crossflow (such as Schaeffler and Jenkins 2004), and control of flow over a hump (such as Greenblatt et al. 2004). For the first case, the actuator is flush mounted on a flat plate (Figure 8.15(a)). The size of the

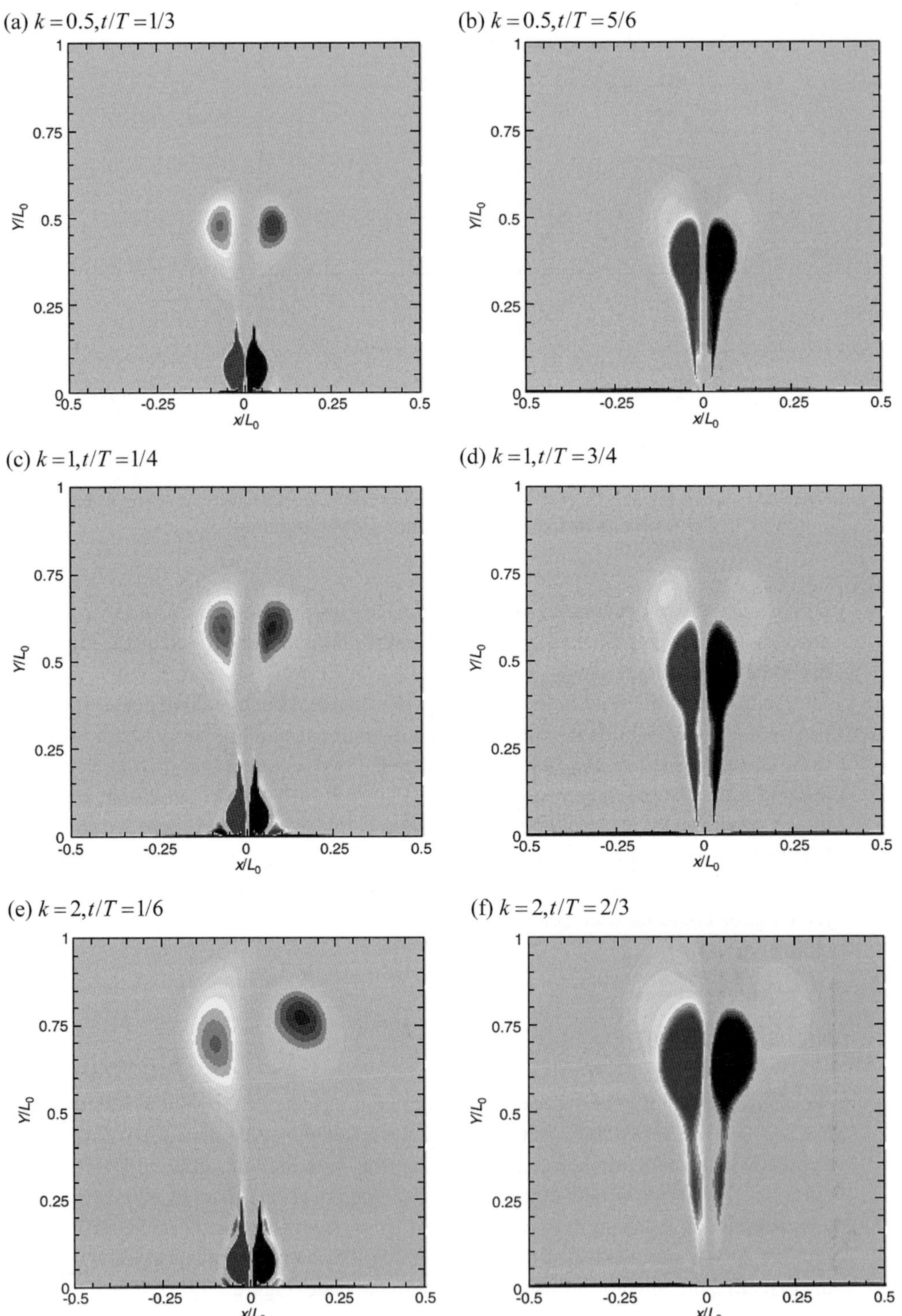

Figure 8.12 Evolution of vortex pair at different phases for different suction duty cycle factors (Zhang and Wang 2007). Reproduced with permission, copyright © The American Institute of Aeronautics and Astronautics.

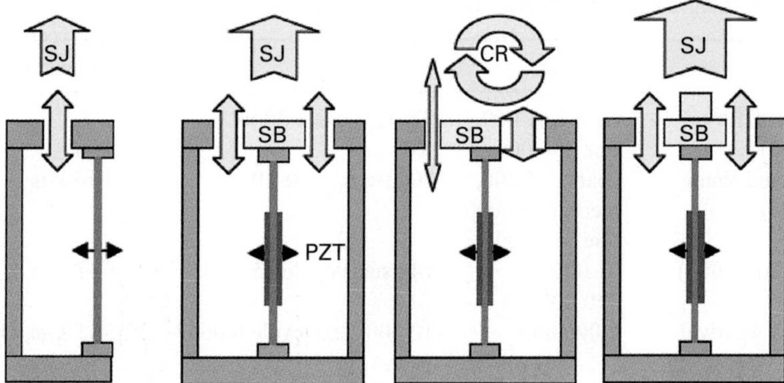

Figure 8.13 Schematics of different synthetic jet actuators. (a) A conventional synthetic jet actuator; (b) a dual synthetic jet actuator; (c) a dual synthetic jet actuator when the slide block shifts to left; and (d) a dual synthetic jet actuator containing a slide block with an extended step (Luo et al. 2006). Reproduced with permission, copyright © The American Institute of Aeronautics and Astronautics.

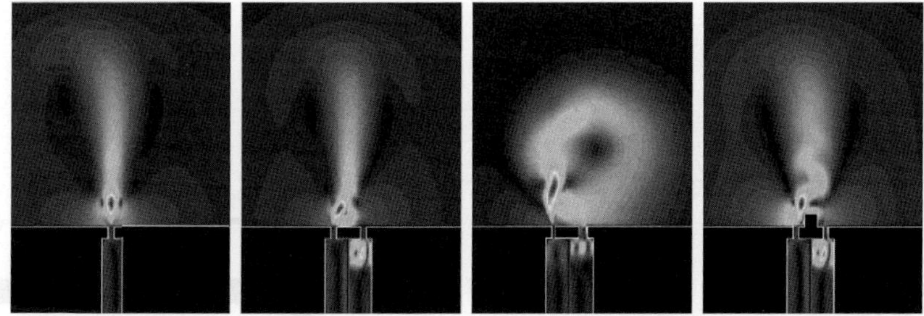

Figure 8.14 Velocity magnitudes of synthetic jets. (a) A conventional synthetic jet actuator; (b) a dual synthetic jet actuator; (c) a dual synthetic jet actuator when the slide block shifts to left by 0.5 mm; and (d) a dual synthetic jet actuator containing a slide block with an 2×2 mm extended step (Luo et al. 2006). Reproduced with permission, copyright © The American Institute of Aeronautics and Astronautics.

exit slot is 1.25 mm $\times$ 35.56 mm. The synthetic jet is driven by a single piezoelectric diaphragm with diameter of 50.8 mm which is mounted on one side of a narrow cavity. The synthetic jet works at 445 Hz, and the maximum velocity at the orifice is about 25–30 m/s. For the second case, the synthetic jet actuator has a circular orifice with a diameter of 6.35 mm, and there is a diaphragm with diameter 15.2 mm in the cavity according to Schaeffler and Jenkins (2004) (Figure 8.15(b)). The free-stream velocity over the plate is 34.6 m/s, and the synthetic jet works at 150 Hz. For the third case, the chord length of the hump is $c = 420$ mm and its maximum thickness is $h = 53.7$ mm with width of $w = 584$ mm (Figure 8.15(c)). The narrow slot of the synthetic jet is located at about $x/c = 65\%$, which is near the separation point. The excitation

Table 8.1 Some representative numerical studies on the three CFD validation cases

Case	Authors	Grid size	Time step	Turbulence model
1	Vatsa and Turkel (2006)	coarse 61 000, fine 250 000	72 steps/cycle 30 µs	SA, SST k-ω
	Zhang and Wang (2007)	coarse 17 800, medium 46 000, fine 315 700	100 steps/cycle 20 µs	Standard k-ω
	Park et al. (2007)	61 000 250 000	144 steps/cycle 15 µs	SA, SST k-ω, k-ε
2	Cui and Agarwal (2005)	4 090 000	10 000 steps/cycle 0.6667 µs	SA, SST k-ω, DES
	Dandois et al. (2006)	9 100 000 1 700 000 220 000	13 333 steps/cycle 0.5 µs 6667 steps/cycle 1 µs 3333 steps/cycle 2 µs	SA, LES
3	Morgan et al. (2006)	50 166	1×10^{-4}s–2.5×10^{-4}s	k-ε
	Rumsey (2007)	Fine 210 000	180 steps/cycle 4×10^{-5}s	SA, SST k-ω, EASM based on k-ω
	Bettini and Cravero (2007)	65 000 104 000	72 steps/cycle 1×10^{-4}s	k-ε, SST k-ω, Reynolds Stress

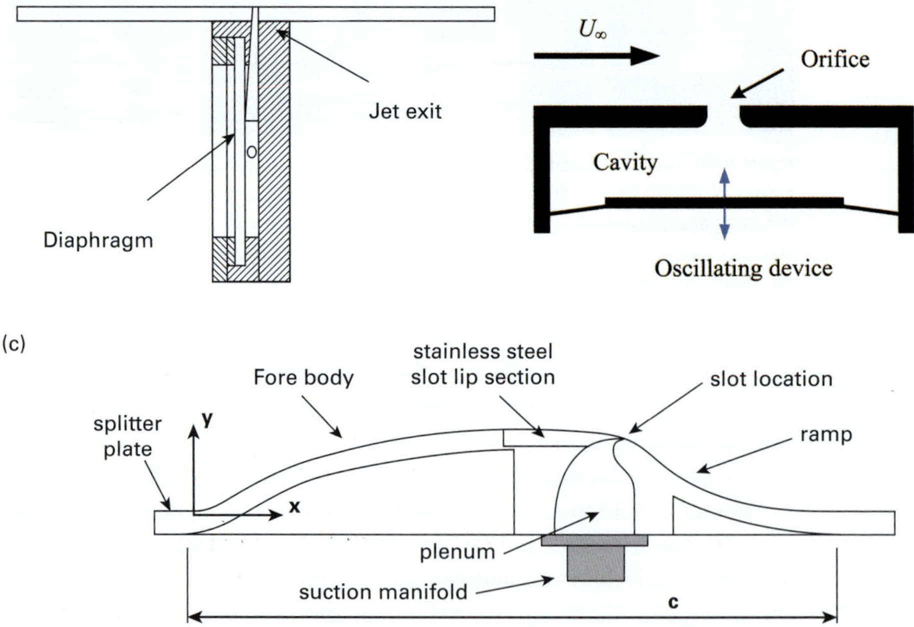

Figure 8.15 Three cases proposed at the workshop on 2004 CFD Validation of Synthetic Jets and Turbulent Separation Control. (a) Synthetic jet into quiescent air (Yao et al. 2004). (b) Synthetic jet in a crossflow. (c) Synthetic jet control of flow over a hump model (Greenblatt et al. 2004). Data kindly supplied by D. Greenblatt. reproduced with permission, copyright © The American Institute of Aeronautics and Astronautics.

frequency is 138.5 Hz, with the peak velocity of 26.6 m/s at the orifice. The free-stream velocity is $Ma = 0.1$.

Different experimental techniques, such as PIV, LDV, and hot-wire, have been used to obtain the dataset for the velocity field, which is used as a reference for the numerical studies. Different numerical methods have been tested, where the influence of the turbulence model, grid size, and time step is studied. Some representative works are summarized in Table 8.1. More details about the CFD validation cases can be found in a review paper by Rumsey et al. (2006).

8.6   Applications of Synthetic Jets

8.6.1   Circular Cylinder

Vortex shedding over a bluff body usually causes serious structural vibration, acoustic noise, and resonance, and significant increases in the mean drag and lift fluctuations. Therefore, the effective control of vortex shedding is important in engineering applications, which can successfully lead to separation delay, drag reduction, and suppression of noise and vibration. The synthetic jet has been used to control the flow around a circular cylinder, which is mainly based on three typical positions, namely the separation point, the front stagnation point, and the rear stagnation point.

Positioned Near the Separation Point
When the synthetic jets are positioned near the separation points, both separation delay and drag reduction can be reduced. Amitay et al. (1997, 1998) found that the synthetic jets positioned at azimuth $100°$ could substantially reduce drag by up to 30%, while the velocity fluctuations and Reynolds stress at $x/D = 3$ downstream of the circular cylinder were also reduced. Tensi et al. (2002) placed two rough strips at $\pm30°$ and found that the separation point was delayed from $95°$ to $135°$ when the synthetic jet was positioned at $112.5°$. It was suggested by Fujisawa and Takeda (2003) that the most effective position of the synthetic jet was near the separation point, and thus the maximum drag reduction of 30% could be achieved.

The synthetic jet can also change the vortex dynamics of the circular cylinder when they are positioned near the separation points, as has been conducted by Liu and Feng (2015). When the excitation frequencies are $f_e/f_0 = 1$ (Figure 8.16(b)) and 4.5 (Figure 8.16(d)), the wake vortex is synchronized by the synthetic jet and it sheds downstream in a symmetric mode. In particular, the lift coefficient is almost zero and does not change with time when the flow becomes symmetric. However, for some other control cases, such as $f_e/f_0 = 3$ (Figure 8.16(c)) and 15 (Figure 8.16(e)), the synthetic jet could not completely dominate the global flow field except for the very near field. Further downstream it shows the asymmetric shedding pattern, which is similar to the natural case (Figure 8.16(a)).

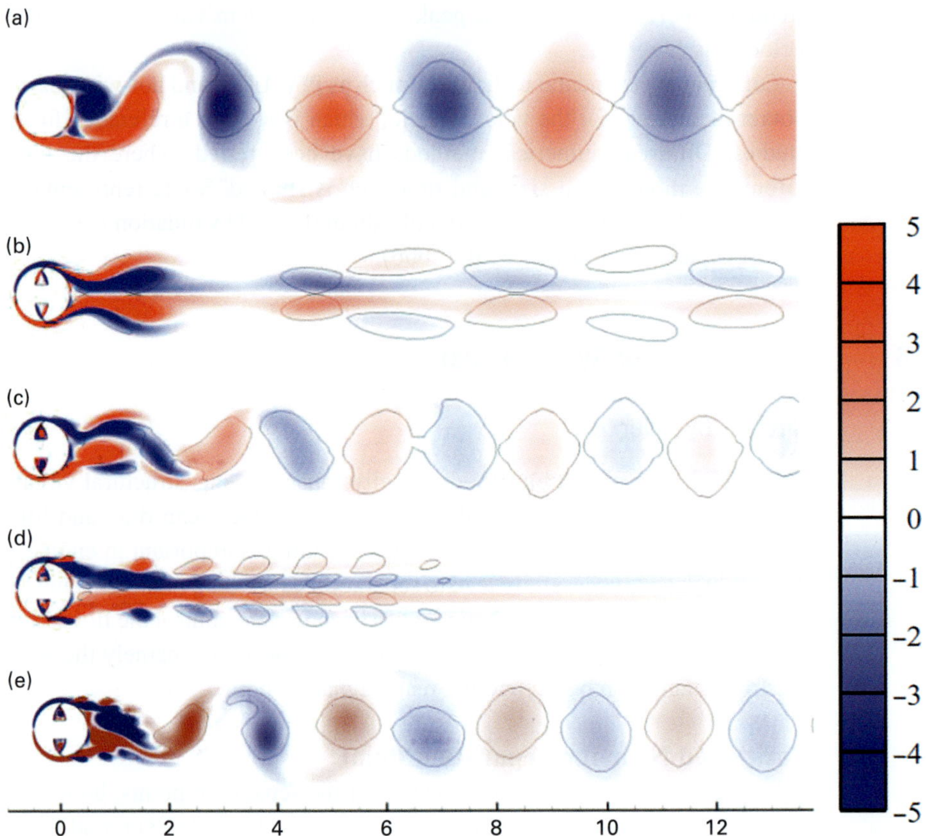

Figure 8.16 Wake patterns illustrated by the instantaneous spanwise vorticity $\omega D/U_\infty$ for (a) natural case and (b–e) control cases at $Re = 500$, $C_\mu = 0.0806$: (b) $f_e^+ = 1$; (c) $f_e^+ = 3$; and (d) $f_e^+ = 4.5$; (e) $f_e^+ = 15$. Solidlines are $\lambda_2 = -0.01$ isolines denoting the individual vortices (Liu and Feng 2015). Reproduced with permission, copyright © 2015 Elsevier Ltd. All rights reserved.

Positioned at the Front Stagnation Point

In other related investigations, the synthetic jet is usually positioned near the separation point. However, Amitay et al. (1997) found that when the synthetic jets were placed at azimuths of 0° and 60°, the interaction of the synthetic jets with the oncoming flow led to the formation of a closed recirculation flow. Wang et al. (2007) and Feng and Wang (2014a) further indicated that the recirculation region formed in the windward side of the cylinder could be divided into a closed envelope and an open envelope, both acting as the virtual aerodynamic shape. Some typical results are shown in Figure 8.17. At small excitation frequencies (Figure 8.17(a)), the velocity defect upstream of the circular cylinder increases as the vortex pair moves upstream and thus forms a new front stagnation point. A closed envelope forms in front of the windward side periodically, displacing the local streamlines and modifying the flow configuration. At high excitation frequencies (Figure 8.17(b)), the interaction between the synthetic jet and the oncoming

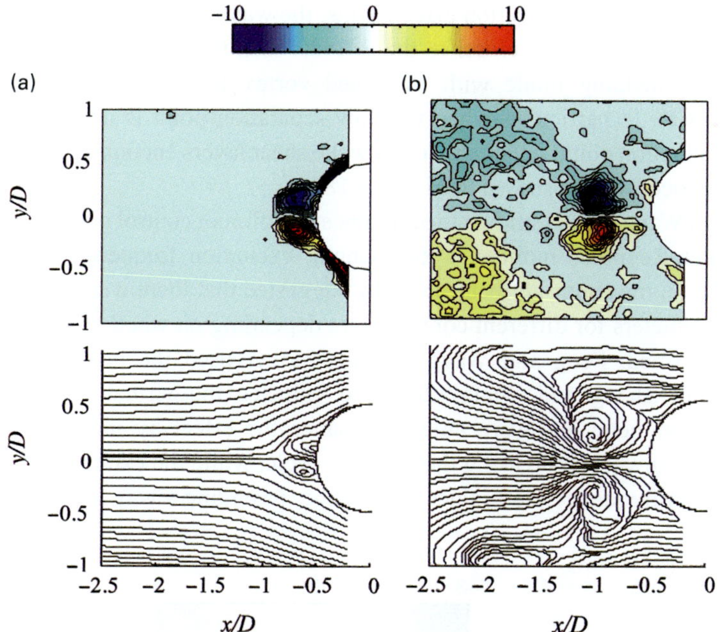

Figure 8.17 Phase-averaged spanwise vorticity and streamline around the windward side of the circular cylinder at $Re = 800$ with synthetic jet control for (a) $L_0/D = 3.3, f_e/f_0 = 1.04, C_\mu = 0.035$ and (b) $L_0/D = 3.3, f_e/f_0 = 6.25, C_\mu = 1.248$ (Feng and Wang 2014a). Reproduced with permission, copyright © 2014 Elsevier Masson. All rights reserved.

flow leads to a further upstream displacement of the front stagnation point and the formation of a quasi-steady open envelope symmetric about the x-axis. The flow field around the circular cylinder could be divided into two regions: an outer flow region and a synthetic jet flow region, but the circular cylinder is mainly influenced by the synthetic jet flow. The perturbations introduced by the synthetic jet are transferred downstream by the periodic closed envelope or the quasi-steady open envelope, which results in different control effects.

When the closed envelope forms, the vortex pair convects downstream from both upper and lower surfaces of the circular cylinder. In the leeward side, they interact with the wake shear layer and trigger its instability. The triggered shear layer develops and its size enlarges. When it becomes larger enough, it detaches from the main shear layer and forms the separated wake vortex shedding downstream in the symmetric mode, as shown in Figure 8.18(b). However, as the wake shear layer is subject to the induction effect of the synthetic jet, it also has inherent instability of asymmetric shedding at the natural frequency. These two effects compete with each other. When the synthetic jet is too weak, the wake vortex still sheds downstream in the asymmetric mode, similar to the natural case shown in Figure 8.18(a). When the two effects are comparable, the vortex shedding mode might vary with the symmetric mode or the asymmetric mode. This will lead to the formation of the bistable state I or II, which differs from the vortex-synchronization state. When the momentum coefficient is high enough, the flow separation is delayed and the distance between upper and lower wake shear layers decreases due to the apparent modification effect. Thus, the wake shear layers interact with each other from positions much closer to the cylinder than the natural case, resulting in the asymmetric shedding mode with shortened vortex formation length, as shown in Figure 8.18(c). In particular, when the flow separation point is delayed very close to the rear stagnation point, upper and lower wake shear layers encounter there, resulting in the new-induced vortex shown in Figure 8.18(d).

Feng and Wang (2013) studied the influence of different control parameters, including the cylinder Reynolds number, stroke length, excitation frequency, and momentum coefficient, on the control effects, and they suggested that there were different essential control parameters for different control cases depending on whether the closed or open envelope formed. Feng and Wang (2014b) further used the novel actuation signal proposed by Zhang and Wang (2007) to enhance the virtual aeroshaping effect. The novel synthetic jets were able to enhance the effect by increasing the suction cycle and meanwhile decreasing the blowing cycle, while the excitation amplitude and the excitation frequency did not need to change, providing a way for efficient control.

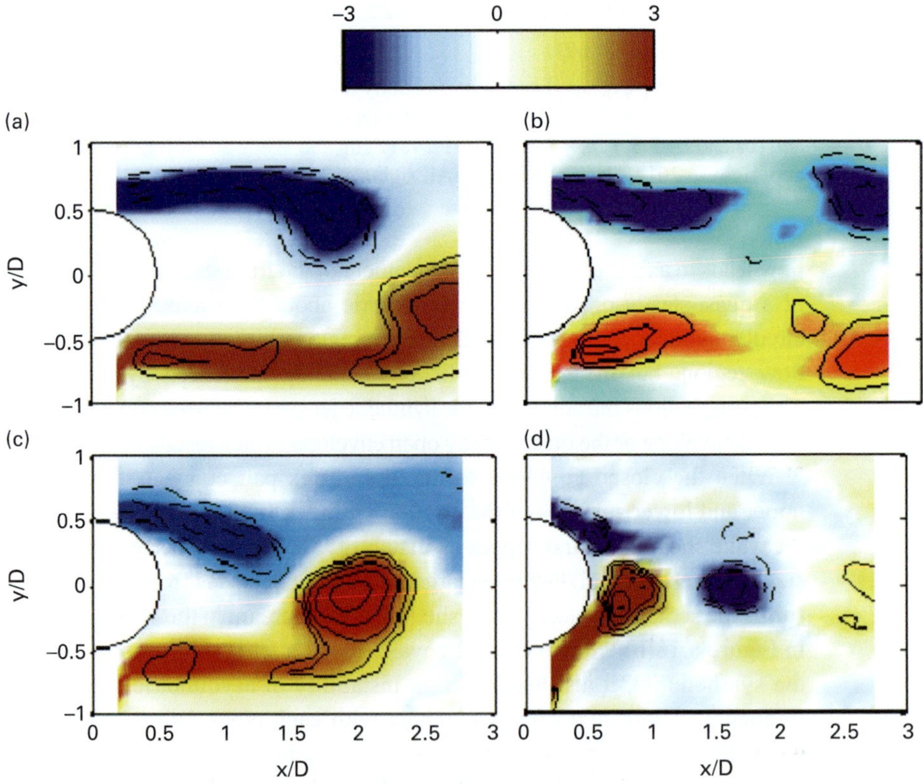

Figure 8.18 Instantaneous reconstructed spanwise vorticity superposed with λ_{ci} contour for the natural case (a) at $Re = 800$ and the control cases at (b) $L_0/D = 3.3$, $f_e/f_0 = 1.25$, $C_\mu = 0.050$, (c) $L_0/D = 3.3$, $f_e/f_0 = 2.08$, $C_\mu = 0.139$, and (d) $L_0/D = 3.3$, $f_e/f_0 = 6.25$, $C_\mu = 1.248$ (Feng and Wang 2014a). Reproduced with permission, copyright © 2014 Elsevier Masson. All rights reserved.

Positioned at the Rear Stagnation Point

Amitay et al. (1997) and Feng et al. (2008) all concluded that the synthetic jet could also control flow separation around a circular cylinder when positioned at the rear stagnation point. This control strategy is similar to other control techniques based on this point, such as splitter plate and base bleed. This field contains abundant flow phenomena, such as the shear layer instability and the interaction between vortex and shear layer. Feng and Wang (2010, 2012) and Feng et al. (2010, 2011) conducted a series of experimental studies into this field to reveal the physics of synthetic jet control, while Ma and Feng (2013) further examined the control effects when the synthetic jets were placed at both the front and rear stagnation points.

It was found that the synthetic jet could control flow separation well during both blowing and suction cycles, however, with different control mechanism. Drag reduction by up to 29% was achieved under this control configuration. Moreover, velocity decomposition, phase averaging, and proper orthogonal decomposition (POD) methods were used to analyze the vortex dynamics. The vortex pattern could be categorized into three groups, namely the asymmetric Kármán vortex mode, vortex synchronization with bistable mode, and vortex synchronization with symmetric mode. The synthetic jet vortex pair is induced from the orifice located at the rear stagnation point. It interacts with the shear layers behind both sides of the circular cylinder as it is convected downstream, resulting in the concentration of vorticities near the tails of the shear layers, which finally roll up into new-induced wake vortices. When the synthetic jet vortex pair is strong enough, it can dominate the global flow field, resulting in the formation of the symmetric shedding mode, as shown in Figure 8.19. Otherwise, the wake vortex might shed with a bistable state varying between the symmetric and asymmetric modes, or still with the asymmetric mode.

8.6.2 Hump and Rump

The synthetic jet can be used to control the flow separation induced by the adverse pressure gradient, thus it has been widely used to control the flow separation over a two-dimensional hump (Suzuki 2006; Wang et al. 2014), ramp (Dandois et al. 2007; Lardeau and Leschziner 2011), and a three-dimensional car model (Brunn and Nitsche 2006; Kourta and Leclerc 2013), etc. The synthetic jet shows a good ability to control the flow separation. However, the control effects are determined by the momentum coefficient and the operating frequency.

One example for the two-dimensional hump by Wang et al. (2014) is shown in Figure 8.20. The streamwise length, height and width of the hump are $c = 140$ mm, $h = 17.9$ mm, and $w = 500$ mm, respectively. A two-dimensional slot with width $d = 1$ mm and length $l = 70$ mm is located at $x/c = 0.65$ in the mid-span region of the hump, which has a 30° angle with the x-axis. The hump model is placed onto a splitter plate of size 2000 mm $\times$ 500 mm $\times$ 15 mm (length $\times$ width $\times$ thickness). The leading edge of the splitter plate has a 4:1 elliptical modification, which is 860 mm upstream of the hump model's leading edge.

Figure 8.21 shows the time-averaged streamlines for natural and control cases at $Re = 700$. For the natural case, the flow separates at $x_s/c = 0.65$ and reattaches at

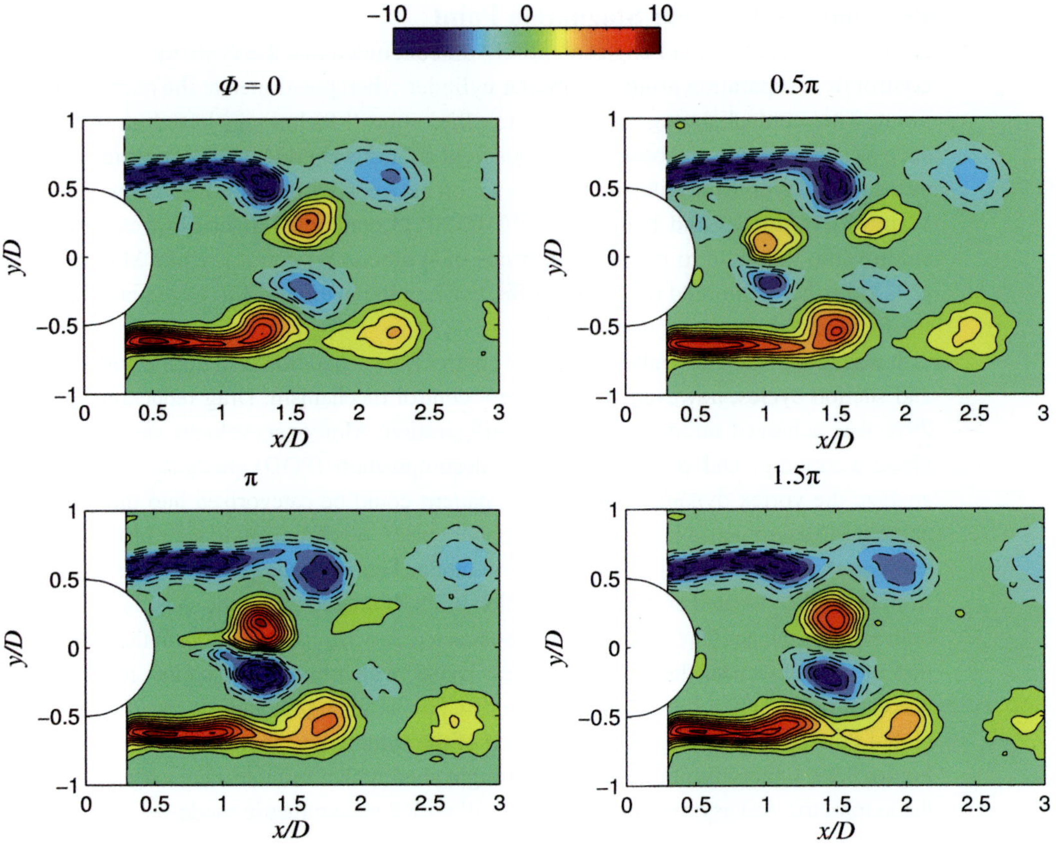

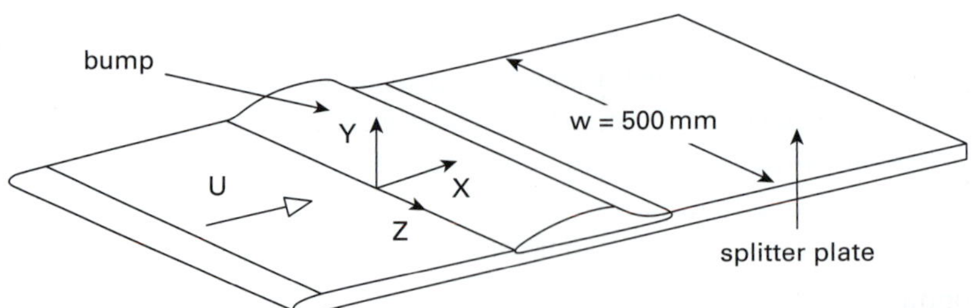

Figure 8.19 Evolution of the phase-averaged spanwise vorticity at four phases of 0, 0.5π, π and 1.5π for $Re = 950$, $L_0/D = 3.3$, $f_e/f_0 = 3.33$, $C_\mu = 0.378$ (Feng and Wang 2010). Reproduced with permission, copyright © 2010 Elsevier Masson. All rights reserved.

Figure 8.20 Schematic of the experimental hump mounted on the splitter plate (Wang et al. 2014). Reproduced with permission, copyright © Taylor Francis, www.tandfonline.com.

$x_r/c = 1.81$ (Figure 8.21(a)). Thus, an extremely narrow and long separation bubble is formed at the leeward side of the hump. For $f_e/f_0 = 1$ and 2, the separation point is delayed to $x_s/c = 0.70$ and 0.72, while the reattachment point is put forward to $x_r/c = 1.55$ and 1.45, corresponding to a reduction in the length of the separation bubble of

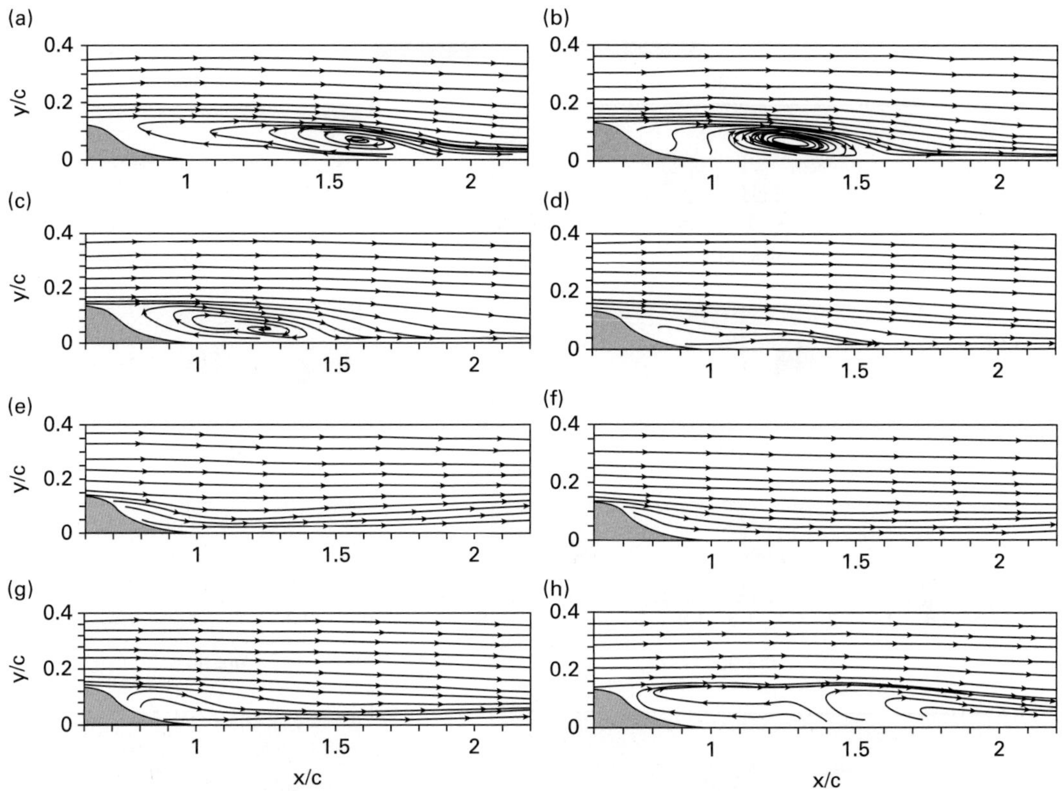

Figure 8.21 Time-averaged streamlines at $Re = 700$. (a) No jet; (b) $f_e/f_0 = 1$; (c) $f_e/f_0 = 2$; (d) $f_e/f_0 = 3$; (e) $f_e/f_0 = 4$; (f) $f_e/f_0 = 5$; and (g) $f_e/f_0 = 6$; (h) $f_e/f_0 = 7$ (Wang et al. 2014). Reproduced with permission, copyright © Taylor Francis, www.tandfonline.com.

about 27% and 37%, respectively (Figure 8.21(b) and (c)). For $f_e/f_0 = 3$, the separation region is further reduced and the time-averaged separation bubble is not observed (Figure 8.21(d)). For $f_e/f_0 = 4$ and 5, the backwards flow is not found in the flow field, validating that the flow has been completely attached (Figure 8.21(e) and (f)). For $f_e/f_0 = 6$, the streamlines are bent close to the leeward side of the hump, indicating that the flow is separated there, but the separation is soon reattached (Figure 8.21(g)). For $f_e/f_0 = 7$, a separation bubble occurs in the region downstream of the hump (Figure 8.21(h)). Thus, it is indicated that the optimal separation control can only be achieved within some excitation frequencies. A similar suggestion has also been made by Dandois et al. (2007), who found that the separation length was reduced by 54% in the low-frequency control case while it was increased by 43% in the high-frequency control case. It was also suggested that when a better control effect was achieved, less control would be acquired by further increasing the excitation frequency (Kotapati et al. 2010).

The change in the flow topology also indicates a variation in the vortex dynamics, as shown in Figure 8.22. For the natural case, a large-scale vortex with negative spanwise

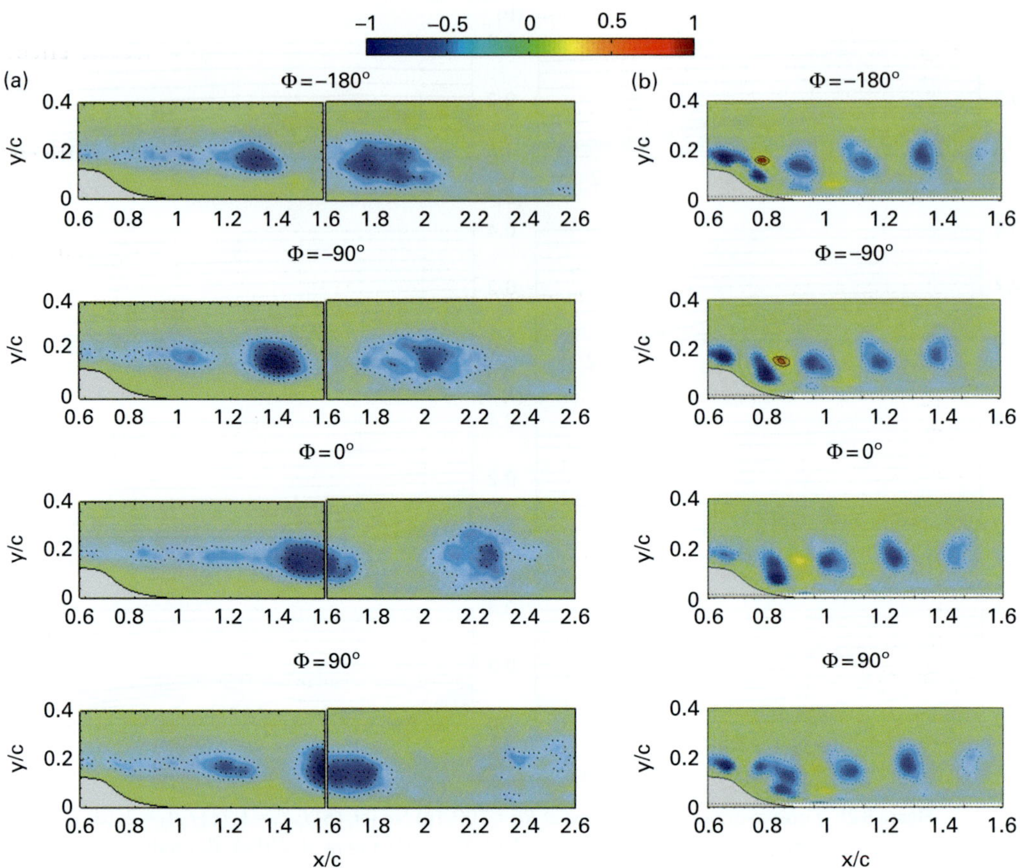

Figure 8.22 Phase-averaged Λ_{ci} for $Re = 700$ for the natural case (a) and control case $F^+ = 4$ (b), where pseudo-color represents the magnitude of Λ_{ci} (Wang et al. 2014). Reproduced with permission, copyright © Taylor Francis, www.tandfonline.com.

vorticity rolls up from the shear layer with shedding frequency of $f_0 = 0.29$ Hz (Figure 8.22(a)). For the control case of $f_e/f_0 = 4$ (Figure 8.22(b)), the large-scale vortices are reduced into small-scale flow structures. As the synthetic jet vortex pair is convected downstream, the negative one draws the separated shear layer and they merge together, while the positive one is restrained by the separated shear layer. After that, the strength of the negative vortex is enhanced and its size is enlarged gradually until a new wake vortex is formed, while the strength of the positive one is reduced. Thus, the vortex formation frequency is identical to the excitation frequency, meaning that the synthetic jet dominates the vortex-shedding frequency and acts as a lock-on phenomenon.

8.6.3   Airfoil

The synthetic jet has been used to control the flow around an airfoil. There have been many different parameters that may determine the control effects, such as the actuation position, the issuing angle, the momentum coefficient, and the excitation frequency. On the other hand, the Reynolds number of the airfoil may also influence the effect since it determines the flow state over the airfoil. In particular, laminar-to-turbulence transition may occur over the upper surface and thus the laminar separation bubble appears at low Reynolds numbers up to 10^6. Thus, the control mechanism and effects for the airfoil at lower Reynolds numbers may be different from that at higher Reynolds numbers.

Usually, there are two main locations where the synthetic jet can be placed, namely the position at or near the leading edge and that near the trailing edge. They usually result in different effects due to the difference in control mechanism. When the synthetic jet is positioned near the leading edge, the synthetic jet can increase the lift coefficient via separation control by imposing momentum into the boundary layer. The lift coefficient may not be changed when the flow is still fully attached, that is before the stall angle, as has been found by many researchers (such as Esmaeili Monir et al. 2014). Thus, it is suitable to control the airfoil with flow separation after stall. One example by Zhang and Wang (2008) is shown in Figure 8.23, where the serious separation region over the airfoil (Figure 8.23(a)) is reduced with the synthetic jet control (Figure 8.23(b)) at an angle of attack $\alpha = 20°$, increasing the lift coefficient by about 40%. When the synthetic jet is positioned near the trailing edge, it can increase the circulation of the airfoil, and thus increase the lift coefficient for all angles of attack (such as Lopez Mejia et al. 2011; Zhang et al. 2012).

At one position, the control effect of the synthetic jet is also determined by the issuing angle, which has been pointed out by Zhang and Wang (2009). They placed the synthetic jet at the upper surface of the airfoil 12% from the leading edge. The issuing angle was

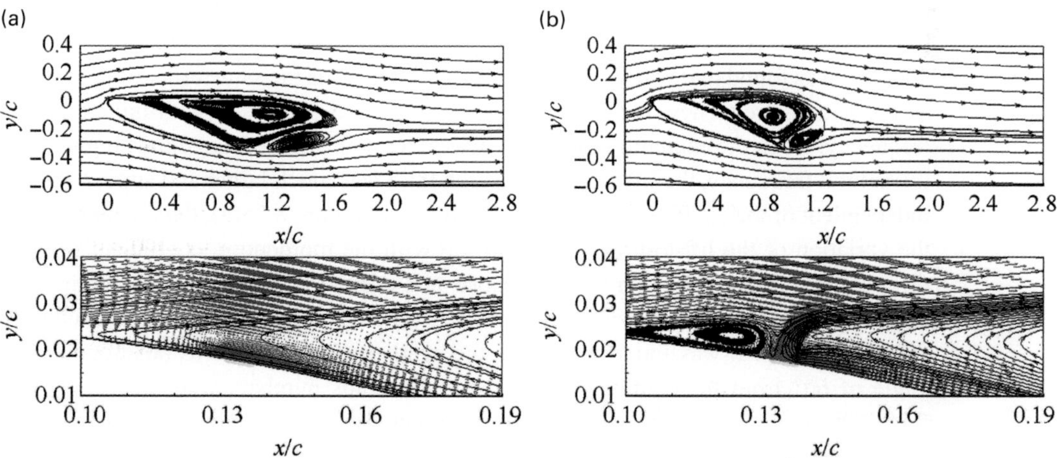

Figure 8.23 Time-averaged streamline around the NACA0015 airfoil without (a) and with (b) synthetic jet control for $\alpha = 20°$, $Re = 8.9 \times 10^5$, $C_\mu = 1\%$, $f_e c/U_\infty = 1$ (Zhang and Wang 2008). Reproduced with permission of Beijing University of Aeronautics and Astronautics.

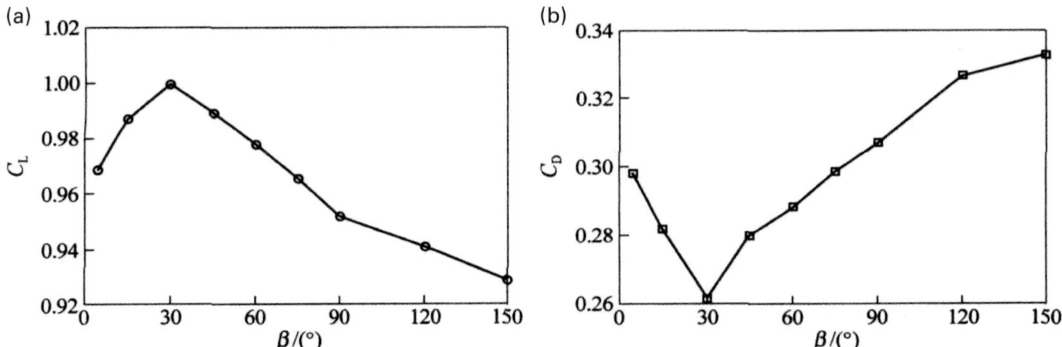

Figure 8.24 Variation of the lift coefficient (a) and drag coefficient (b) with the jet angle β for $\alpha = 20°$, $Re = 8.9 \times 10^5$, $C_\mu = 1\%$, $f_e c/U_\infty = 1$ (Zhang and Wang 2009). Reproduced with permission of Beijing University of Aeronautics and Astronautics.

related to the tangential line of the surface pointing to the downstream position. The angle of attack was fixed at $\alpha = 20°$, where the lift and drag coefficients for the natural case were 0.68 and 0.22, respectively. Figure 8.24 shows that for all control cases with different jet angles, the synthetic jet could increase the lift coefficient (Figure 8.24 (a)) and also the drag coefficient (Figure 8.24(b)). In particular, the optimal control case is obtained when the jet angle is 30°, where the lift coefficient is maximum while the drag coefficient is minimum, leading to a maximum lift-to-drag ratio. Duvigneau and Visonneau (2006) also obtained a similar optimal jet angle of about 25° and many researchers adopted the jet angle of 30° for flow control, such as You and Moin (2008).

The issue of parameter influence on the effect is also important in the control of the flow around an airfoil using synthetic jets. A series of investigations have been conducted about the influence of the excitation frequency and the momentum coefficient (such as Smith et al. 1998; Amitay et al. 2001; Amitay and Glezer 2002; Glezer et al. 2005). It was found that when the excitation frequency was the same order of magnitude as the natural frequency, the lift-to-pressure drag ratio decreased with the excitation frequency; when the excitation frequency was at least an order of magnitude higher than the natural frequency, the lift-to-pressure drag ratio was much larger and appeared to be independent of the excitation frequency, as shown in Figure 8.25(a). On the other hand, the variation of the lift-to-pressure drag ratio with the momentum coefficient showed different trends depending on the jet location and excitation frequency. One example is presented here when the synthetic jet was excited at the dimensionless excitation frequency of 10. It was found that when the synthetic jet control was positioned at an angle of 60° from the leading edge and the Reynolds numbers were $Re = 3.1 \times 10^5$, 5.25×10^5, and 7.25×10^5, the lift-to-drag ratio increased to a nominal value of 3.6 and then it was independent of the Reynolds number and the momentum coefficient, as shown in Figure 8.25(b).

Similar to the control of flow around an airfoil, the synthetic jet has also been used to control the flow around a finite and swept-back wing (Elimelech et al. 2011), a pitching

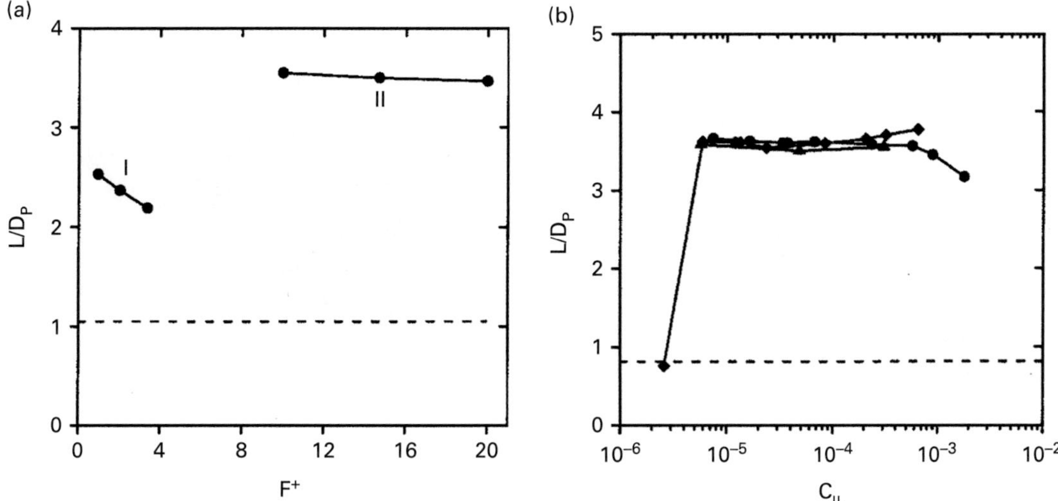

Figure 8.25 Variation of the lift-to-drag ratio with the excitation frequency (a) for the jet angle of 60°, $C_\mu = 0.35\%$, $Re = 3.1 \times 10^5$ and the momentum coefficient (b) for the jet angle of 60°, $F^+ = 10$, $Re = 3.1 \times 10^5$ (●), 5.25×10^5 (◆), and 7.25×10^5 (▲) (Amitay et al. 2001). Reproduced with permission, copyright © The American Institute of Aeronautics and Astronautics.

airfoil (Rehman and Kontis 2006), and the tip vortex (Margaris and Gursul 2006), where it also showed a significant control effect.

8.6.4   Vehicle

The synthetic jet has also shown potential applications in vehicles, including the unmanned underwater vehicle (UUV) (Mohseni 2006), the miniducted-fan unmanned aerial vehicle (UAV) (Fung and Amitay 2002), the Stingray UAV (Amitay et al. 2004), the Cessna aircraft model (Ciuryla et al. 2007), and the F16 Aircraft, etc. These studies were conducted either in the laboratory by wind tunnel or water tunnel test or in real flight condition, which highlighted the control ability of the synthetic jet.

The synthetic jet could be used as a propeller thruster for maneuvering and propulsion of UUV, as has been proved by Mohseni 2006 (Figure 8.26(a)). Fung and Amitay (2002) used synthetic jets to control a miniducted-fan UAV by placing them at four stator vanes, as shown in Figure 8.26(b). It was found that the flow field could be modified, and thus synthetic jet control was suggested to be used instead of moving control surfaces. The research group from Georgia Institute of Technology conducted a series of investigations into the control of a Stingray UAV by using synthetic jets, such as Amitay et al. (2004). The synthetic jet actuators were embedded inside the wing of the UAV and the slot was along the leading edge, as shown in Figure 8.26(c). It was found that the synthetic jets were able to create significant control forces and moments when the flow was separated from the leading edge of UAV. Ciuryla et al. (2007) conducted a wind tunnel study into the separation and roll control of a Cessna 182 model with synthetic

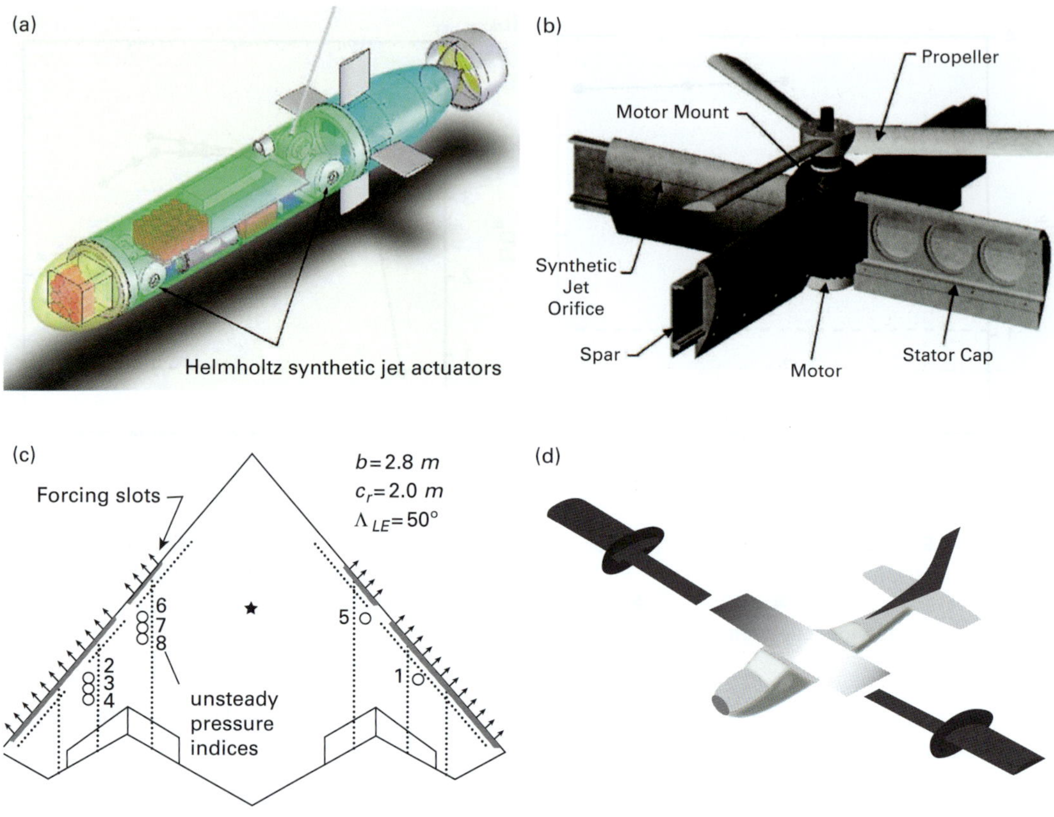

Figure 8.26 Examples of synthetic jet applications in different vehicles. (a) Unmanned underwater vehicle (UUV) (Mohseni 2006), reproduced with permission, copyright © 2006 Elsevier Ltd. All rights reserved. (b) Miniducted-fan unmanned aerial vehicle (UAV) (Fung and Amitay 2002), reproduced with permission, copyright © The American Institute of Aeronautics and Astronautics. (c) Stingray UAV (Amitay et al. 2004), reproduced with permission, copyright © The American Institute of Aeronautics and Astronautics. (d) Cessna model (Ciuryla et al. 2007), reproduced with permission, copyright © The American Institute of Aeronautics and Astronautics; (e) F16 Aircraft.

jets embedded within the outer portion of the wing, as shown in Figure 8.26(d). It was found that the stall angle could be delayed by 2° and the maximum lift coefficient could be increased by up to 15%. Shaw et al. (2006) conducted a flight test on the control of the wake from a pod of an F16 aircraft by placing the synthetic jets at the trailing edge of the pod, as shown in Figure 8.26(e). It was found that both the wake location and the shedding frequency could be controlled at Mach number 0.5, leading to a suppression of the intensity of the fluctuating pressures.

8.6.5   Inlet Duct

The synthetic jet is also capable of controlling the inside flow, such as the flow inside an inlet duct. In particular, the aggressive S-shaped inlet duct is an economic choice for modern aircraft. However, the fluids inside it undergo serious flow separation due to the strong adverse pressure gradient.

A typical S-inlet duct is shown in Figure 8.27. The duct has a length-to-width ratio of 1.5 with a rectangular cross-section of 114.3 mm wide and 88.9 mm high at the inlet and a square cross-section of 114.3 mm wide and 114.3 mm high at the outlet. Narrow slots with 1 mm width are arranged on the lower surface of the first turn. They act as the orifice of the synthetic jet with an oblique angle of 50° to the downstream surface. There are three different control configurations: the jet slots are along the spanwise direction of the inlet with the length of $0.98D$, $0.34D$ in the middle, and two $0.32D$ off the centerline of the inlet, which are denoted by up, up-mid and up-side actuations, respectively.

It was found by Chen and Wang (2012) that the synthetic jet was able to increase the total pressure recovery, decrease the velocity of crossflow and unify the flow in the outlet

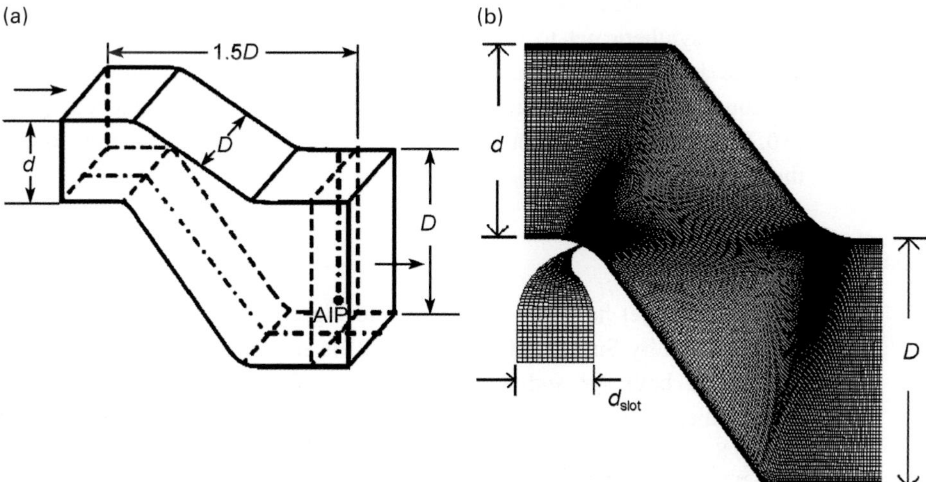

Figure 8.27 (a) Sketch of the computational inlet duct; and (b) grid in the symmetry plane where the synthetic jet actuator is shown (Chen and Wang 2012). Reproduced with permission, copyright © Springer 2012.

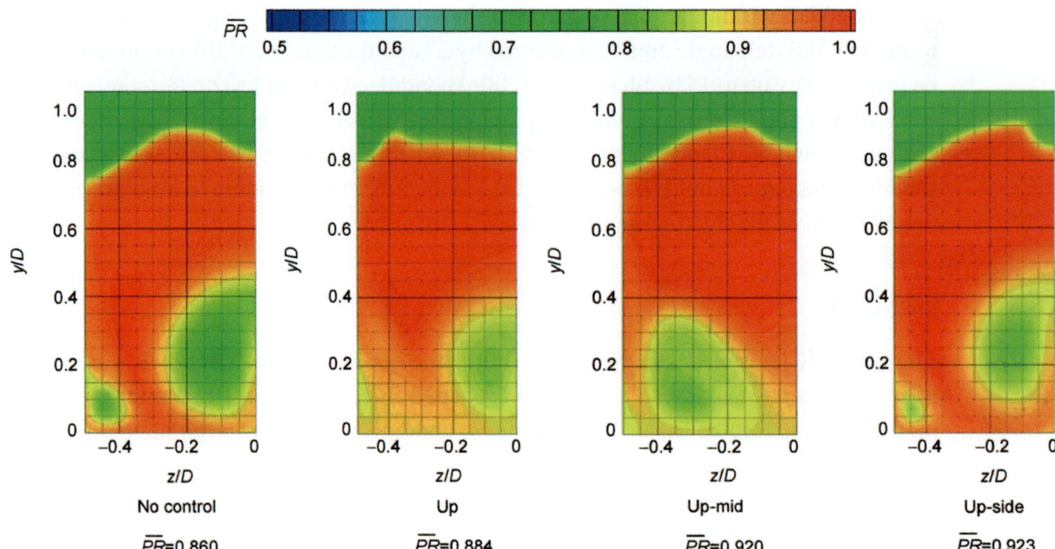

Figure 8.28 Distribution of total pressure recovery coefficient in the outlet interface of inlet duct at $x/D = 1.5$ (Chen and Wang 2012). Reproduced with permission, copyright © Springer 2012.

interface if the synthetic jet was applied in certain positions. Figure 8.28 shows the distribution of total pressure recovery coefficient in the outlet interface of inlet duct at $x/D = 1.5$. It is obvious that the quality of flow is better when the synthetic jet is across the spanwise direction, and it is more energetic and uniform for up-mid case. The flow becomes worse for the off-centerline control case due to the occurrence of a strong secondary vortex.

There have been few studies about the control of flow in an inlet. Some researchers used the synthetic jet to control a simplified duct. Amitay et al. (2002) placed the actuator arrays in the separation region of the inlet diffuser along the flow direction, and found that the flow attached completely at $Ma < 0.2$ and attached partly at $Ma = 0.2$ to 0.3. The control effect could be further enhanced by increasing the blowing velocity of the synthetic jet.

8.6.6   Vectoring

The synthetic jet has been used as vectoring control for a continuous jet. A general configuration by Smith and Glezer (2002) is shown in Figure 8.29, where the issuing orifices of both jet and synthetic jet are the rectangles parallel to each other. The schlieren images for the unforced and forced continuous jet are shown in Figure 8.29(a) and Figure 8.29(b), respectively. The centerline velocity of the unforced jet is 7 m/s, and the synthetic jet is operating at $Re_{U0} = 380$. The laminar jet is symmetric about the jet centerline, which is vectored toward the synthetic jet at a mean angle of approximately 30°.

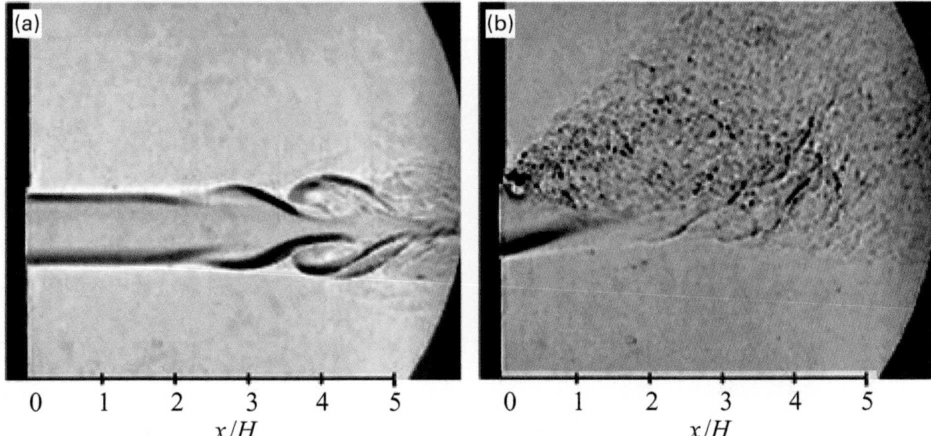

Figure 8.29 Schlieren images of the continuous jet without (a) and with (b) synthetic jet control (Smith and Glezer 2002). Reproduced with permission, copyright © Cambridge University Press 2002.

There are still many factors that can influence the vectoring angle, such as the velocity ratio λ between the synthetic jet and the primary jet, the excitation frequency, and the distance between the jet and the synthetic jet. One study case was conducted by Guo et al. (2003), where the results are shown in Figure 8.30. For all control cases, the vectoring angle of the continuous jet increases with an increase in one of these parameters until the maximum vectoring angle around 30°, and then decreases with it.

8.6.7   Heat Transfer

In many fields heat transfer needs to be enhanced for heat removal. Many techniques have been proposed, and one way is impinging-jet cooling. In comparison, the synthetic jet can be used as a cooling technique due to the strong entrainment capacity of the vortical structures by impinging on a fixed wall. Many researchers have studied the impingement of the synthetic jet onto a fixed wall, mainly concerning two issues: heat transfer and the flow field. The related works have been well reviewed by Xu and Wang (2013).

Figure 8.31 shows the variation of the stagnation Nusselt number with the orifice-to-surface distance at different synthetic jet Reynolds numbers and operating frequencies. For all cases, the stagnation Nusselt number increases gradually as the orifice becomes closer to the wall until a maximum and then decreases. The best cooling performance is obtained for the distance in the range of 3 to 11. A higher stagnation Nusselt number is obtained at a higher Reynolds number for the same excitation frequency at the same orifice-to-surface distance, as shown in Figure 8.31(a), since the Reynolds number determines the strength of the synthetic jet vortex. On the other hand, the high frequency synthetic jet removes heat better than the low frequency jet for smaller distances,

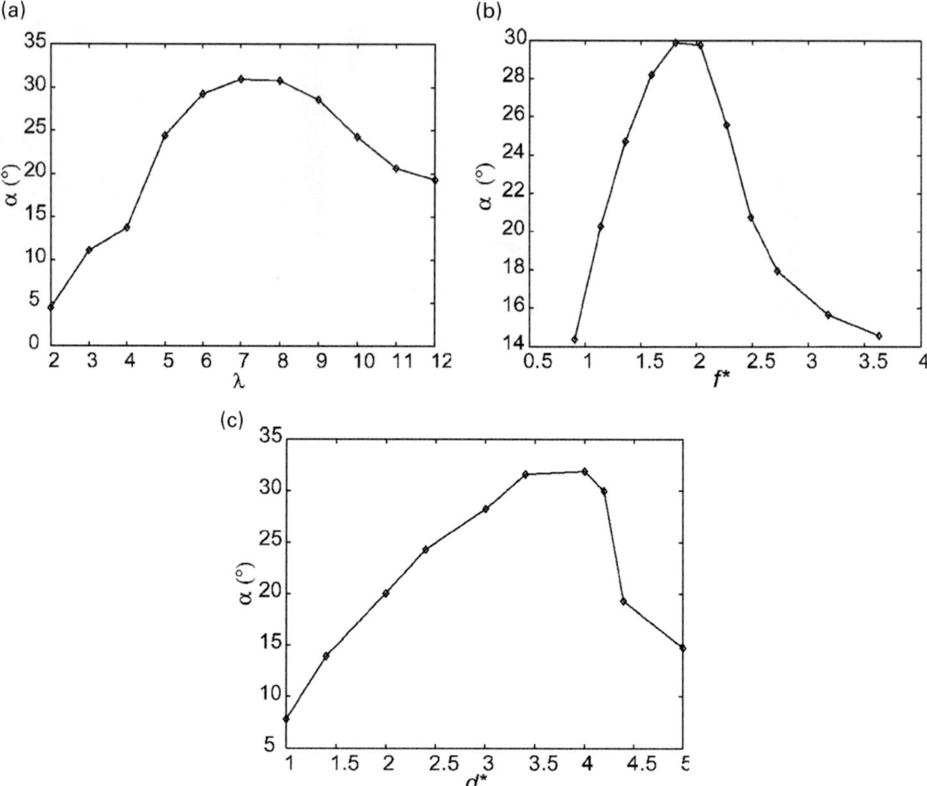

Figure 8.30 Variation of the vectoring angle of the continuous jet with different parameters: (a) the velocity ratio of the synthetic jet to the primary jet at f_e = 1120 Hz, U_{jet} = 7.0 m/s, d^* = 3.556; (b) the excitation frequency at Re_{U0} = 383, U_{jet} = 7 m/s; and (c) the distance between the jet and the synthetic jet at Re_{U0} = 383, U_{jet} = 7 m/s, f_e = 1120 Hz (Guo et al. 2003). Reproduced with permission, copyright © The American Institute of Aeronautics and Astronautics.

whereas the low frequency jet is more effective at larger distances, as shown in Figure 8.31(b). In particular, synthetic jets cool the surface better than the corresponding continuous jet.

Analyzing the flow evolution during impingement helps us to understand the heat transfer mechanism. Figure 8.32 shows the phase-averaged impingement process of the vortex ring. It can be seen that when the primary vortex ring is approaching the wall, the previous vortex with weaker vortical strength is pushed away. Then the primary vortex ring expands along the wall, while the secondary vortex with its rotation direction opposite to the primary vortex is induced near the wall surface. Thus, one can find three pairs of vortical structures in the near-wall field, including the primary vortex, secondary vortex, and previous vortical structure clustering the wall. This study was conducted by Xu et al. (2013) at low Reynolds number of 126. At higher Reynolds number, the secondary vortex rolls up from the wall to pair with the primary vortex, as has been found by Cheng et al. (2010) and Couch and Krueger (2011).

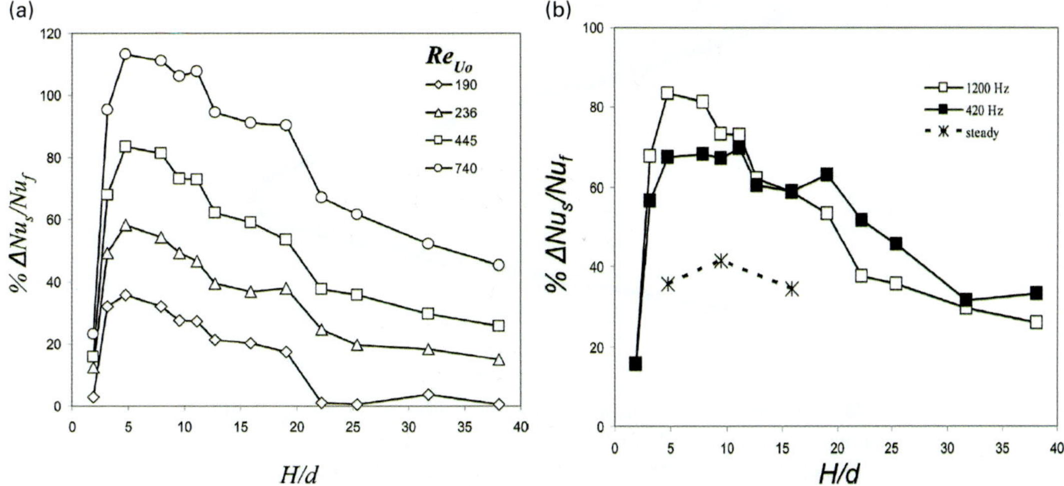

Figure 8.31 Variation of the stagnation Nusselt number with the orifice-to-wall distance H/d for different Reynolds numbers at $f = 1200$ Hz (a) and for different excitation frequencies at $Re_{U0} = 445$ (b). The results of those induced by a continuous jet are also plotted in (b) (Pavlova and Amitay 2006). Reproduced with permission of ASME Internationals.

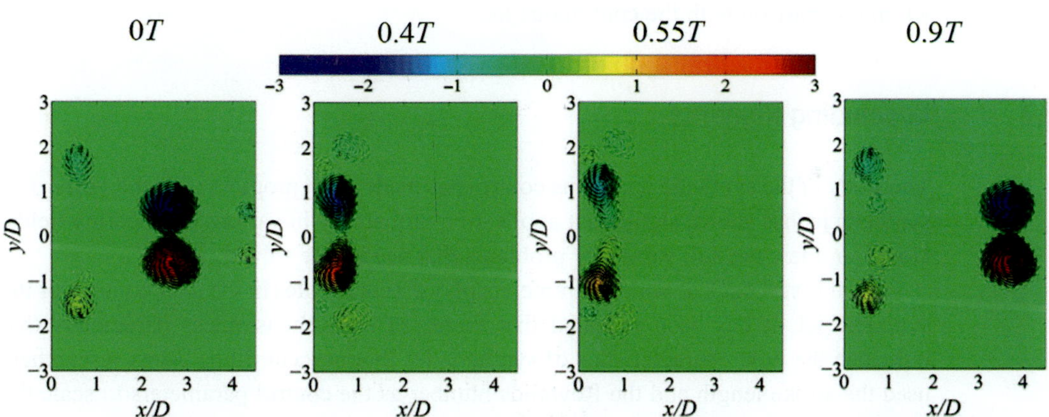

Figure 8.32 Phase-averaged λ_{ci} contours to show the impinging process of the synthetic jet vortex pair. Threshold is $|\lambda_{ci}| \geq 0.1$ and contour spacing is 0.02 (Xu et al. 2013). Reproduced with permission, copyright © Springer 2013.

During the impingement process, a wall jet is forming along the wall. Typical mean radial velocity profiles are plotted in Figure 8.33. Here, the mean radial velocity is normalized by the local maximum velocity, while the streamwise location is normalized by the half-width of the wall jet. It is shown that the nondimensional velocity profiles demonstrate self-similar behavior in the radial region $y/D \geq 2$, which is consistent with the theoretical solution of the laminar wall jet (the dashed line). On the other hand, Krishnan and Mohseni (2010) suggested that at a relatively high jet Reynolds number of

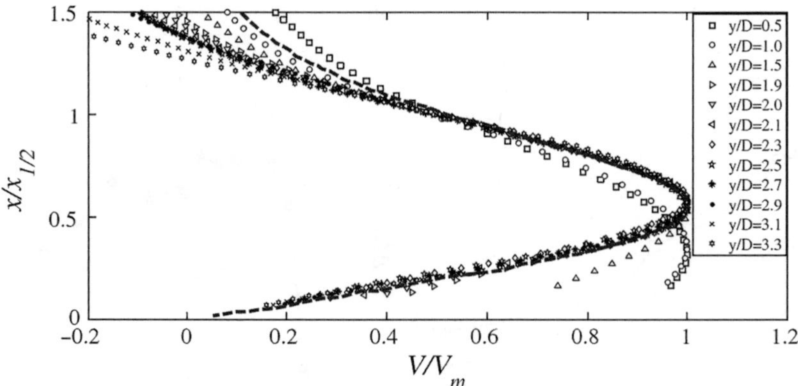

Figure 8.33 Self-similar curves of mean radial velocities. $x_{1/2}$ is the wall jet half-width, and V_m is the local maximum mean radial velocity. Dashed line represents theoretical solution of laminar wall jet (Xu et al. 2013). Reproduced with permission, copyright © Springer 2013.

$1632 \leq Re_{U0} \leq 2285$, the wall jet could only be described by an entire empirical wall jet solution.

From the study of the flow physics, it is suggested that the existence of the vortical structure and the induced wall jet is the key factor for higher heat transfer of the synthetic jet, in comparison with the continuous jet.

8.7   Concluding Remarks

The study of the synthetic jet and its control has made great progress over the years. The synthetic jet has shown significant application probability in various engineering fields. However, there are still some key problems that need to be solved.

First, it is still unclear which parameters play essential roles in the flow control for the synthetic jet. It has been revealed that there are six dimensionless parameters that influence the flow characteristics of synthetic jet in quiescent fluids. Most researchers used the stroke length and the Reynolds number as the control parameters to scale the vortex evolution and velocity field. However, some other parameters, such as Strouhal number, Stokes number, and geometric parameters, can also determine the development of the synthetic jet. Some of them may be related to each other. However, there is still the question of which parameter is the most important. Furthermore, when the synthetic jets are used for flow control, there are other factors that should be considered, such as the actuation location, the momentum coefficient, and the excitation frequency, etc. This makes the influence of parameters for the synthetic jet and its applications a complicated problem. However, for the purpose of flow control, it is required that we not only obtain the most effective control but also the best efficiency. In order to achieve this, we have to analyze the role of the individual parameters in flow control and deduce the optimal control parameters, and thus obtain the optimal effects.

Second, in the applications of the synthetic jet, we can see significant control effects. However, we do not understand how the disturbance induced by the synthetic jets can influence the global flow field for some cases. Some researchers have tried to reveal the control mechanism based on the velocity fields, vorticity fields or pressure distributions, etc. However, there is not a unified understanding.

Thirdly, the goal for the synthetic jet is its actual application in engineering, which is a great challenge for researchers. The actual applications need to take into account more factors rather than fluids mechanics. Take the widely used piezoceramic synthetic jet actuator as an example, the working principle suggests that it works well at the Helmholtz frequency of the cavity resonance, where it may encounter the problem of fatigue damage. In order to make a more reliable and robust control system, the support of other techniques, such as material, electronics, mechanism and manufacture, etc. are required.

References

Amitay, M. and Glezer, A. Role of actuation frequency in controlled flow reattachment over a stalled airfoil. *AIAA Journal*, 2002, 40(2): 209–216

Amitay, M., Honohan, A., Trautman, M., and Glezer, A. Modification of the aerodynamic characteristics of bluff bodies using fluidic actuators. AIAA Paper 1997–2004

Amitay, M., Pitt, D., and Glezer, A. Separation control in duct flows. *Journal of Aircraft*, 2002, 39(4): 616–620

Amitay, M., Smith, B. L., and Glezer, A. Aerodynamic flow control using synthetic jet technology. AIAA Paper 1998–0208

Amitay, M., Smith, D. R., Kibens, V., Parekh, D. E., and Glezer, A. Aerodynamic flow control over an unconventional airfoil using synthetic jet actuators. *AIAA Journal*, 2001, 39(3): 361–370

Amitay, M., Washburn, A. E., Anders, S. G., and Parekh, D. E. Active flow control on the Stingray uninhabited air vehicle: transient behavior. *AIAA Journal*, 2004, 42(11): 2205–2215

Bettini, C. and Cravero, C. Computational analysis of flow separation control for the flow over a wall-mounted hump using a synthetic jet. AIAA Paper 2007–0516

Brunn, A. and Nitsche, W. Active control of turbulent separated flows over slanted surfaces. *International Journal of Heat and Fluid Flow*, 2006, 27(5): 748–755

Cater, J. E. and Soria, J. The evolution of round zero-net-mass-flux jets. *Journal of Fluid Mechanics*, 2002, 472: 167–200

Chen, Z. J. and Wang, J. J. Numerical investigation on synthetic jet flow control inside an S-inlet duct. *Science China Technological Sciences*, 2012, 55(9): 2578–2584

Cheng, M., Lou, J., and Luo, L. S. Numerical study of a vortex ring impacting a flat wall. *Journal of Fluid Mechanics*, 2010, 660: 430–455

Cicca, G. M. D. and Iuso, G. On the near field of an axisymmetric synthetic jet. *Fluid Dynamics Research*, 2007, 39(9–10): 673–693

Ciuryla, M., Liu, Y., Farnsworth, J., Kwan, C., and Amitay, M. Flight control using synthetic jets on a Cessna 182 model. *Journal of Aircraft*, 2007, 44(2): 642–653

Couch, L. D. and Krueger, P. S. Experimental investigation of vortex rings impinging on inclined surfaces. *Experiments in Fluids*, 2011, 51(4): 1123–1138

Crook, A. and Wood, N. J. Measurements and visualisations of synthetic jets. AIAA 2001–0145

Cui, J. and Agarwal, R. K. 3-D CFD validation of an axisymmetric jet in cross-flow (NASA Langley Workshop Validation: Case 2). AIAA paper 2005–1112

Dandois, J., Garnier, E., and Sagaut, P. Unsteady simulation of synthetic jet in a crossflow. *AIAA Journal*, 2006, 44 (2): 225–238

Dandois, J., Garnier, E., and Sagaut, P. Numerical simulation of active separation control by a synthetic jet. *Journal of Fluid Mechanics*, 2007, 574: 25–58

Deng, X., Xia, Z. X., Luo, Z. B., and Li, Y. J. Vector-adjusting characteristic of dual-synthetic-jet actuator. *AIAA Journal*, 2015, 53(3): 794–797

Duvigneau, R. and Visonneau, M. Optimization of a synthetic jet actuator for aerodynamic stall control. *Computers & Fluids*, 2006, 35(6): 624–638

Elimelech, Y., Vasile, J., and Amitay, M. Secondary flow structures due to interaction between a finite-span synthetic jet and a 3-D cross flow. *Physics of Fluids*, 2011, 23(9): 094104

Esmaeili Monir, H., Tadjfar, M., and Bakhtian, A. Tangential synthetic jets for separation control. *Journal of Fluids and Structures*, 2014, 45: 50–65

Feng, L. H. and Wang, J. J. Circular cylinder vortex-synchronization control with a synthetic jet positioned at the rear stagnation point. *Journal of Fluid Mechanics*, 2010, 662: 232–259

Feng, L. H. and Wang, J. J. Synthetic jet control of separation in the flow over a circular cylinder. *Experiments in Fluids*, 2012, 53(2): 467–480

Feng, L. H. and Wang, J. J. Particle image velocimetry study of parameter influence for synthetic-jet application. *Flow Measurement and Instrumentation*, 2013, 34: 53–67

Feng, L. H. and Wang, J. J. Modification of a circular cylinder wake with synthetic jet: vortex shedding modes and mechanism. *European Journal of Mechanics B/ Fluids*, 2014a, 43: 14–32

Feng, L. H. and Wang, J. J. The virtual aeroshaping enhancement by synthetic jets with variable suction and blowing cycles. *Physics of Fluids*, 2014b, 26(1): 014105

Feng, L. H., Wang, J. J., and Pan, C. Effect of novel synthetic jet on wake vortex shedding modes of a circular cylinder. *Journal of Fluids and Structures*, 2010, 26(6): 900–917

Feng, L. H., Wang, J. J., and Pan, C. Proper orthogonal decomposition analysis of vortex dynamics of a circular cylinder under synthetic jet control. *Physics of Fluids*, 2011, 23(1): 014106

Feng, L. H., Wang, J. J., and Xu, C. J. Experimental verification of a novel actuator signal for efficient synthetic jet (in Chinese). *Journal of Experiments in Fluid Mechanics*, 2008, 22(1): 6–10

Fujisawa, N. and Takeda, G. Flow control around a circular cylinder by internal acoustic excitation. *Journal of Fluids and Structures*, 2003, 17(7): 903–913

Fung, P. and Amitay, M. Control of a miniducted-fan unmanned aerial vehicle using active flow control. *Journal of Aircraft*, 2002, 39(4): 561–571

Gharib, M., Rambod, E., and Shariff, K. A universal time scale for vortex ring formation. *Journal of Fluid Mechanics*, 1998, 360: 121–140

Gilarranz, J. L., Traub, L. W., and Rediniotis, O. K. A new class of synthetic jet actuators-Part I: design, fabrication and bench top characterization. *Journal of Fluids Engineering*, 2005, 127 (2): 367–376

Glezer, A. The formation of vortex rings. *Physics of Fluids*, 1988, 31(12): 3532–3542

Glezer, A. and Amitay, M. Synthetic jets. *Annual Review of Fluid Mechanics*, 2002, 34: 503–529

Glezer, A., Amitay, M., and Honohan, A. M. Aspects of low- and high-frequency actuation for aerodynamic flow control. AIAA Journal, 2005, 43(7): 1501–1511

Greenblatt, D., Paschal, K. B., Yao, C. S., Harris, .J, Schaeffler, N. W., and Washburn, A. E. A separation control cfd validation test case. Part 1: baseline & steady suction. AIAA Paper 2004–2220

Guo, D. H., Cary, A. W., and Agarwal, R. K. Numerical simulation of vectoring of a primary jet with a synthetic jet. *AIAA Journal*, 2003, 41(12): 2364–2370

Holman, R., Utturkar, Y., Mittal, R., Smith, B. L., and Cattafesta L. Formation criterion for synthetic jets. *AIAA Journal*, 2005, 43(10): 2110–2116

Ingard, U. and Labate, S. Acoustic circulation effects and the nonlinear impedance of orifices. *Journal of the Acoustical Society of America*, 1950, 22(2): 211–218

James, R. D., Jacobs, J. W., and Glezer, A. A round turbulent jet produced by an oscillating diaphragm. *Physics of Fluids*, 1996, 8(9): 2484–2495

Kotapati, R. B., Mittal, R., Marxen, O., Ham, F., You, D., and Cattafesta III, L. N. Nonlinear dynamics and synthetic-jet-based control of a canonical separated flow. *Journal of Fluid Mechanics*, 2010, 654: 65–97

Kourta, A. and Leclerc, C. Characterization of synthetic jet actuation with application to Ahmed body wake. *Sensors and Actuators A:* Physical, 2013, 192: 13–26

Krishnan, G. and Mohseni, K. An experimental study of a radial wall jet formed by the normal impingement of a round synthetic jet. *European Journal of Mechanics B/ Fluids*, 2010, 29(4): 269–277

Langley Research Center Workshop "CFD Validation of Synthetic Jets and Turbulent Separation Control," URL: http://cfdval2004.larc.nasa.gov

Lardeau, S. and Leschziner, M. A. The interaction of round synthetic jets with a turbulent boundary layer separating from a rounded ramp. *Journal of Fluid Mechanics*, 2011, 683: 172–211

Lebedeva, I. V. Experimental study of acoustic streaming in the vicinity of orifices. *Soviet Physics – Acoustics*, 1980, 26(4): 331–333

Liu, Y. G. and Feng, L. H. Suppression of lift fluctuations on a circular cylinder by inducing the symmetric vortex shedding mode. *Journal of Fluids and Structures*, 2015, 54: 743–759

Lopez Mejia, O. D., Moser, R. D., Brzozowski, D. P., and Glezer, A. Effects of trailing-edge synthetic jet actuation on an airfoil. *AIAA Journal*, 2011, 49(8): 1763–1777

Luo, Z. B., Deng, X., Xia, Z. X., Wang, L., and Gong, W. J. Flow field and heat transfer characteristics of impingement based on a vectoring dual synthetic jet actuator. *International Journal of Heat and Mass Transfer*, 2016, 102: 18–25

Luo, Z. B. and Xia, Z. X. Advances in synthetic jet technology and applications in flow control. *Advances in Mechanics*, 2005, 35(2): 221–234 (in Chinese)

Luo, Z. B. and Xia, Z. X. PIV Measurements and mechanisms of adjacent synthetic jets interactions. *Chinese Physics Letters*, 2008, 25(2): 612–615

Luo, Z. B., Xia, Z. X., and Liu, B. New generation of synthetic jet actuators. *AIAA Journal*, 2006, 44(10): 2418–2419

Luo, Z. B., Xia, Z. X., and Xie, Y. G. Jet vectoring control using a novel synthetic jet actuator. *Chinese Journal of Aeronautics*, 2007, 20(3): 193–201

Ma, L. Q. and Feng, L. H. Experimental investigation on control of vortex shedding mode of a circular cylinder using synthetic jets placed at stagnation points. *Science China Technological Sciences*, 2013, 56(1): 158–170

Margaris, P. and Gursul, I. Wing tip vortex control using synthetic jets. *Aeronautical Journal*, 2006, 110(1112): 673–681

Mednikov, E. P. and Novitskii, B. G. Experimental study of intense acoustic streaming. *Soviet Physics-Acoustics*, 1975, 21(2): 152–154

Mohseni, K. Pulsatile vortex generators for low-speed maneuvering of small underwater vehicles. *Ocean Engineering*, 2006, 33(16): 2209–2223

Morgan, P. E., Rizzetta, D. P., and Visbal, M. R. High-order numerical simulation of turbulent flow over a wall-mounted hump. *AIAA Journal*, 2006, 44(2): 239–251

Park, S. H., Yu, Y. H., and Byun, D. Y. RANS simulations of a synthetic jet in quiescent air. AIAA Paper 2007–1131

Pavlova, A. and Amitay, M. Electronic cooling using synthetic jet impingement. *Journal of Heat Transfer*, 2006, 128(9): 897–907

Rehman, A. and Kontis, K. Synthetic jet control effectiveness on stationary and pitching airfoils. *Journal of Aircraft*, 2006, 43(6): 1782–1789

Rumsey, C. L. Reynolds-averaged Navier-Stokes analysis of zero efflux flow control over a hump model. *Journal of Aircraft*, 2007, 44 (2): 444–452

Rumsey, C. L., Gatski, T. B., Sellers III, W. L., Vasta, V. N., and Viken, S. A. Summary of the 2004 Computational Fluid Dynamics Validation Workshop on Synthetic Jets. *AIAA Journal*, 2006, 44 (2): 194–207

Schaeffler, N. W. and Jenkins, L. N. The isolated synthetic jet in crossflow: a benchmark for flow control simulation. AIAA Paper 2004–2219

Shan, R. Q. and Wang, J. J. Experimental studies of the influence of parameters on axisymmetric synthetic jets. *Sensors and Actuators A: Physical*, 2010, 157(1): 107–112

Shaw, L. L., Smith, B. R., and Saddoughi S. Full scale flight demonstration of active flow control of a pod wake. AIAA Paper 2006–3185

Shuster, J. M. and Smith, D. R. Experimental study of the formation and scaling of a round synthetic jet. *Physics of Fluids*, 2007, 19(4): 045109

Smith, B. L. and Glezer, A. Jet vectoring using synthetic jets. *Journal of Fluid Mechanics*, 2002, 458: 1–34

Smith, B. L. and Glezer, A. The formation and evolution of synthetic jets. *Physics of Fluids*, 1998, 10(9): 2281–2297

Smith, B. L. and Swift, G. W. Synthetic jet at large Reynolds number and comparison to continuous jets. AIAA Paper 2001–3030

Smith, D. R., Amitay, M., Kibens, V., Parekh, D., and Glezer A. Modification of lifting body aerodynamics using synthetic jet actuators. AIAA Paper 1998–0209

Suzuki, T. Effects of a synthetic jet acting on a separated flow over a hump. *Journal of Fluid Mechanics*, 2006, 547: 331–359

Tensi, J., Boué, I., Paillé, F., and Dury, G. Modification of the wake behind a circular cylinder by using synthetic jets. *Journal of Visualization*, 2002, 5(1): 37–44

Toyoda, K. and Hiramoto, R. Manipulation of vortex rings for flow control. *Fluid Dynamics Research*, 2009, 41(5): 051402

Utturkar, Y., Holman, R., Mittal, R., Carroll, B., Sheplak, M., and Cattafesta, L. A jet formation criterion for synthetic jet actuators. AIAA Paper 2003–0636

Vatsa, V. N. and Turkel, E. Simulation of synthetic jets using unsteady Reynolds-averaged Navier-Stokes equations. *AIAA Journal*, 2006, 44(2): 217–224

Wang, J. J., Ba, Y. L., and Feng, L. H. Experimental investigation on laminar separation control for flow over a two-dimensional bump. *Journal of Turbulence*, 2014, 15(4): 221–240

Wang, J. J., Feng, L. H., and Xu, C. J. Experimental investigations on separation control and flow structure around a circular cylinder with synthetic jet. *Science in China E: Technological Sciences*, 2007, 50(5): 550–559

Wang, J. J., Shan, R. Q., Zhang, C., and Feng, L. H. Experimental investigation of a novel two-dimensional synthetic jet. *European Journal of Mechanics – B/ Fluids*, 2010, 29(5): 342–350

Xu, Y., Feng, L. H., and Wang, J. J. Experimental investigation of a synthetic jet impinging on a fixed wall. *Experiments in Fluids*, 2013, 54(5): 1512

Xu, Y. and Wang, J. J. Recent development of vortex ring impinging onto the wall. *Science China Technological Sciences*, 2013, 56(10): 2447–2455

Yao, C., Chen, F. J., Neuhart, D., and Harris J. Synthetic jet flow field database for CFD validation. AIAA Paper 2004–2218

You, D. and Moin, P. Active control of flow separation over an airfoil using synthetic jets. *Journal of Fluids and Structures*, 2008, 24(8): 1349–1357

Zhang, P. F. and Wang, J. J. Novel signal wave pattern for efficient synthetic jet generation. *AIAA Journal*, 2007, 45(5): 1058–1065

Zhang, P. F. and Wang, J. J. Effect of orifice inclined angle on flow control of the stalled airfoil with synthetic jet actuator. *Acta Armamentarii*, 2009, 30(12): 1658–1662 (in Chinese)

Zhang, P. F. and Wang, J. J. Numerical simulation on flow control of stalled NACA0015 airfoil with synthetic jet actuator in recirculation region. *Journal of Beijing University of Aeronautics and Astronautics* 2008, 34(4): 443–446 (in Chinese)

Zhang, P. F., Wang, J. J., and Feng, L. H. Review of zero-net-mass-flux jet and its application in separation flow control. *Science in China Series E: Technological Sciences*, 2008, 51(9): 1315–1344

Zhang, P. F., Yan, B., and Dai, C. F. Lift enhancement method by synthetic jet circulation control. *Science China Technological Sciences*, 2012, 55(9): 2585–2592

Zhong, S., Jabbal, M., Tang, H., Garcillan, L., Guo, F. S., Wood, N., and Warsop, C. Towards the design of synthetic-jet actuators for full-scale flight conditions, Part 1: the fluid mechanics of synthetic-jet actuators. *Flow, Turbulence and Combustion*, 2007, 78(3–4): 283–307

Zhou, J., Tang, H., and Zhong, S. Vortex roll-up criterion for synthetic jets. *AIAA Journal*, 2009, 47(5): 1252–1262

9   Plasma Actuator

9.1   Background

Plasma is one of the four fundamental states of matter, with the others being solid, liquid, and gas (Wu and Li 2015). With a strong electromagnetic field, plasma can be created with positive or negative charged particles, namely ions. The ions move in the presence of the electromagnetic field, which will also induce motion of the surrounding air due to viscous effects. Thus, with a specified design, the plasma actuator may produce ionization of the flowing air and thus add localized momentum to the flow through the collision process of migrating charged particles with the neutral species of the gas (Sosa and Artana 2006).

The idea to use plasma for flow control has been proposed for several decades. However, particular attention to plasma actuators for flow control started in the 1990s (Roth et al. 1995). The advantages of plasma control include electronic design with no moving parts, extremely fast response and real-time control ability, low power consumption, low mass, simplicity to use, etc. Thus, the plasma actuator becomes one of the most popular techniques for flow control and there have been many researches in this field. One can also refer to the review papers for recent progress on plasma flow control (such as Moreau 2007; Corke et al. 2010; Wang et al. 2013; Wu and Li 2015).

In this chapter we will introduce the main characteristics of plasma actuators and their conventional applications in various fields. In addition, some novel flow control conceptions based on plasma actuators will be presented.

9.2   Classification of Plasma Actuators

There are different kinds of plasma actuators, based on their configurations, such as dielectric barrier discharge (DBD) plasma actuator, surface corona discharge actuator, plasma spark-jet actuator. Among these, the DBD plasma actuator is the most frequently used technique for flow control. In this section we will first introduce the dominant features of these actuators.

9.2.1 DBD Plasma Actuator

A typical DBD plasma actuator consists of an exposed electrode and an embedded electrode, separated by a dielectric sheet (Figure 9.1). When the electrodes are supplied with high voltage and frequency, the air over the embedded electrode is ionized and thus a wall jet forms.

The two main features of the DBD plasma actuator are that it can induce a wall jet and a starting vortex. As we can see from Figure 9.2(a), when the velocity profile is normalized, the flow induced by the DBD plasma actuator shows the wall jet characteristics. This is consistent with the laminar solution. Accompanying the wall jet, the flow separates from the wall and rolls up into a vortex (Figure 9.2(b)). This vortex forms when the actuator is turned on, thus it is usually called the starting vortex. If the plasma actuator is controlled in a periodic manner, the wall jet and the starting vortex can also be induced periodically. It is easy to implement unsteady flow control by using the DBD plasma actuator.

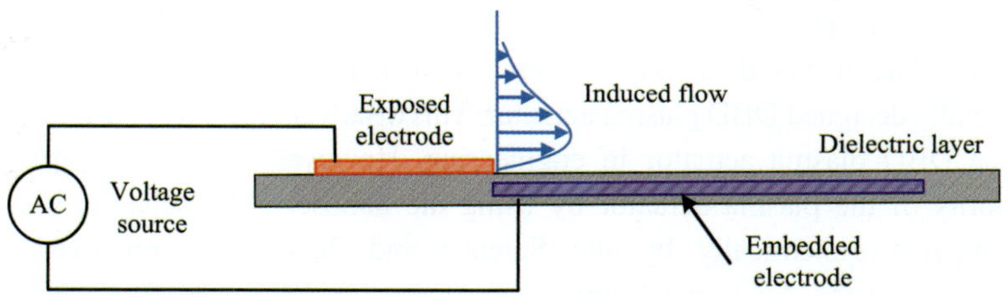

Figure 9.1 Schematic of the DBD plasma actuator.

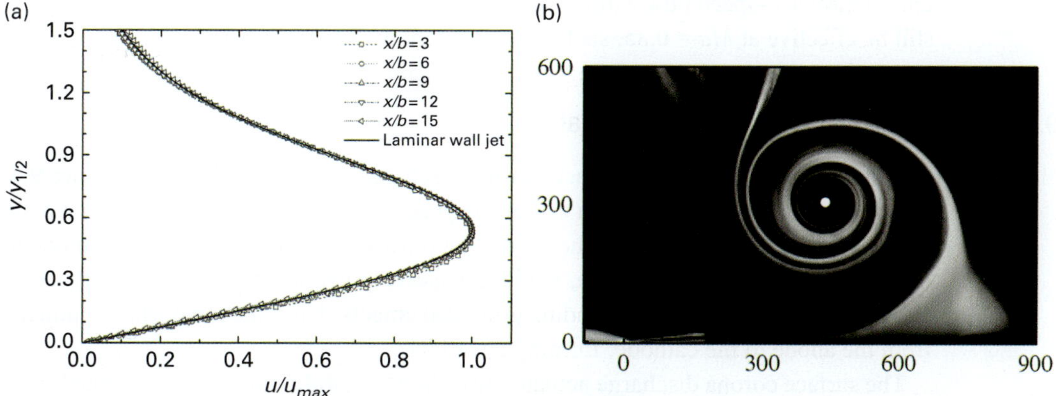

Figure 9.2 (a) Velocity profiles (Zhang et al. 2010c), reproduced with permission of Beijing University of Aeronautics and Astronautics. (b) The starting vortex induced by a DBD plasma actuator in quiescent flow (Whalley and Choi 2012). Reproduced with permission, copyright © Cambridge University Press 2012.

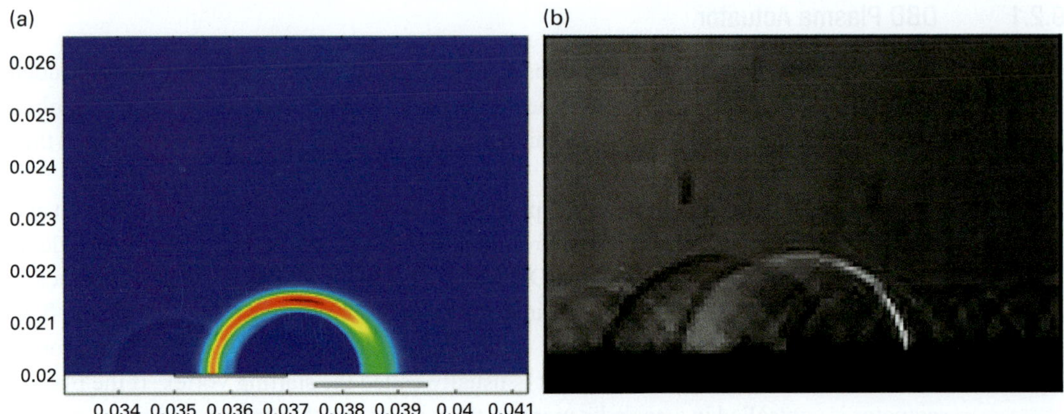

Figure 9.3 Calculated pressure perturbation (a) and experimental schlieren image (b) at 5 μs after the discharge (Zhu et al. 2013). Reproduced with permission, copyright © IOP Publishing. All rights reserved.

Most of the studies have mainly focused on the microsecond discharge. The microsecond plasma actuator can induce a wall jet with maximum velocity of several meters per second. However, the maximum velocity can hardly reach 20 m/s at present, even with a specially designed DBD plasma actuator. This disadvantage may restrict the applications of the DBD plasma actuator in engineering. However, one can improve the control authority of the plasma actuator by using the nanosecond discharge, which has been investigated systemically by the Science and Technology on Plasma Dynamics Laboratory at the Air Force Engineering University (Wu et al. 2012, 2013; Zhu et al. 2013; Zhao et al. 2015a, 2015b, 2015c, 2015d). The nanosecond plasma actuator induces a local compression wave or even shock wave due to fast heating near the exposed electrode edge, as shown in Figure 9.3. In general, the microsecond plasma actuator can only control low-speed flow within $Ma \leq 0.3$, while the nanosecond plasma actuator might still be effective at $Ma = 0.85$, such as in the work by Roupassov et al. (2009).

9.2.2   Surface Corona Discharge Actuator

A typical schematic for the surface corona discharge actuator is shown in Figure 9.4. It usually contains an anode and a cathode exposed on the surface. Either wire or a plate can be used as the anode and cathode. A high positive potential is used at the anode to obtain a homogeneous discharge, while a negative potential is used at the cathode to avoid the influence of any surrounding grounded objects. Thus, the ionic wind is induced from the anode to the cathode, forming a wall jet.

The surface corona discharge actuator also shows its ability to control separated flow, and one example is shown in Figure 9.5. We can see that the large-scale separation region is reduced significantly by plasma control at a low speed of $U_\infty = 0.4$ m/s. However, the wind tunnel test by Moreau et al. (2006) indicated that the surface corona discharge could still be effective at $U_\infty = 25$ m/s.

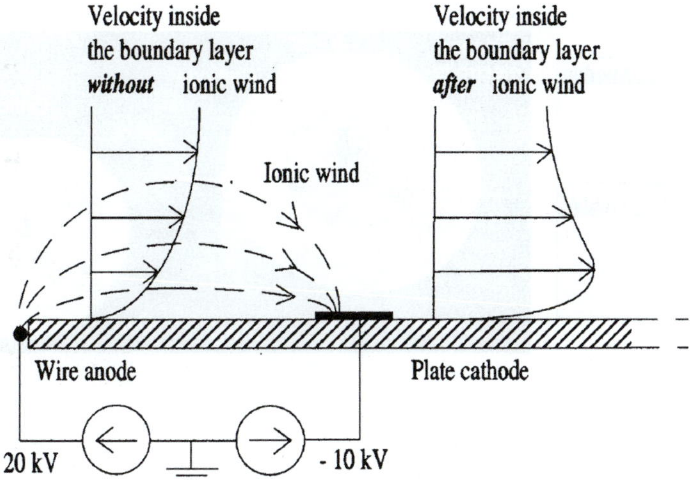

Figure 9.4 Schematic of the surface corona discharge actuator (Léger et al. 2001).

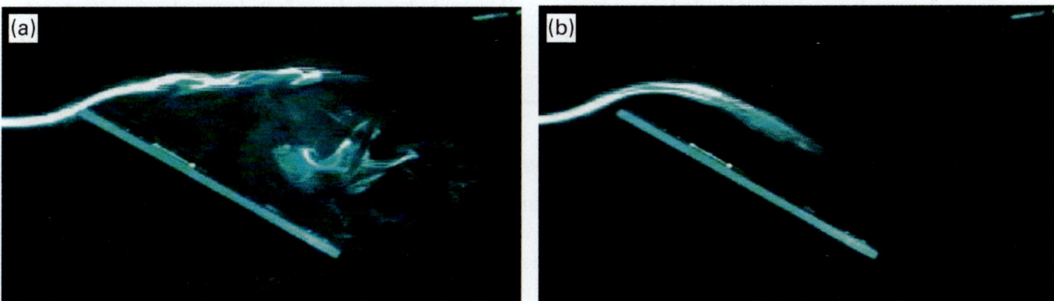

Figure 9.5 Flow visualization of an inclined flat plate without (a) and with (b) corona discharge control at $U_\infty = 0.4$ m/s (Moreau et al. 2006). Reproduced with permission, copyright © 2006 Elsevier Ltd. All rights reserved.

9.2.3 Plasma Spark-Jet Actuator

The Johns Hopkins University Applied Physics Laboratory have started to conduct a series of investigations on a new system of plasma actuator for high-speed flow control, which is called the plasma spark-jet actuator. It produces a synthetic jet with high exhaust velocity and holds the promise of manipulating high-speed flow without moving aerodynamic control surfaces. A schematic of the spark-jet device is shown in Figure 9.6(a). The device is made from an electrical insulator. Three electrodes are fashioned into the device: an anode, a cathode and a grid, as indicated by Cybyk et al. (2003). The working process includes three stages: energy deposition, discharge and recovery, as shown in Figure 9.6(c). When the grid is pulsed, a streamer discharge initiates between the cathode and grid, and current flows from the cathode to the anode, thereby heating the bulk chamber gas.

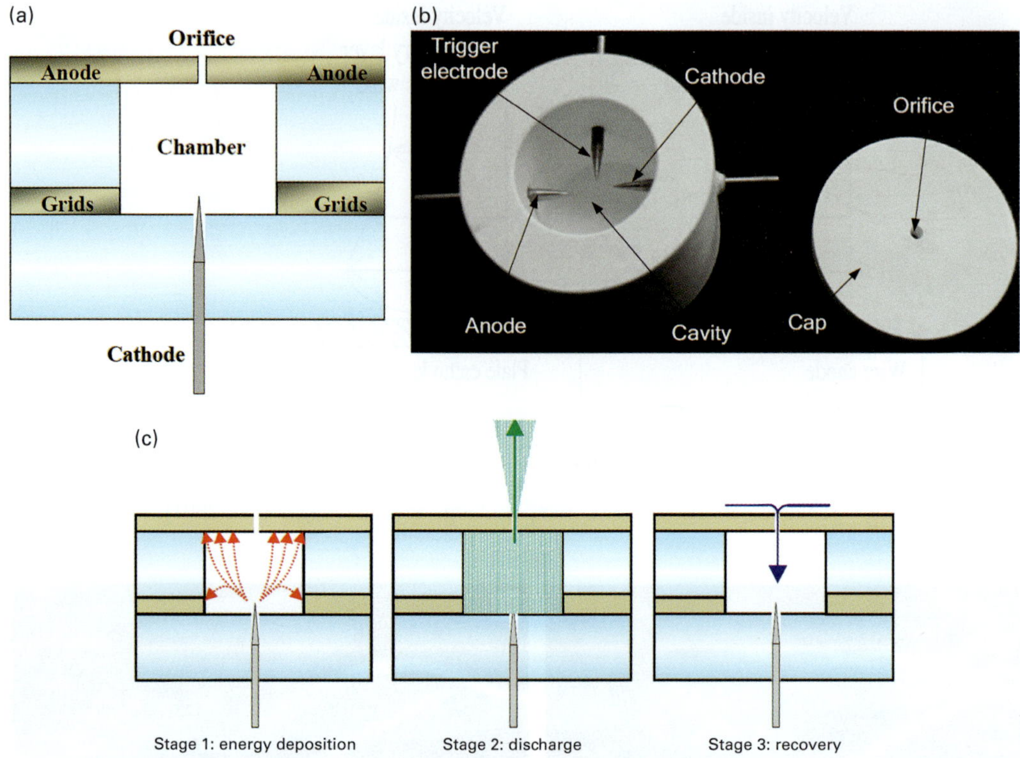

Figure 9.6 (a) Schematic of the plasma spark-jet actuator with an anode, a cathode and a grid (Cybyk et al. 2003). (b) Three electrodes with an anode, a cathode and a trigger electrode (Wang et al. 2014). (c) Stages of the spark-jet operating cycle (Cybyk et al. 2003). Reproduced with permission, copyright © The American Institute of Aeronautics and Astronautics.

A recent design by Wang et al. (2014) added a trigger electrode, as shown in Figure 9.6(b). Then, a short duration high-voltage pulse is flashed onto the trigger electrode to initiate a streamer discharge between the trigger electrode and cathode. Thus, a more energetic anode-to-cathode spark discharge is induced immediately by the streamer discharge. The energy deposited heats the chamber gas to high temperature and high pressure. Then, the gas expels from the device orifice at high speed, leading to a drop of chamber pressure and temperature, and thus the exiting velocity decreases. Finally, the chamber cools rapidly, and fresh air is drawn from outside the device into the chamber. Then, the device is ready for operation again. This working process is similar to that of a conventional synthetic jet, thus the device is also called the plasma synthetic jet.

The schlieren images of a three-electrode spark-jet designed by Zong et al. (2015) are shown in Figure 9.7. We can see the formation of shock wave and jet front after the device is actuated (Figure 9.7(a)). They are convecting downstream and enlarging its range rapidly in the following evolution. Their propagation velocities can be calculated,

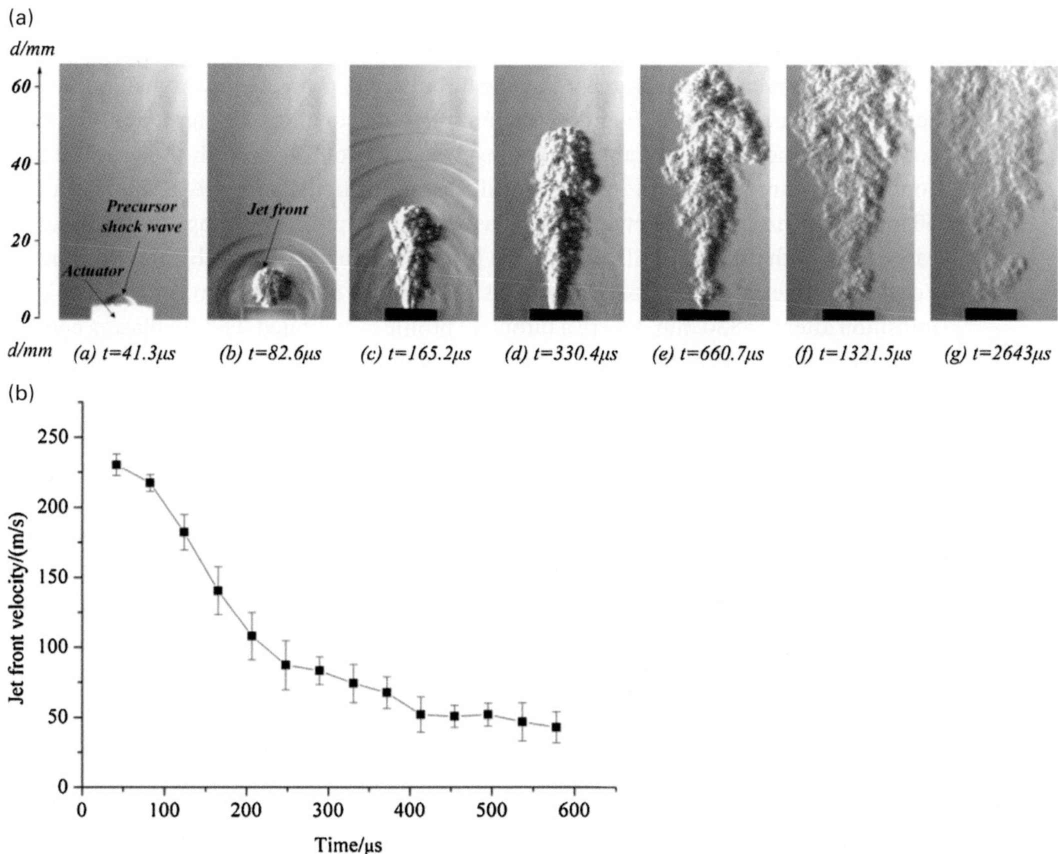

(a) $t=41.3\mu s$　　(b) $t=82.6\mu s$　　(c) $t=165.2\mu s$　　(d) $t=330.4\mu s$　　(e) $t=660.7\mu s$　　(f) $t=1321.5\mu s$　　(g) $t=2643\mu s$

Figure 9.7 (a) Schlieren images of the spark-jet at atmospheric condition. (b) Variation of jet front propagation velocity with time (Zong et al. 2015). Reproduced with permission, copyright © 2014 Elsevier B.V.

as shown in Figure 9.7(b). It is indicated that the maximum jet front velocity is about 230 m/s. It decreases with time, but still keeps velocity at around 50 m/s until 500 μs. These features enable the plasma spark-jet actuator to control the supersonic or hypersonic flow.

9.3　Conventional Applications

9.3.1　Boundary Layer

The plasma actuator can be used for boundary layer control, and one of first studies was conducted by Grundmann and Tropea (2007, 2008). An actuator used as the turbulator was placed at around 400 mm from the leading edge, as shown in Figure 9.8. It was powered at an AC peak-to-peak voltage of 9.2 kV and a frequency of 6 kHz. Two actuators were placed at 500 mm and 600 mm, respectively, which were both operated with an AC voltage of 10 kV and frequency of 6 kHz. The free-stream velocity was fixed at 6 m/s.

The boundary layer was triggered by the turbulator to promote flow transition, which acted as the baseline case. Two downstream plasma actuators were turned on for flow control. The mean velocity and fluctuating velocity profiles for both the baseline and control cases are shown in Figure 9.9 at different positions. At $x = 650$ mm, which is 50 mm downstream of the actuator, the Blasius velocity profile develops for the baseline case, while both the near wall velocity and its fluctuation are reduced by plasma control. For the baseline case, the transition process is underway at $x = 730$ mm, and it completes transition after $x = 830$ mm, where a turbulent profile is exhibited. For the plasma control case, the mean velocity profile still corresponds to the Blasius case while the velocity fluctuation is reduced in comparison with the baseline case. It starts to change to a turbulent profile from $x = 830$ mm and also becomes turbulent at $x = 850$ mm; the velocity fluctuations at these two locations show similar distributions to the baseline case.

The variation of the shape factor along the streamwise direction for the baseline and control cases is shown in Figure 9.10. For the baseline case, the flow transition is completed at around $x = 710$ mm, while it could be delayed by approximately 100 mm by using plasma control. In the subsequent study by Grundmann and Tropea (2008), it was indicated that the actuator in an unsteady mode was also effective but with less consumed energy. Duchmann et al. (2014) further indicated that the plasma control could delay transition of approximately $3\%c$ of a trapezoidal wing under free-flight conditions, which led to a reduction of the drag by about 1.9%.

Though most experiments were conducted at low speed, Im et al. (2010) used plasma actuators to control the turbulent boundary layer upstream of a ramp model at $Ma = 4.7$, as shown in Figure 9.11. The model had a sharp leading edge with a spanwise length of 40 mm. A compression ramp was located 120 mm downstream of the leading edge. Sandpaper (180 Grit, 20 mm long, 40 mm wide) was attached to the top of the model 10 mm downstream from the leading edge to ensure a turbulent boundary layer. The plasma actuator was parallel to the flow, with the

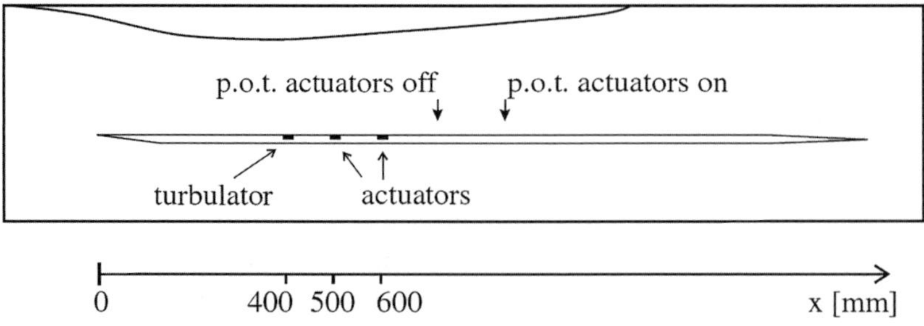

Figure 9.8 Schematic of the flat plate with plasma actuators (Grundmann and Tropea 2007). Reproduced with permission, copyright © Springer 2007.

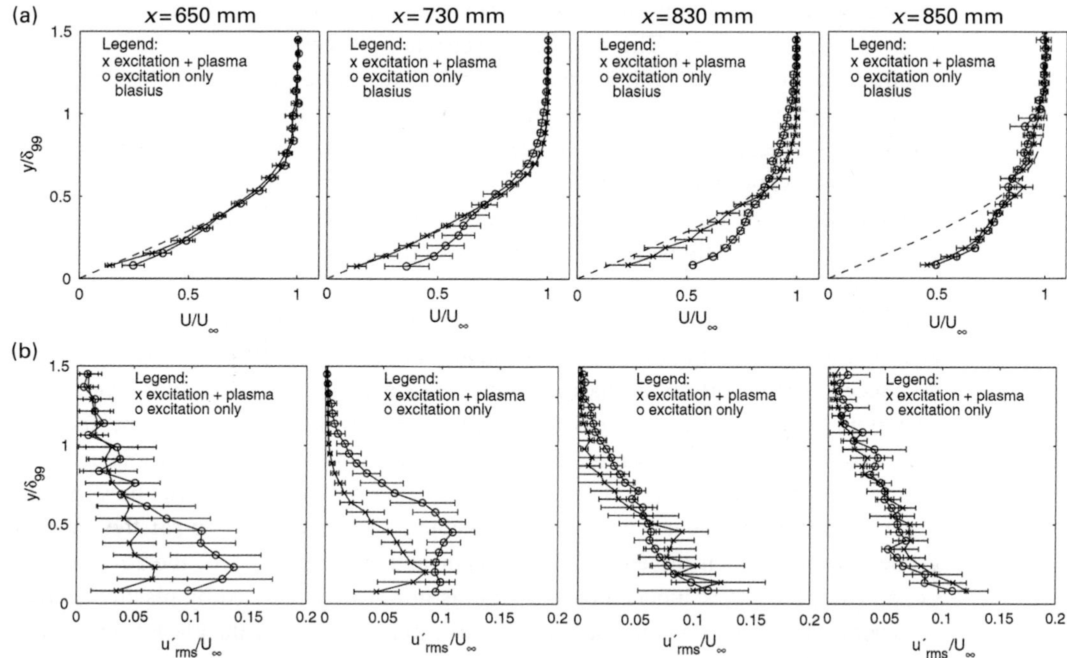

Figure 9.9 Profiles of the mean velocity (a) and velocity fluctuation (b) for the two cases (Grundmann and Tropea 2007). Reproduced with permission, copyright © Springer 2007.

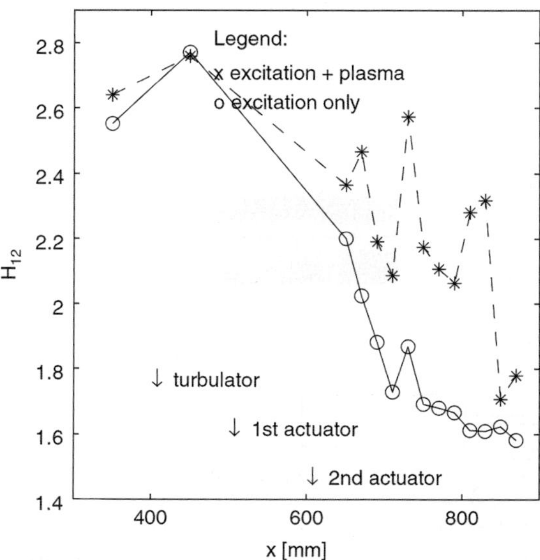

Figure 9.10 Development of the shape factor of the boundary layer for the two cases (Grundmann and Tropea. 2007). Reproduced with permission, copyright © Springer 2007.

exposed electrode 75 mm long and 7 mm wide and the embedded electrode 85 mm long and 25 mm wide. The exposed electrode was centered over the embedded electrode, which were both in the center of the model with their leading edge located 40 mm downstream of the model leading edge. The actuator was driven at 6 kV peak-to-peak voltage and 28 kHz, approximately 6.8 W of delivered power. The field of view was illuminated by a laser sheet located between 85 mm and 140 mm downstream of the leading edge and 3.5 mm off the centerline. Flow visualization was achieved by using Rayleigh scattering from condensed CO_2 particles.

Typical flow patterns for the turbulent boundary layer without and with plasma control are shown in Figure 9.12. It is found that the thickness of the boundary layer is reduced significantly with plasma control. For example, the boundary layer thickness with plasma forcing is less than half of that without forcing at the location indicated by the arrow, which is 25 mm upstream of the end of the plasma actuator. The control mechanism is related to the generation of the spanwise flow by the plasma actuator. However, the turbulent boundary layer appears to recover quickly in the region between the end of the actuator and the beginning of the compression ramp. Thus, the shock induced by the ramp is not greatly affected by plasma control. In general, this study demonstrates an effective manipulation of the plasma actuator in supersonic flow.

9.3.2 Airfoil

There have been numerous studies on the control of flow around stationary and oscillating airfoils by using plasma actuators, and typical investigations by Benard et al. (2009)

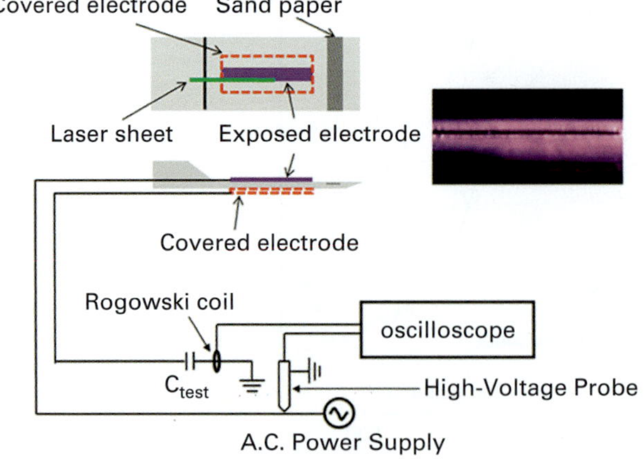

Figure 9.11 Schematic of the flat plate with the plasma actuator (Im et al. 2010). Reproduced with permission from AIP Publishing.

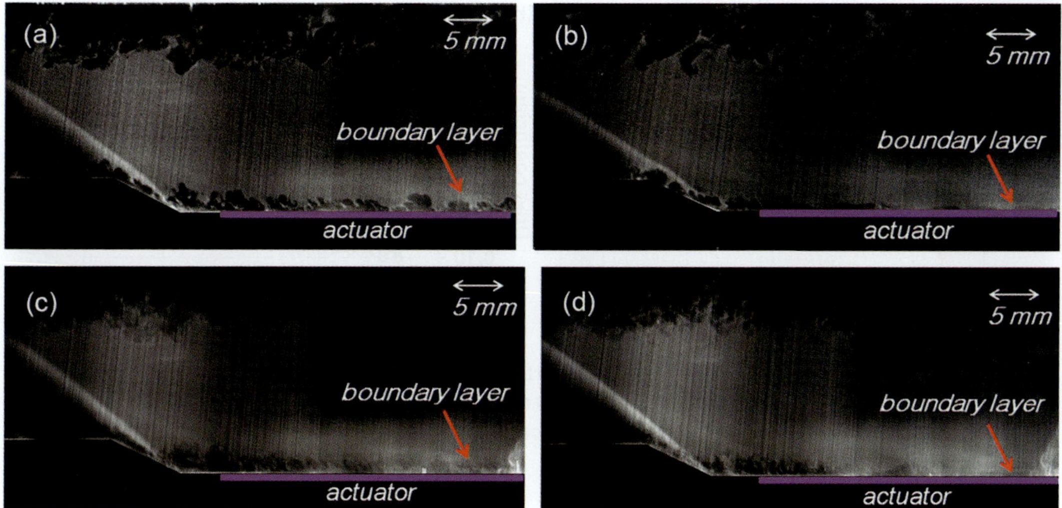

Figure 9.12 Instantaneous (a, b) and ten-frame averaged (c, d) CO_2 Rayleigh scattering images for the natural (a, c) and control (b, d) cases. The flow is from right to left (Im et al. 2010). Reproduced with permission from AIP Publishing.

are introduced here. They used a NACA 0015 airfoil with chord length of $c = 200$ mm and spanwise length of 300 mm. The 20 mm-wide exposed electrode of the plasma actuator was placed at the leading edge of the airfoil. The 15 mm-wide grounded electrode was mounted below the suction side with a gap of 5 mm to the exposed electrode. The actuator was powered at a high-voltage amplitude of 11 to 20 kV and frequency of 0.5 to 1.5 kHz, corresponding to the induced time-averaged velocity of about 2 to 4 m/s. The experiment was conducted at $U_\infty = 20$ m/s, corresponding to $Re = 2.6 \times 10^5$.

One example of the control effects are shown in Figure 9.13. The excitation frequency was fixed at 1 kHz while the voltage amplitude was ranged from 12 to 20 kV. It is obvious that the plasma control could increase the lift coefficient while decreasing the drag coefficient after stall. The control effects increase with the applied voltage amplitude. It is suggested that the plasma control achieve such goals by delaying flow separation point or reducing flow separation region. Thus, the aerodynamic forces in the natural and control cases are very similar before stall angle. Benard et al. (2009) further indicated that an unsteady actuation could lead to an increase in the aerodynamic performances of the airfoil compared to a steady actuation.

Similar effects by placing the plasma actuator near the airfoil leading edge have also been indicated by Sosa et al. (2007). The plasma actuators have also been placed at other positions on the airfoil suction surface. Mabe et al. (2009) placed the plasma actuator at 5% and 75%c from the airfoil leading edge, and they found that the former control case could result in a better effect.

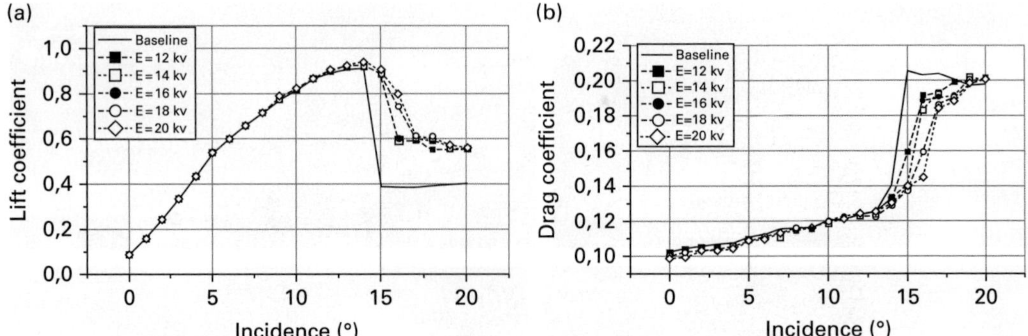

Figure 9.13 Aerodynamic forces on the airfoil without and with plasma actuator control with frequency of 1 kHz but different voltages at $U_\infty = 20$ m/s. (a) Lift coefficient; and (b) drag coefficient (Benard et al. 2009). Reproduced with permission, copyright © 2019 Elsevier Ltd. All rights reserved.

9.3.3   Straight Wing

Researchers have also tested the plasma control effects on straight wings. Similar to the effects for airfoils, plasma control could also improve the aerodynamic performance of the straight wing. Also, there are some new applications, such as the control of the tip vortex and momentum control.

The tip vortex forms from the wing tip due to the difference between the pressures on upper and lower wing surfaces. The tip vortex may influence both the drag and lift coefficients of the wing, and thus studies has been conducted to manipulate the tip vortex. Boesch et al. (2010) numerically studied the control of the tip vortex by using plasma actuators, as shown in Figure 9.14. A straight wing with a semicircular rounded tip was made based on a NACA 4418 airfoil, with chord length of $c = 15.24$ cm and a half span of 31.88 cm. Two plasma actuators were placed on the suction and pressure surfaces of the straight wing near the tip, with the induced flow towards and outwards the wing tip, respectively. The study was conducted at zero angle of attack and $U_\infty = 15$ m/s, corresponding to $Re = 1.5 \times 10^5$.

The effect of the plasma actuator with 400 mN/m actuation is shown in Figure 9.15, in comparison with the natural case. The induced flow over both the suction and pressure surfaces can be clearly seen. Thus, the vortex core is more diffused with the plasma actuator, with a larger-diameter and lower-vorticity distribution, though the total circulation of the tip vortex is not decreased. The lift coefficient is increased to 0.2740 by about 11.9% from the original 0.2448 for the natural case.

Boesch et al. (2010) also studied the influence of the actuation strength and the actuation manner, such as only suction side actuation and only pressure side actuation, on the tip vortex development and the related lift and drag coefficients. It is possible to alter the lift coefficient on either side of the straight wing by manipulating the tip vortex, which provides a potential application for roll control.

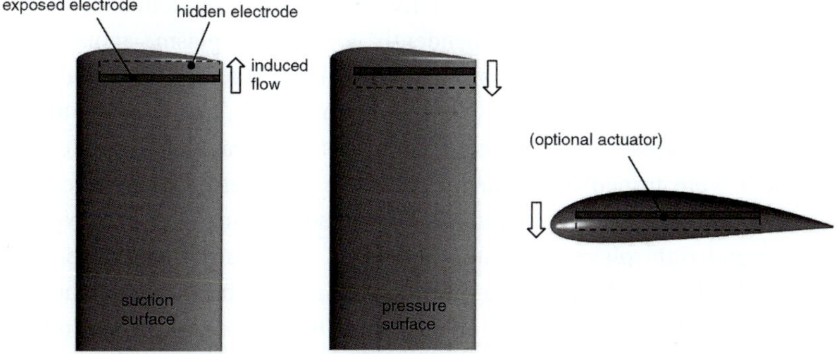

Figure 9.14 Schematic of the straight wing and the plasma actuator arrangement (Boesch et al. 2010). Reproduced with permission.

(a) (b)

(c) (d)

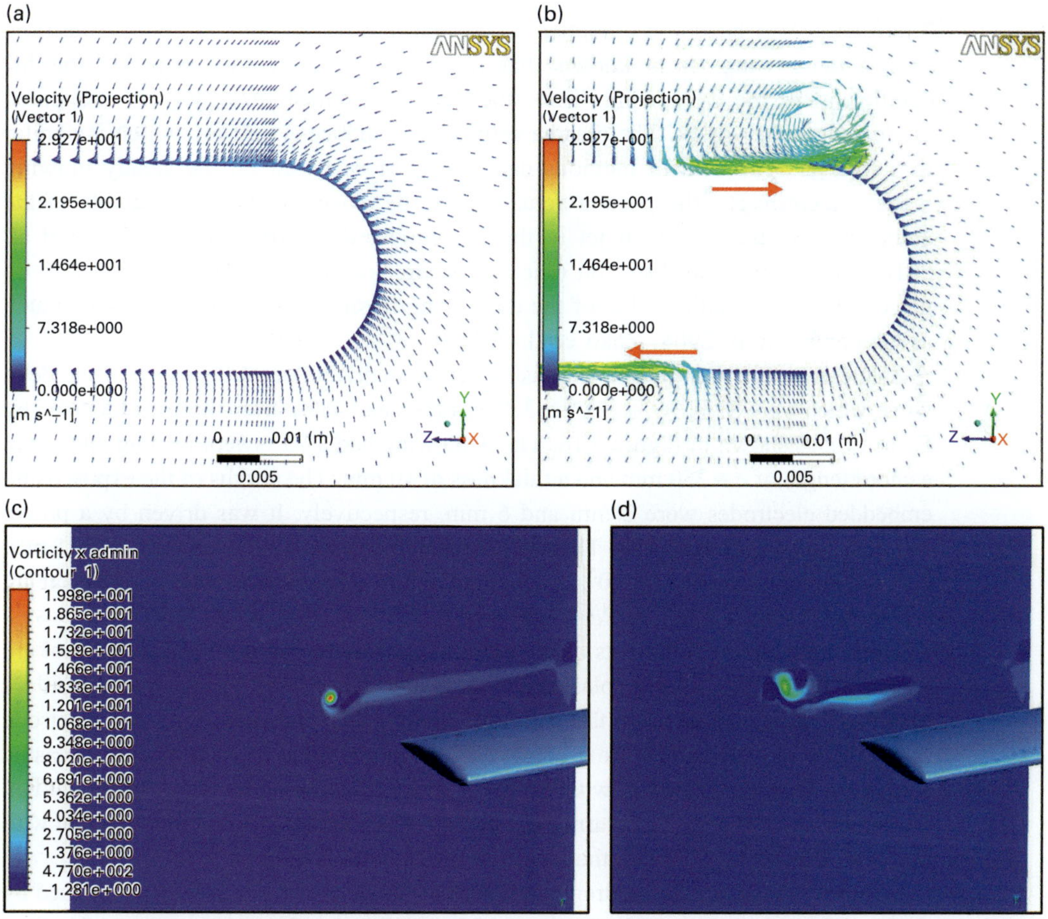

Figure 9.15 Transverse velocity vectors at midchord (a, b) and streamwise vorticity contours one chord downstream of trailing edge (c, d) for the natural and control cases (Boesch et al. 2010). Reproduced with permission.

Vorobiev et al. (2008) experimentally validated the roll control ability of the plasma actuator. Two separate and independently controllable plasma actuators were installed on the left and right sides of a straight wing. Each actuator was operated at a constant mean dissipated power of 65 W. They tested four cases, including the actuators off, both actuators operated together, and each actuator operated independently. When both actuators were turned on, an increase in the lift coefficient by about 0.09 could be found, while there was no additional roll momentum increment as expected. However, the left and right plasma actuators produced a similar lift coefficient when operated independently, which was about half of that induced by the case where both operated together. In addition, they could induce equal but opposite-directed roll moment. Thus, plasma actuators show significant ability for momentum control, such as roll, pitch and yaw control. These findings showed a potential application for the proposed concept to replace conventional movable control surfaces on future aircraft.

9.3.4   Delta Wing

There have also been some studies to use the plasma actuator to control the flow over a delta wing. It is known that the leading-edge vortex provides the lift coefficient for the delta wing. Thus, the essential control strategy is to strengthen the leading-edge vortex. The plasma actuator can induce a wall jet, which is similar to a free jet. The control methods can follow those based on the steady blowing model. Accordingly, the plasma actuators can be placed at the leeward side of the delta wing to induce a wall jet in the streamwise direction (such as Zhang et al. 2010a) or vortex axial direction (such as Sidorenko et al. 2013). They can also be placed along the leading edge of the delta wing to simulate spanwise blowing (such as Greenblatt et al. 2008; Zhao et al. 2015b).

Zhang et al. (2010a) placed six plasma actuators on the leeward surface of the delta wing, which were 8%c, 16%c, 24%c, 32%c, 40%c, and 48%c from the apex (denoted by E1 to E6), as shown in Figure 9.16. The delta wing had a 75° swept leading edge with a chord length of c = 250 mm and a thickness of 10 mm. The widths of the exposed and embedded electrodes were 2 mm and 6 mm, respectively. It was driven by a power source with 5.4 kV and 20 kHz. The force measurement was conducted at U_∞ = 16.5 m/s, corresponding to Re = 2.82 × 10^5, and the smoke flow visualization was conducted at U_∞ = 5.35 m/s, corresponding to Re = 9.08 × 10^4.

The lift and drag coefficients of the delta wing without and with plasma control are shown in Figure 9.17. As the plasma actuators placed at 8%c, 16%c, 24%c, 40%c, are turned on, they can increase the lift coefficient, which increases with the electrode downstream movement (Figure 9.17(a)). The optimum value appears when the plasma actuator was placed at 40%c, with the maximum lift increased by about 10.6%. When the actuator placed at 48%c is turned on, the lift enhancement is not as much as that with that placed at 40%c, although the length of the actuator is greater.

It is also indicated that the drag coefficient of the delta wing is greatly influenced by the position of the actuator (Figure 9.17(b)). With the 8%c actuator on, the drag coefficients of the delta wing becomes much greater than the natural case. With the

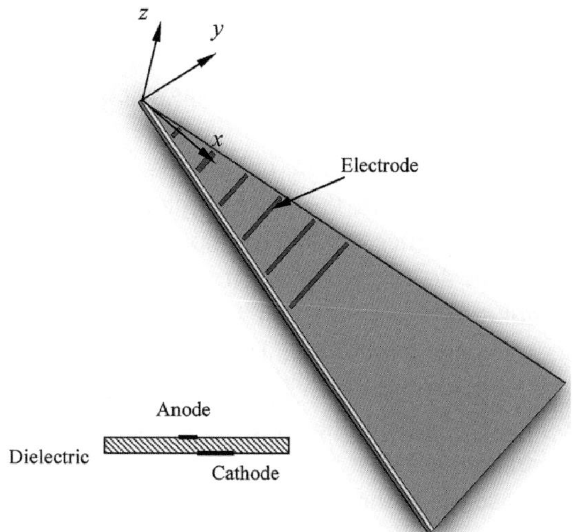

Figure 9.16 Schematic of the delta wing with the plasma actuator (Zhang et al. 2010a). Reproduced with permission, copyright © The American Institute of Aeronautics and Astronautics.

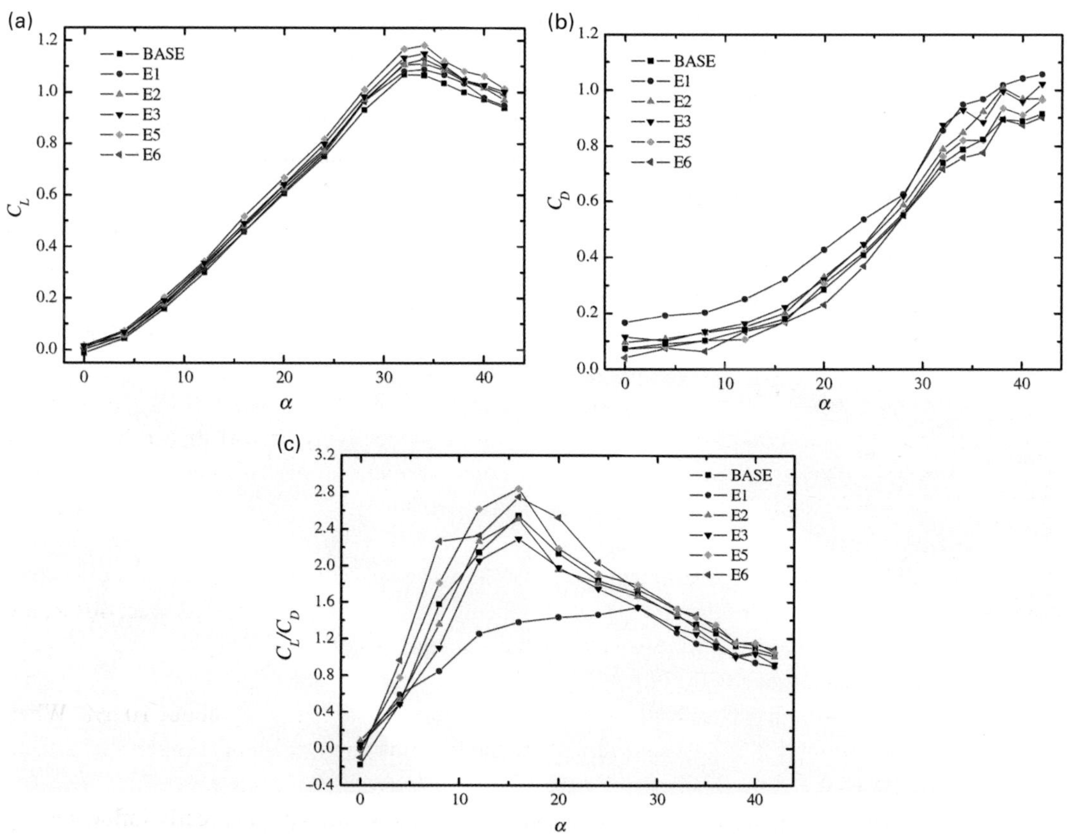

Figure 9.17 Aerodynamic forces on the delta wing with plasma actuator control. (a) Lift coefficient; (b) drag coefficient; and (c) lift-to-drag ratio (Zhang et al. 2010a). Reproduced with permission, copyright © The American Institute of Aeronautics and Astronautics.

16%c or 24%c actuator on, the drag coefficient is still greater than the natural case, but smaller than the control case with 8%c actuator. For the 40%c and 48%c actuators, the drag coefficients are very close to the natural case at most angles of attack and even less at small angles of attack.

The lift-to-drag ratio is therefore influenced by plasma control, as shown in Figure 9.17(c). It may be decreased when the plasma actuators are placed before 24%c, and be increased when placed after 40%c. The maximum lift-to-drag ratio can be increased by about 11.7% by the 40%c plasma actuator.

Figure 9.18 shows the smoke flow visualization for the natural and one typical control case in the poststall region at $\alpha = 38°$. From the flow visualization, the leading-edge vortex can be clearly observed. The breakdown point can also be judged as the chordwise location where the dark core begins to expand. It is indicated that the vortex breakdown point is delayed from 0.68c for the natural case to 0.76c for the control case, resulting in lift and lift-to-drag ratio performance enhancement of the delta wing.

Zhao et al. (2015b) used nanosecond plasma actuators to control the flow around a delta wing by placing them along the leading edge, as shown in Figure 9.19. The delta wing had a leading-edge sweep angle of 47°, chord length of $c = 240$ mm, and thickness of 8 mm. The force measurement was conducted at $U_\infty = 50$ m/s, corresponding to $Re = 7.2 \times 10^5$, and the PIV measurement was conducted at $U_\infty = 20$ m/s, corresponding to $Re = 2.9 \times 10^5$. The power supply for the plasma actuator was a nanosecond pulse generator capable of producing pulses of up to 80 kV of magnitude with rising time of 10–30 ns.

The lift and drag coefficients of the delta wing without and with plasma control at $U_\infty = 50$ m/s are shown in Figure 9.20. During the experiment, the applied peak-to-peak voltage was fixed at 12 kV, while the excitation frequency f_e was varied from 200 Hz, 400 Hz, 600 Hz, 1000 Hz, corresponding to the reduced excitation frequency $f_e c/U_\infty = 1, 2, 3$, and 5, respectively. The plasma control does not change the lift and drag coefficients before

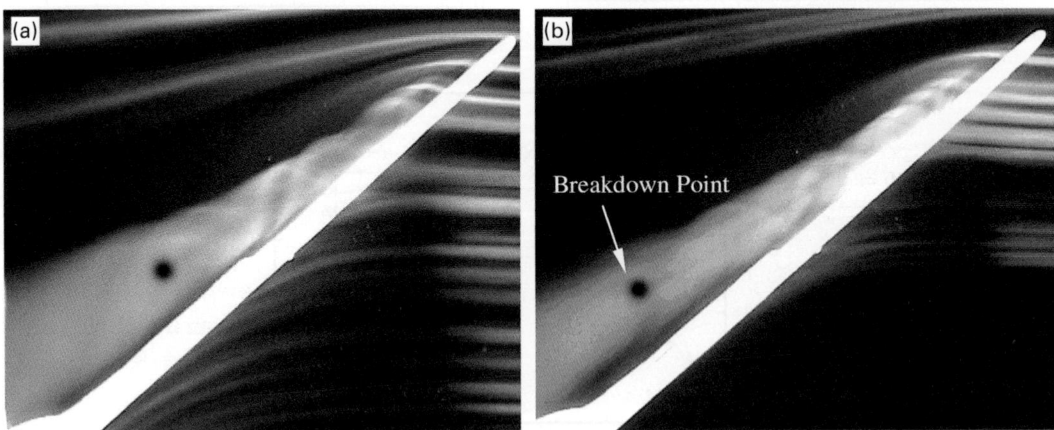

Figure 9.18 The leading-edge vortex breakdown point for the natural case (a) and the control case (b) with 40%c plasma actuator in the poststall region at $\alpha = 38°$ (Zhang et al. 2010a). Reproduced with permission, copyright © The American Institute of Aeronautics and Astronautics.

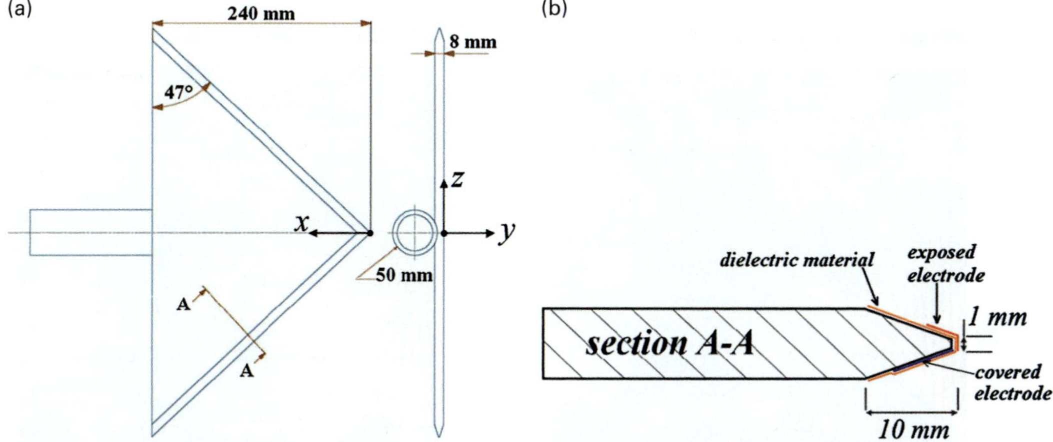

Figure 9.19 Schematic of the delta wing (a) and the arrangement of the nanosecond plasma actuator (b) (Zhao et al. 2015b). Reproduced with permission, copyright © Springer 2015.

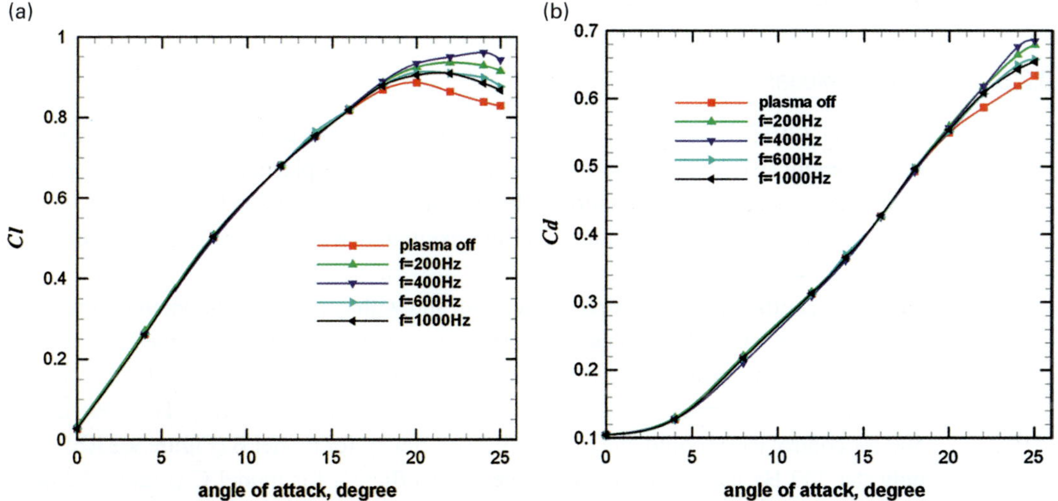

Figure 9.20 Aerodynamic forces on the delta wing without and with plasma actuator control at U_∞ = 50 m/s. (a) Lift coefficient; and (b) drag coefficient (Zhao et al. 2015b). Reproduced with permission, copyright © Springer 2015.

stall angle, but can increase the lift and drag coefficients in the poststall region and postpone the stall angle. The most effective control frequency is $f_e c/U_\infty = 2$, with the maximum lift coefficient increased by about 14.7% and the stall angle delayed by about 4°.

Velocity distributions at $x/c = 0.35$ for the natural and one typical control cases are shown in Figure 9.21. It is indicated that the plasma control enhances the strength of the leading-edge vortex. This is evidence to interpret the lift improvement by plasma forcing.

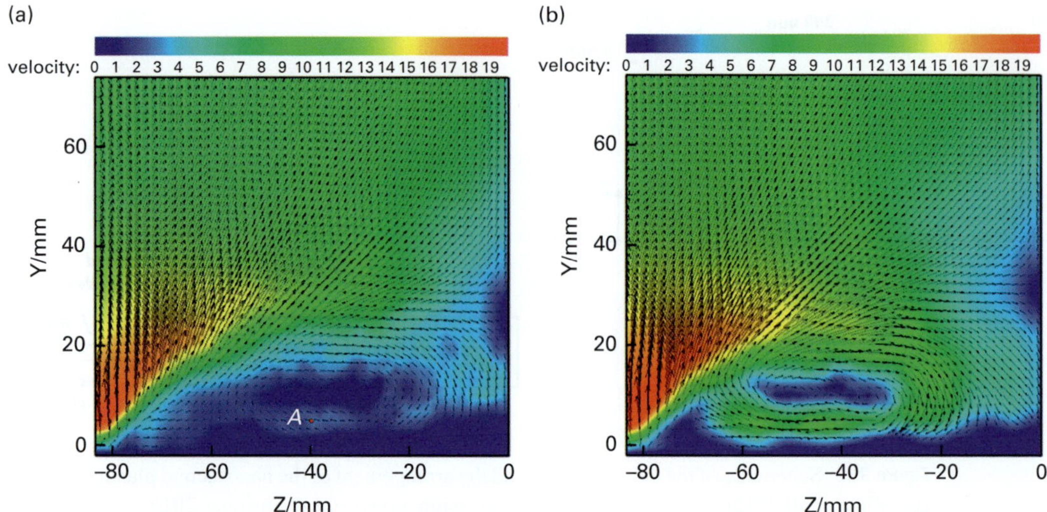

Figure 9.21 Mean velocity field of the delta wing without (a) and with (b) plasma actuator control at $V_{p\text{-}p} = 10.5$ kV, $f_e = 200$ Hz, $x/c = 0.35$, $\alpha = 22°$, $U_\infty = 20$ m/s (Zhao et al. 2015b). Reproduced with permission, copyright © Springer 2015.

9.3.5   Aircraft Model

Zhao et al. (2015a) used the nanosecond plasma actuator to control four aircraft models with different sweep angles, as shown in Figure 9.22. The plasma actuators were placed along the leading edge of the model. The free-stream velocity was fixed at $U_\infty = 30$ m/s. The applied peak-to-peak voltage and the excitation were varied to see the different effects.

The lift coefficients of the aircrafts without and with control are shown in Figure 9.23. It is indicated that all the plasma control cases for model A could increase the lift coefficient in the near and post stall angles of attack, and the lift coefficient further increases with a decrease in the excitation frequency (Figure 9.23(a)). The 200 Hz and 500 Hz control cases could increase the lift coefficient for model B (Figure 9.23(b)), and only the 200 Hz case could increase the lift coefficient for model C (Figure 9.23(c)). However, the lift coefficients are almost the same with plasma actuator control for model D (Figure 9.23(d)). Thus, it is indicated that the effects of the plasma actuator are highly dependent on the configuration of the aircraft.

9.3.6   Bluff Body

Vortex shedding over a bluff body usually causes serious structural vibration, acoustic noise and resonance, enhanced mixing, and significant increase in the mean drag and lift fluctuations. Therefore, the effective control of vortex shedding is important in engineering applications, which can successfully lead to separation delay, drag reduction, and suppression of noise and vibration. There have been many researchers who have

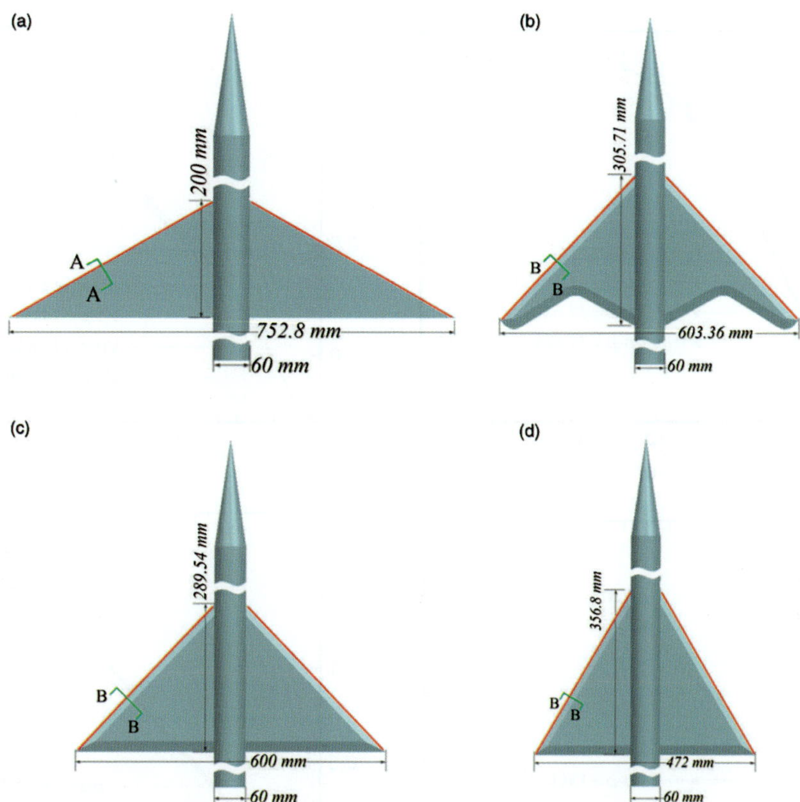

Figure 9.22 Schematics of four aircraft models with different sweep angles. (a) Model A with sweep angle of 30°; (b) model B with sweep angle of 47°; (c) model C with sweep angle of 47°; and (d) model D with sweep angle of 60° (Zhao et al. 2015a). Reproduced by permission of SAGE Publications Ltd.

used plasma actuators to control the flow around bluff bodies. The objective of bluff body control may be different according to practical requirements. However, a commonly essential factor is the vortex dynamics, which could well account for the variations in the aerodynamics and aeroacoustics characteristics. Thus, it is an important issue that if the external perturbations can change the vortex shedding mode around a bluff body.

McLaughlin and his co-workers conducted a series of experiments to study the control of flow around a circular cylinder with DBD plasma actuators. McLaughlin et al. (2004) placed a pair of plasma actuators at azimuth ±90°from the front stagnation point of the cylinder at $Re = 7.4 \times 10^3$. They found that the plasma control could delay flow separation and the wake vortex was also locked-on by the excitation frequency. Munska and McLaughlin (2005) further presented a lock-on regime at a much larger range of Reynolds numbers from $Re = 2.9 \times 10^4$ to 8.8×10^4. McLaughlin et al. (2006) indicated that asymmetric forcing could significantly mitigate the effects of the Kármán vortex at lower Reynolds numbers but at higher energy cost. Two different control strategies were investigated by Rizzetta and Visbal

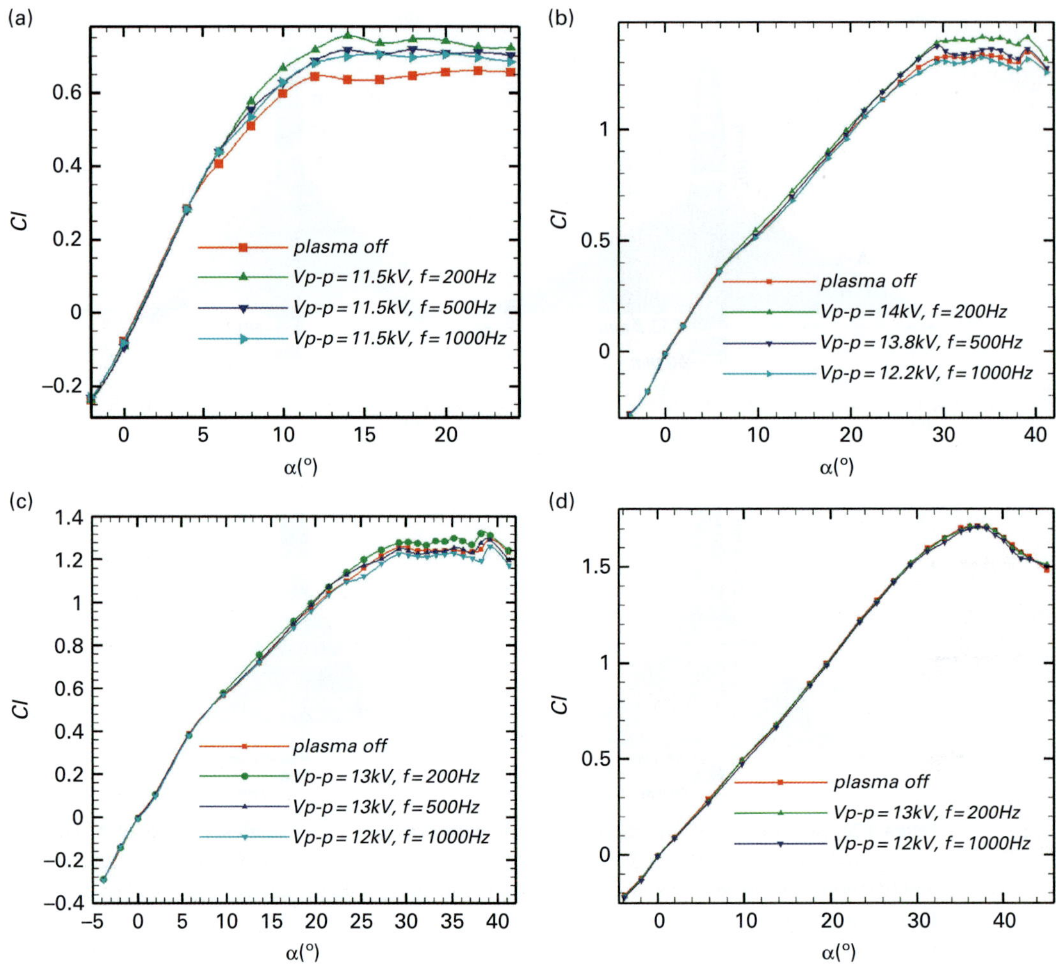

Figure 9.23 Lift coefficients of four aircraft models with different sweep angles. (a) Model A with sweep angle of 30°; (b) model B with sweep angle of 47°; (c) model C with sweep angle of 47°; and (d) model D with sweep angle of 60° (Zhao et al. 2015a). Reproduced by permission of SAGE Publications Ltd.

(2009) at $Re = 1 \times 10^4$, consisting of larger actuators that produced a wall-jet-like flow and smaller actuators that perturbed the unstable shear layers near the separation location. Plasma control was found to result in at least a 50% decrease in the drag for both cases. Sung et al. (2006), on the other hand, indicated that an opposite electrode configuration with the exposed electrode placed downstream of the embedded electrode led to a greater flow separation, which could be advantageous in enhancing mixing or flow vectoring.

Thomas et al. (2008) placed the plasma actuators at ±90° and ±135° with respect to the cylinder front stagnation point, as shown in Figure 9.24. Using either steady or unsteady actuation, plasma control was able to substantially reduce the flow separation region. The Kármán vortex shedding appeared to be eliminated and the turbulence levels in the

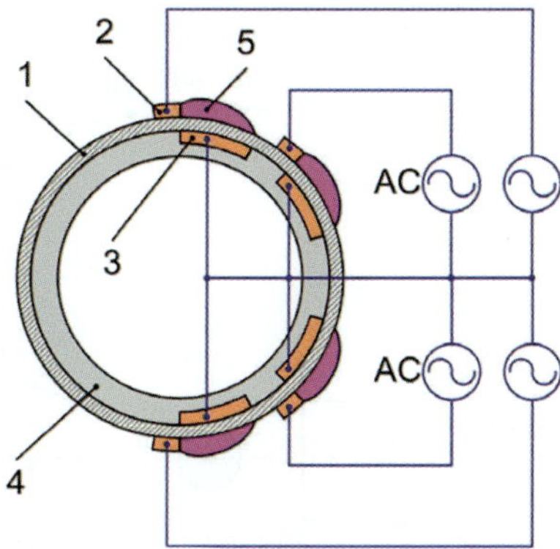

Figure 9.24 Schematic of a circular cylinder with plasma control (Thomas et al. 2008). Reproduced with permission, copyright © The American Institute of Aeronautics and Astronautics.

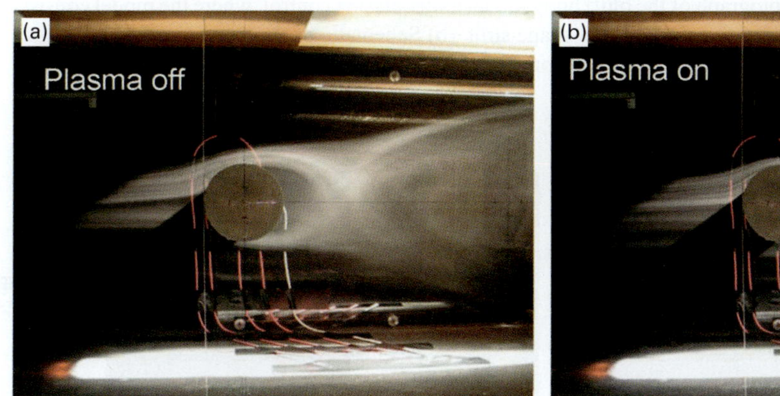

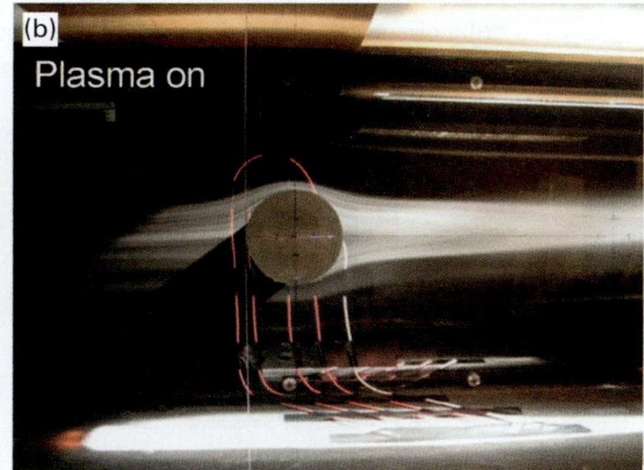

Figure 9.25 Smoke visualization for the flow around a circular cylinder without (a) and with (b) plasma control at $Re = 3.3 \times 10^4$, $C_\mu = 1.1\%$ (Thomas et al. 2008). Reproduced with permission, copyright © The American Institute of Aeronautics and Astronautics.

wake were decreased significantly, as shown in Figure 9.25. As a result, the near-field sound pressure levels were found to be reduced by 13.3 dB in comparison with the natural case.

Huang and his co-workers were also able to reduce the flow-induced broadband noise radiating from a bluff body by DBD plasma actuators (Vinogradov and Huang 2011; Li et al. 2010; Huang et al. 2010). For example, Li et al. (2010) placed the plasma actuators on the ±90° from the front stagnation point of a circular cylinder, as shown in Figure 9.26(a). An oblique I-shape strut was placed downstream of the cylinder. The cylinder model was $D = 100$ mm in diameter and 500 mm in length. The width of the I-shape oblique strut was 70 mm. The free-stream velocities were varied from 20 to 40 m/s, corresponding to $Re = 1.7 \times 10^5$ to 3.4×10^5 based on the cylinder diameter. The noise level was measured using a phased microphone array and far-field microphones in the anechoic chamber, as shown in Figure 9.26(b).

(a)

(b)

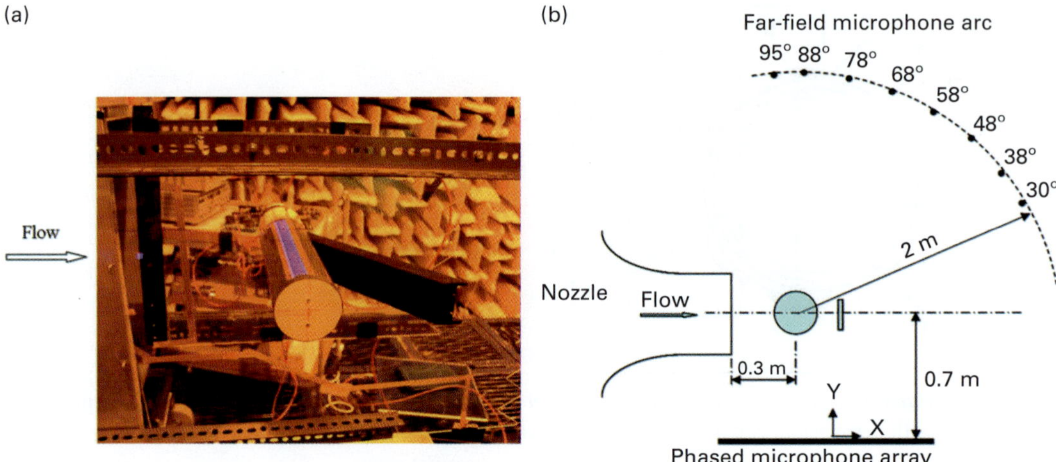

Figure 9.26 (a) Photograph of the bluff body model under plasma control, where the model consists of a circular cylinder and an oblique I-shape strut. (b) Schematic of the anechoic chamber noise measurements setup (Li et al. 2010). Reproduced with permission, copyright © Springer-Verlag 2010.

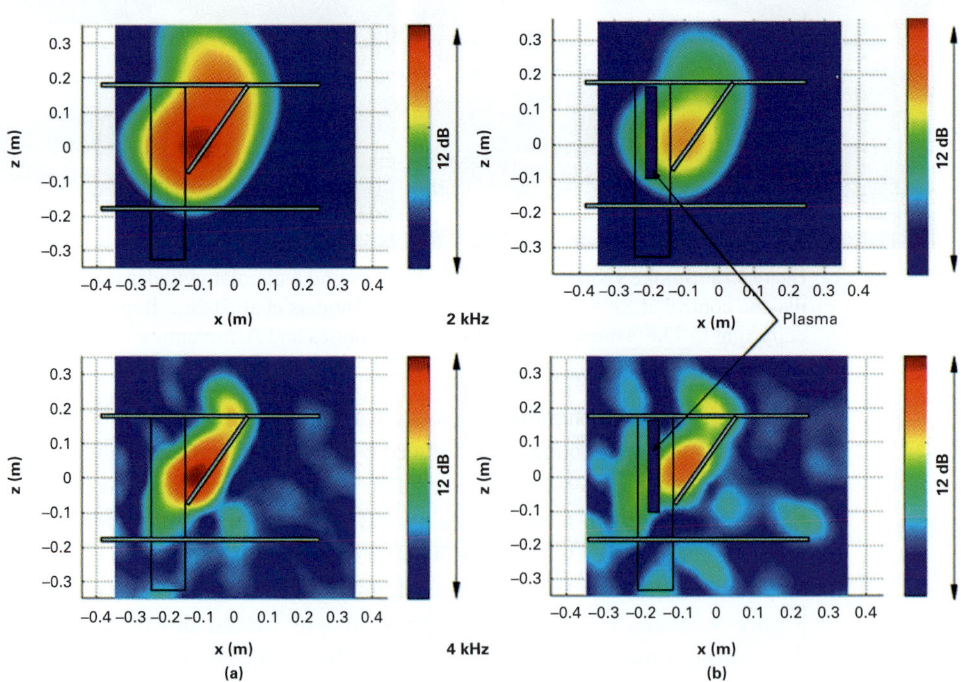

Figure 9.27 Noise source maps at 2 and 4 kHz from the phased microphone array test for the natural (a) and control (b) cases at $U_\infty = 30$ m/s (Li et al. 2010). Reproduced with permission, copyright © Springer-Verlag 2010.

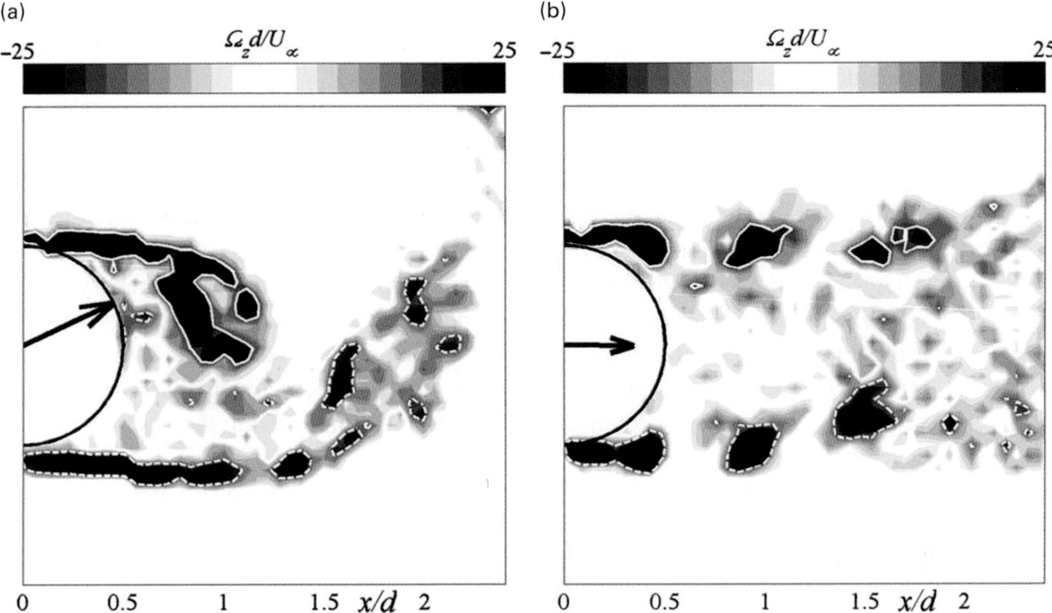

Figure 9.28 Instantaneous spanwise vorticity for the flow around a circular cylinder without (a) and with (b) plasma control at $Re = 1.5 \times 10^4$, $f_e^+ = 1.0$, $C_e = 0.17\%$ (Jukes and Choi 2009b). Reproduced with permission from AIP Publishing.

Figure 9.27 shows the noise source maps at two frequencies of 2 and 4 kHz for $U_\infty = 30$ m/s. It is indicated that the major noise sources are located in the gap area between the cylinder and the strut for both cases. The noise level was reduced by a considerable amount when the plasma was activated. In particular, reductions of 3.8 dB and 2.2 dB were obtained in the frequency ranging between 80 and 4 kHz at $U_\infty = 20$ and 30 m/s, respectively.

Jukes and Choi (2009a, 2009b, 2009c) conducted a series of experimental studies on the behavior of unsteady flow separation over a circular cylinder with DBD plasma actuators positioned near the separation point at $Re = 1.5 \times 10^4$. For a typical control case at forcing frequency of $f_e^+ = f_e D/U_\infty = 1.0$ and forcing coefficient of $C_e = F_e/0.5U_\infty^2 D = 0.17\%$, a chain of small-scale spanwise vortices appeared in the shear layers as they separated from the upper and lower surfaces of the cylinder, as shown in Figure 9.28(b). These vortices were created each time the plasma was activated and they traveled parallel to the oncoming flow, indicating that the vortex lock-on phenomenon occurred and the vortex shedding mode was changed. Such a pattern is completely different from that for the natural case, as shown in Figure 9.28(a).

The cylinder drag was reduced by up to 32% and the lift fluctuations were reduced by up to 72% at the maximum forcing parameters of $f_e^+ = 2.0$ and $C_e = 0.32\%$. Thus, plasma control exhibits a potential application to reduce the vortex-induced-vibration by varying the vortex shedding mode. The power saving ratio, defined as a ratio of the power saved by drag reduction to the fluidic power introduced by the plasma, could be as much as 1170, as indicated by Jukes and Choi (2009c).

The most frequently used plasma actuator mentioned above is the DBD one. However, some novel conceptions have also been proposed based on the main characteristics of the DBD plasma actuator.

Sosa et al. (2011) placed four DBD plasma actuators on the cylinder surface with plasma forcing in the anticlockwise direction to simulate the inverse Magnus effect of a rotating cylinder, as shown in Figure 9.29(a). Such a configuration demonstrated the ability to induce significant transverse forces and the optimal frequencies were found to be close to the natural frequency.

Gregory and his co-workers (Gregory et al. 2008; Bhattacharya and Gregory 2015a, 2015b) conducted a series of studies to control the flow around a circular cylinder by using a three-dimensional plasma actuator, which could produce spatially distributed forcing. One example is shown in Figure 9.29(b). It was found that the spanwise variation in the plasma forcing had a substantial impact on the spanwise wake development.

Kozlov and Thomas (2011a, 2011b) used a new conception of plasma vortex generator, which introduced streamwise vorticity similar to a mechanical vortex generator, to control the flow around a circular cylinder (Figure 9.29(c)). The flow separation was delayed because of the enhanced momentum transfer by the counter-rotating vortices. Consequently, the Kármán vortex shedding was strongly suppressed. The plasma vortex generator was also found to be effective in controlling the flow around cylinders in a tandem configuration.

9.4   Novel Plasma Actuators

It has been indicated that the two main features of the DBD plasma actuator are that it can induce a wall jet and a starting vortex, which are similar to that induced by other devices or methods. Thus, some novel control conceptions based on the plasma actuator have been proposed to simulate some of the traditional control functions. Here, we only present the main principle and the effect of some typical novel control conceptions, such as plasma synthetic jet, plasma Gurney flap, plasma circular control, plasma vortex generator. Readers can refer to a recent review paper by Wang et al. (2013) for more details.

9.4.1   Plasma Synthetic Jet

The plasma synthetic jet can be achieved by a pair of plasma actuators, with two wall jets moving oppositely, as shown in Figure 9.30(a). When they meet near the centerline, the horizontal wall jets roll up to form a vertical jet outwards from the wall. Note that when the plasma actuator is turned on, it will first produce a starting vortex. Thus, when the plasma actuator is actuated periodically, vortices will also be induced periodically, which is similar to the conventional synthetic jet. This manner of operation can be achieved with a pulsed actuation signal, as shown in Figure 9.30(b).

Figure 9.31 shows the instantaneous spanwise vorticity and streamline induced by the plasma actuator at two phases. We can detect a pair of vortical structures at around $x = 8$ mm

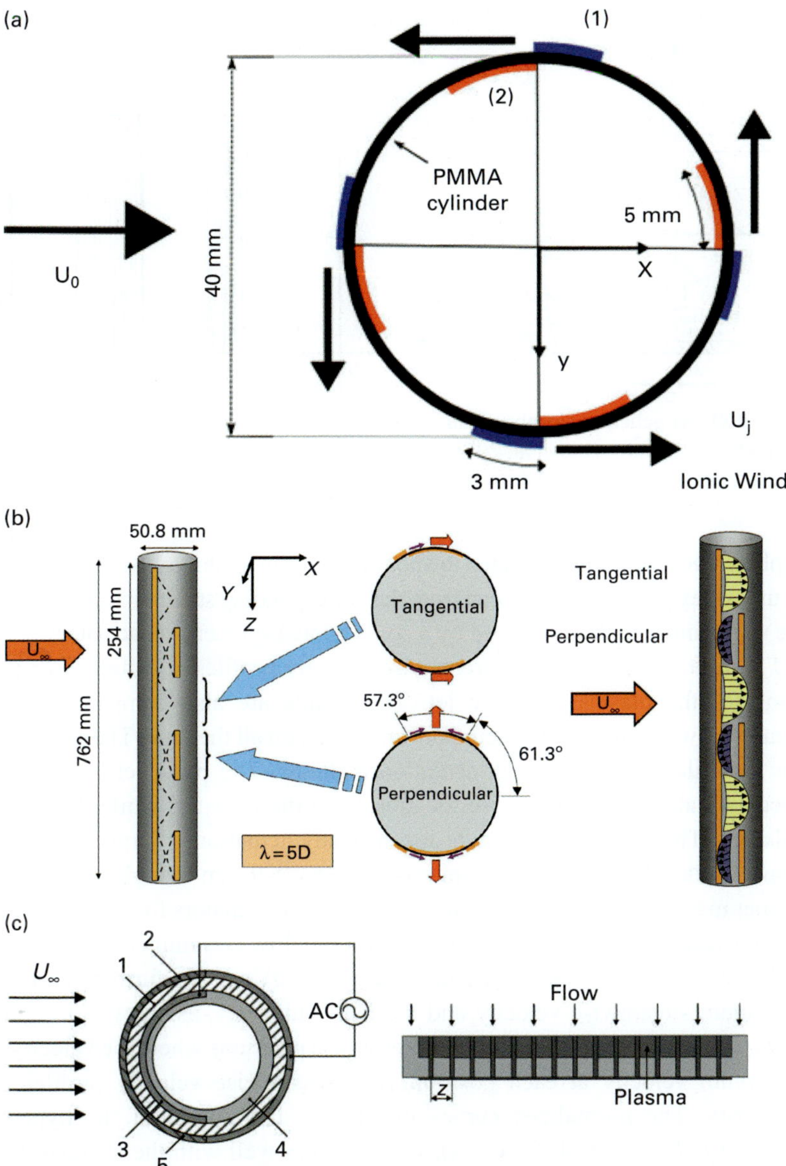

Figure 9.29 Some new conceptions for the control of flow around a circular cylinder by plasma actuators. (a) Four plasma actuators mounted on the cylinder surface forcing flow in anticlockwise direction (Sosa et al. 2011), reproduced with permission, copyright © Springer-Verlag 2011. (b) Three-dimensional plasma actuator to produce the spatially-distributed forcing (Gregory et al. 2008), reproduced with permission; and (c) Schematic of plasma vortex generator for circular cylinder control (Kozlov and Thomas 2011a). Reproduced with permission, copyright © The American Institute of Aeronautics and Astronautics.

(a)

(b)

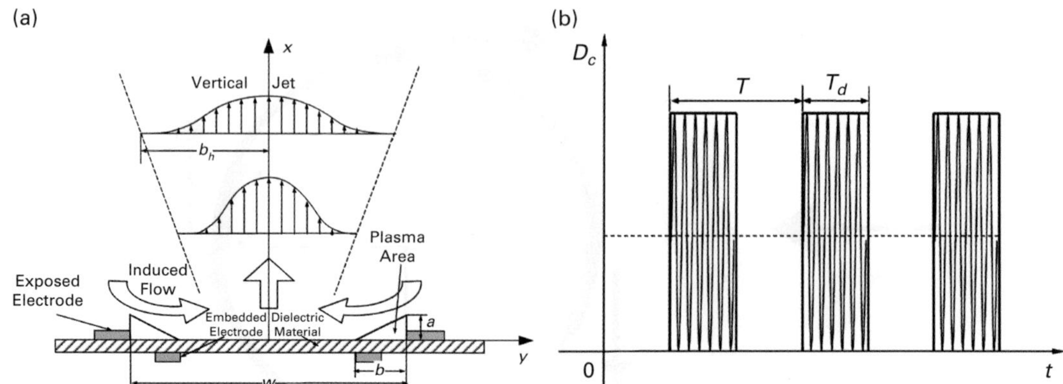

Figure 9.30 (a) Schematic of the plasma synthetic jet actuator; and (b) excitation signal for synthetic jet actuator (Liu et al. 2011). Reproduced with permission, copyright © The American Institute of Aeronautics and Astronautics.

at phase 0.8π, which is convected to about $x = 25$ mm at phase 1.8π. The entrainment effect of the vortex pair can also be seen from the corresponding streamlines. These characteristics are very similar to those induced by a conventional synthetic jet actuator.

Figure 9.32(a) shows the streamwise velocity profiles at different streamwise positions for the plasma synthetic jet. The results are also compared with the steady actuation, where the plasma actuators are turned on all the time. The streamwise velocity has a similar profile for both actuation methods. The peak velocity at the centerline decreases along the streamwise location from the actuator, while the influence region enlarges. The velocity magnitude for unsteady actuation is larger than that for steady actuation in a further downstream position of $x \geq 9$ mm though the magnitude for the former may be smaller than the latter between two actuators for $x \leq 6$ mm. It is indicated that unsteady actuation is more able to achieve flow entrainment of larger regions.

The velocity profiles shown in Figure 9.32(a) can be further normalized by the maximum streamwise velocity and the half-width $b_{0.5}$, as shown in Figure 9.32(b,c). Here, $b_{0.5}$ is defined as the distance between the positions where the velocity is half of the maximum velocity at each position. The streamwise velocity profiles exhibit self-similarity. The normalized curves match the distribution of the hyperbolic cosine function $U/U_{max} = \cosh^{-2}(Ay/b_{0.5})$, which agrees well with the characteristics of traditional synthetic jets.

Similar to the conventional synthetic jet actuator, the plasma synthetic jet can produce a vortex ring or vortex pair periodically. An annular one can be made by an annular exposed electrode and an annular embedded electrode, while a linear one can be made by two linear exposed electrodes and one or two linear embedded electrodes.

It is indicated that the plasma synthetic jet can simulate the features of the traditional synthetic jet, thus it can also be used for related flow control, such as in the work by Feng et al. (2011) who placed the plasma synthetic jet at the rear stagnation point of the circular cylinder, as shown in Figure 9.33(a). The plasma actuator was actuated at an excitation frequency the same as the natural frequency of the Karman vortex. It is found

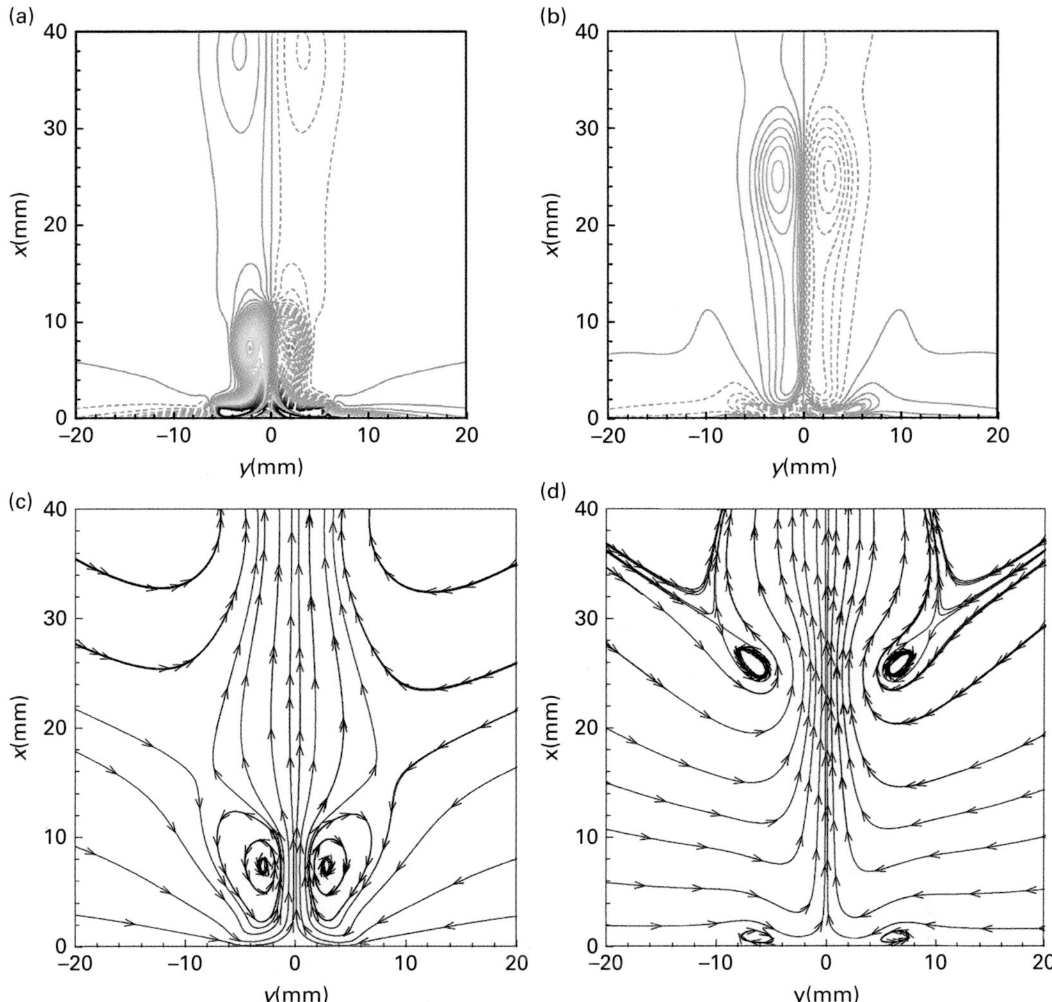

Figure 9.31 Instantaneous spanwise vorticity (a, b) and streamline (c, d) of the plasma synthetic jet at two different phases of 0.8π (a, c) and 1.8π (b, d) (Liu et al. 2011). Reproduced with permission, copyright © The American Institute of Aeronautics and Astronautics.

that as the vortex pair induced by the plasma actuator is convected downstream, it introduces high-momentum fluids into the recirculation region, reducing the velocity defect there. Therefore, the flow topology in the near wake is changed. A new recirculation region is induced by the periodic plasma forcing which is located just upstream of the original recirculation region induced by the inherent flow separation of the near wake, as shown in Figure 9.33(b). The effects are similar to those induced by the transitional synthetic jets by Feng and Wang (2010).

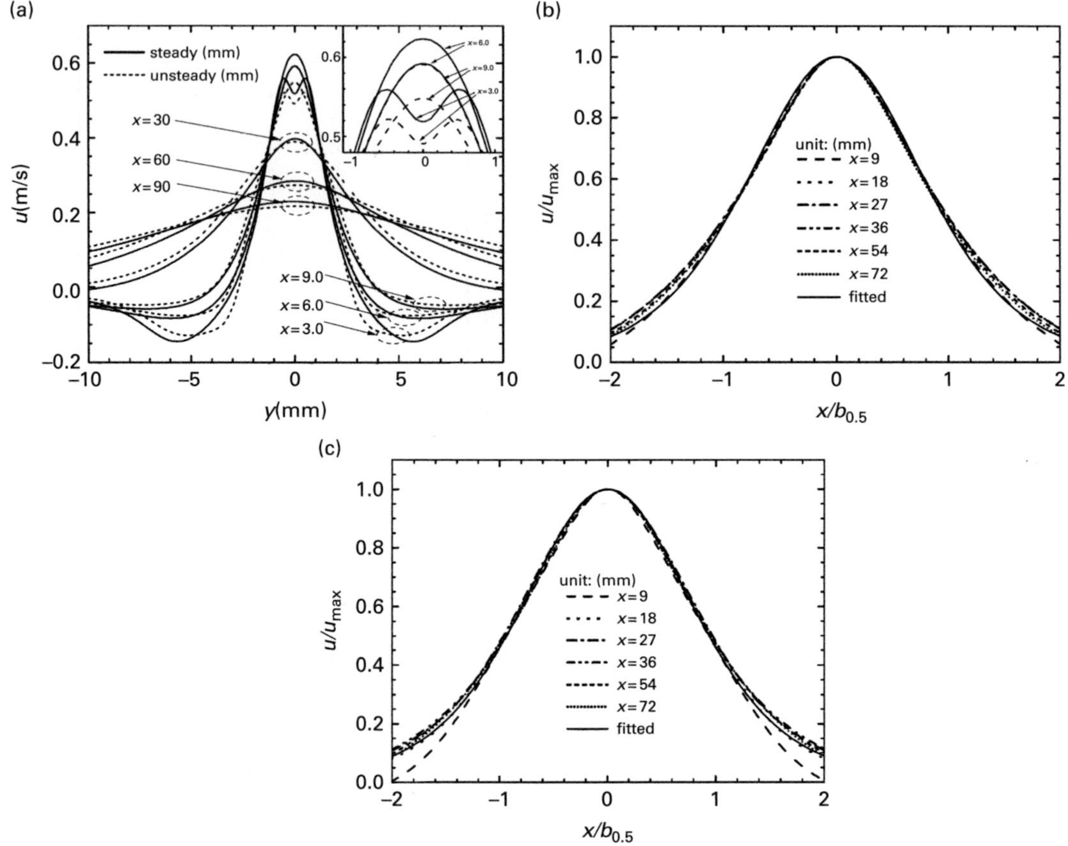

Figure 9.32 (a) Streamwise velocity profiles at different streamwise positions of the plasma synthetic jet with steady and unsteady actuation. Self-similarity of the streamwise velocity profiles for the steady (b) and unsteady (c) actuation (Liu et al. 2011). Reproduced with permission, copyright © The American Institute of Aeronautics and Astronautics.

9.4.2   Plasma Gurney Flap

Based on the features of the induced wall jet, the DBD plasma actuator can simulate the control function of the mechanical Gurney flap. The conception was first proposed by Zhang et al. (2009), who attached the plasma actuator onto the cut section of the airfoil trailing edge, as shown in Figure 9.34(a). The plasma actuator could induce a wall jet to alter the airfoil trailing-edge Kutta condition and to produce enough pressure difference between the upper and lower surfaces. The numerical simulation indicated that such an arrangement could increase the lift coefficient of the airfoil for all angles of attack by shifting the lift curve upwards, which is similar with that induced by a mechanical Gurney flap.

It is difficult to implement the conception proposed by Zhang et al. (2009), as the airfoil trailing edge is generally too small to contain a plasma actuator. Thus, in the following study, Feng et al. (2012) placed the plasma actuator onto the mechanical

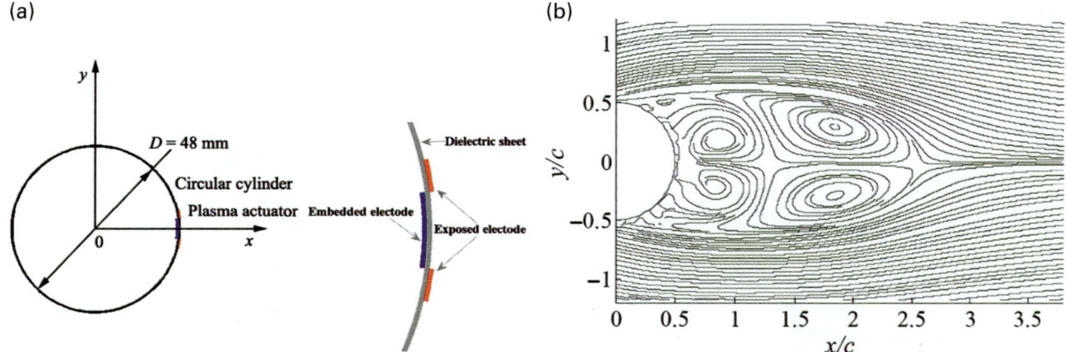

Figure 9.33 (a) Schematic of the circular cylinder with plasma synthetic jet positioned at the rear stagnation point. (b) Mean streamline of the flow around a circular cylinder controlled at $f_e/f_0 = 1$, $C_\mu = 0.052$ (Feng et al. 2011).

Gurney flap attached to an airfoil, as shown in Figure 9.34(b). The plasma induced a wall jet along the Gurney flap surface. The experimental investigation indicated that the plasma control could act to increase the equivalent height of the mechanical Gurney flap, and thus further increase the lift coefficient of the airfoil installed Gurney flap.

Feng et al. (2014) placed two DBD plasma actuators onto the pressure (lower) surface of the airfoil near the trailing edge to produce a wall-normal jet, as shown in Figure 9.34(c). First, they tested two configurations, with the distance from the downstream edge of the second electrode to the airfoil trailing edge 1 mm and zero, respectively. They found that the latter case could lead to a greater lift enhancement, suggesting that the right plasma actuator played a more important role. Accordingly, they proposed another novel conception for the plasma Gurney flap. Only one DBD plasma actuator was positioned onto the airfoil pressure surface to produce a horizontal wall jet opposite to the free stream, as shown in Figure 9.34(d). The force measurement indicated that this conception could significantly improve the lift coefficient.

Feng et al. (2015) conducted a detailed experimental study into the control effect and mechanism of the plasma Gurney flap shown in Figure 9.34(d) on an airfoil by using force and PIV measurements. A NACA 0012 airfoil with chord length of $c = 100$ mm and span of 250 mm was used for flow control. A DBD plasma actuator, which consisted of a 2.5 mm wide exposed electrode and a 6 mm wide embedded electrode, was placed on the pressure side of the airfoil near the trailing edge. The downstream edge of the exposed electrode was 1 mm upstream from the trailing edge. The plasma actuator was powered sinusoidally at an AC peak-to-peak voltage of 9.8 kV with a frequency of 18 kHz. Experiments were conducted at $U_\infty = 3.0$ m/s, 4.3 m/s and 5.3 m/s, corresponding to Reynolds numbers $Re = 2 \times 10^4$, 2.8×10^4 and 3.5×10^4, and momentum coefficients of plasma forcing $C_\mu = 1.9\%$, 0.9% and 0.6%, respectively.

Typical aerodynamic forces on the airfoil without and with plasma control are shown in Figure 9.35. It is indicated that the plasma Gurney flap increases the lift coefficient for all angles of attack, though the drag coefficient is also increased. The maximum lift coefficient could be increased by about 23%. As a result, the lift-to-drag ratio is

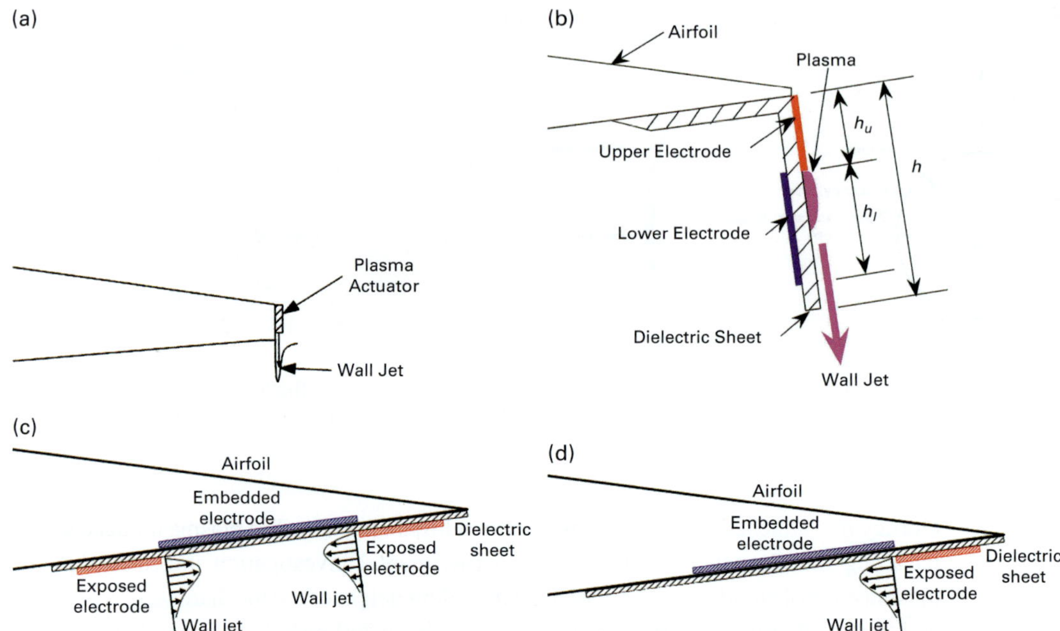

Figure 9.34 Schematic of different conceptions of the plasma Gurney flap. (a) The plasma actuator is attached to the cut section of the airfoil trailing edge (Zhang et al. 2009), reproduced with permission, copyright © The American Institute of Aeronautics and Astronautics. (b) The plasma actuator is attached to the mechanical Gurney flap (Feng et al. 2012), reproduced with permission, copyright © Springer Nature 2012. (c) A pair of plasma actuators is attached to the suction surface of the airfoil close to the trailing edge, which could induce a normal wall jet. (d) One plasma actuator is attached to the suction surface of the airfoil close to the trailing edge, which could induce a wall jet against the free stream (Feng et al. 2014).

increased by plasma control until the stall angle. Such characteristics are very similar to those induced by a 3%c mechanical Gurney flap.

The increase in the lift coefficient ΔC_L by plasma control at zero angle of attack as a function of the plasma momentum coefficient C_μ is shown in Figure 9.35(d). The corresponding values for mechanical Gurney flap from different references are also given as a function of height h/c. It is indicated that ΔC_L increases nearly linearly with the jet momentum coefficient C_μ, giving $\Delta C_L = 12C_\mu$. It is very similar to that induced by a mechanical Gurney flap, namely $\Delta C_L = 12h/c$. Thus, it can be concluded that the lift enhancement by plasma control with a 1% momentum coefficient is equivalent to a 1% chord Gurney flap.

Figure 9.36 compares the time-averaged flow field near the trailing edge of the airfoil without and with plasma control. For the no control case, a separation bubble forms over the suction surface near the trailing edge (Figure 9.36(a)). With plasma control (Figure 9.36(b)), the plasma wall jet interacts with the free stream to form a recirculation region over the pressure surface. The recirculation region stretches downstream beyond the trailing edge, effectively extending the airfoil chord length.

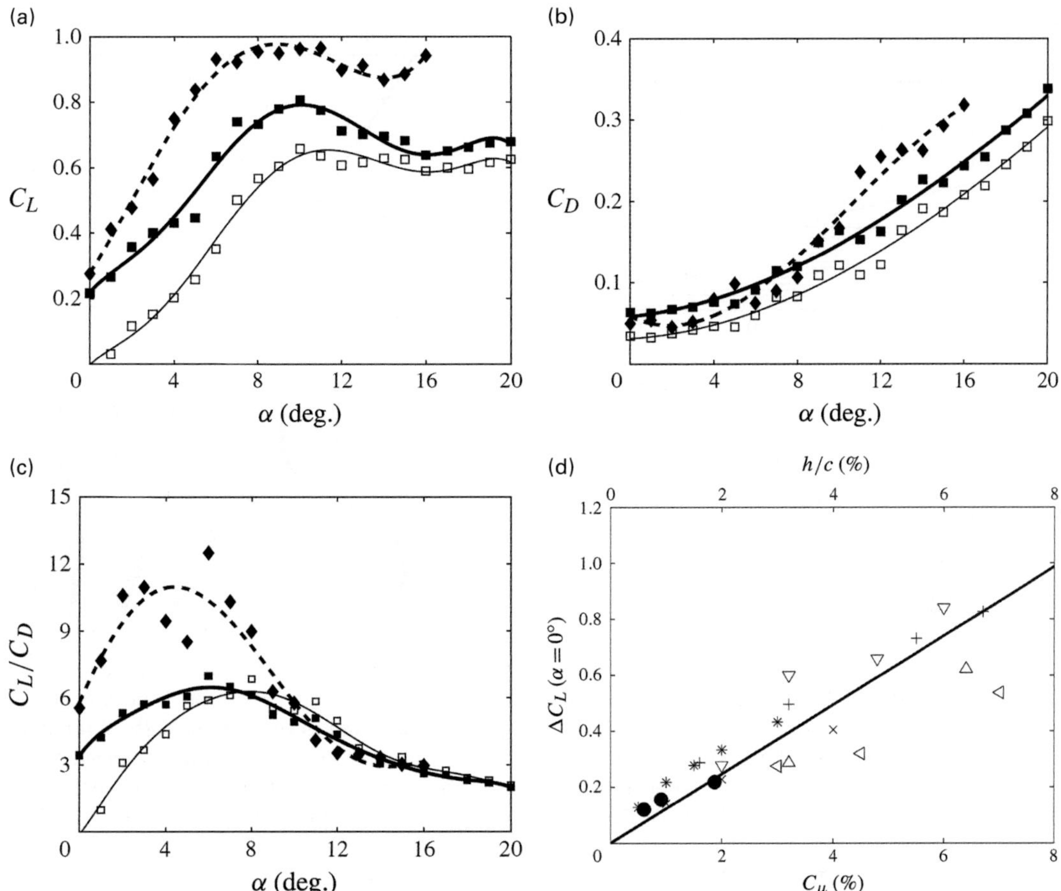

Figure 9.35 (a) Lift coefficient, (b) drag coefficient and (c) lift-to-drag ratio, versus angle of attack at $Re = 2 \times 10^4$: $\square$, plasma off; $\blacksquare$, plasma on with $C_\mu = 1.9\%$; $\blacklozenge$, 3%c mechanical Gurney flap. The thin solid, thick solid, and thick dashed lines denote the fitting curves of without plasma, with plasma, and with Gurney flap, respectively. (d) Increase in lift coefficient by plasma forcing at zero angle of attack as a function of moment coefficient C_μ, denoted by $\bullet$. The corresponding values for mechanical Gurney flap from related references are given as a function of height h/c, denoted by other symbols (Feng et al. 2015). Reproduced with permission, copyright © Cambridge University Press 2015.

This is similar to the flow field induced by the mechanical Gurney flap (Figure 9.36(c)). Figure 9.36(d) further indicates that the formation of the recirculation region by the plasma Gurney flap reduces the velocity over the pressure surface while increasing the velocity over the suction surface, similar to that induced by a mechanical Gurney flap. According to the Bernoulli equation, the pressure over the lower surface and the suction over the upper surface should be increased, resulting in an increase in the pressure difference over the two surfaces. Thus, the lift coefficient of the airfoil can be improved

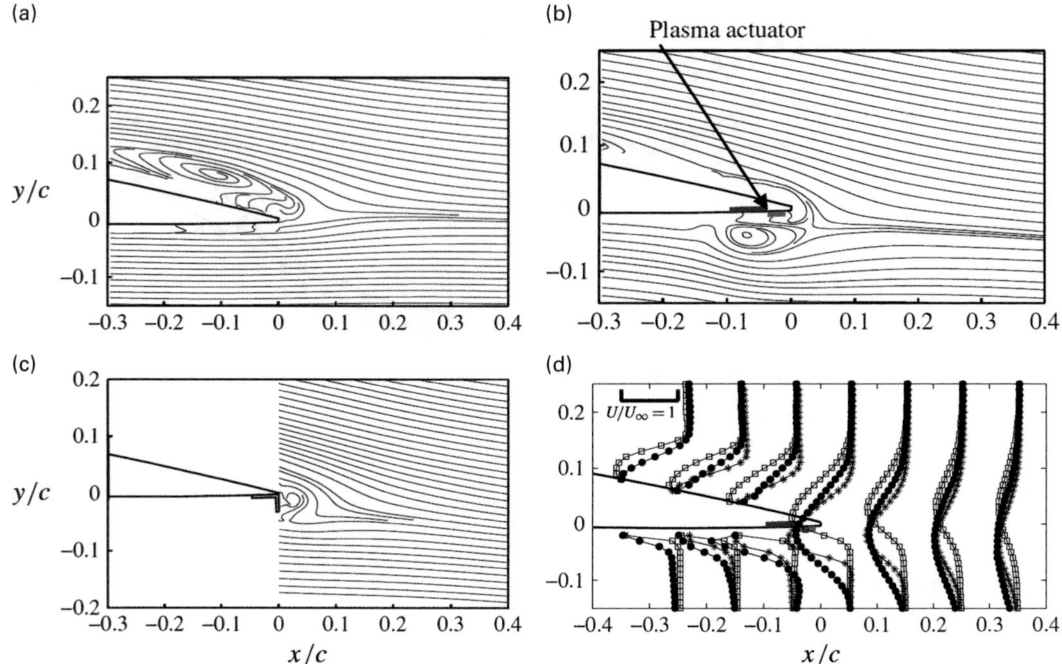

Figure 9.36 Flow field around the airfoil at $\alpha = 6°$, $Re = 2 \times 10^4$, $C_\mu = 1.9\%$. (a) Time-averaged streamlines without plasma control; (b) time-averaged streamlines with plasma control; (c) time-averaged streamlines with 3%c mechanical Gurney flap; (d) comparison of time-averaged streamwise velocity: -□-, without plasma control; -●-, with plasma control; - * -, 3%c mechanical Gurney flap (Feng et al. 2015). Reproduced with permission, copyright © Cambridge University Press 2015.

greatly by plasma Gurney flap, which is consistent with the force measurement results shown in Figure 9.35.

9.4.3   Plasma Circulation Control

Zhang et al. (2010b) first proposed to implement circulation control by using the plasma actuator. The schematic is shown in Figure 9.37(a), where a DBD plasma actuator is attached to the surface of the rounded trailing edge. With plasma turned on, the ambient neutral air is absorbed into the plasma region and forms a high-velocity wall jet compared with the surrounding boundary-layer flow, which attaches to the trailing-edge surface due to the Coanda effect. The lift augmentation efficiency could reach $\Delta C_L/C_\mu = 134.86$, in comparison with the maximum lift augmentation efficiency of 80 for conventional jet circulation control. Their results indicated that the efficiency of plasma circulation control for lift augmentation was more significant than that of conventional jet circulation control.

In the subsequent study, Feng et al. (2013) further proposed the conception of plasma circulation control that could be used on the airfoil with sharp trailing edge.

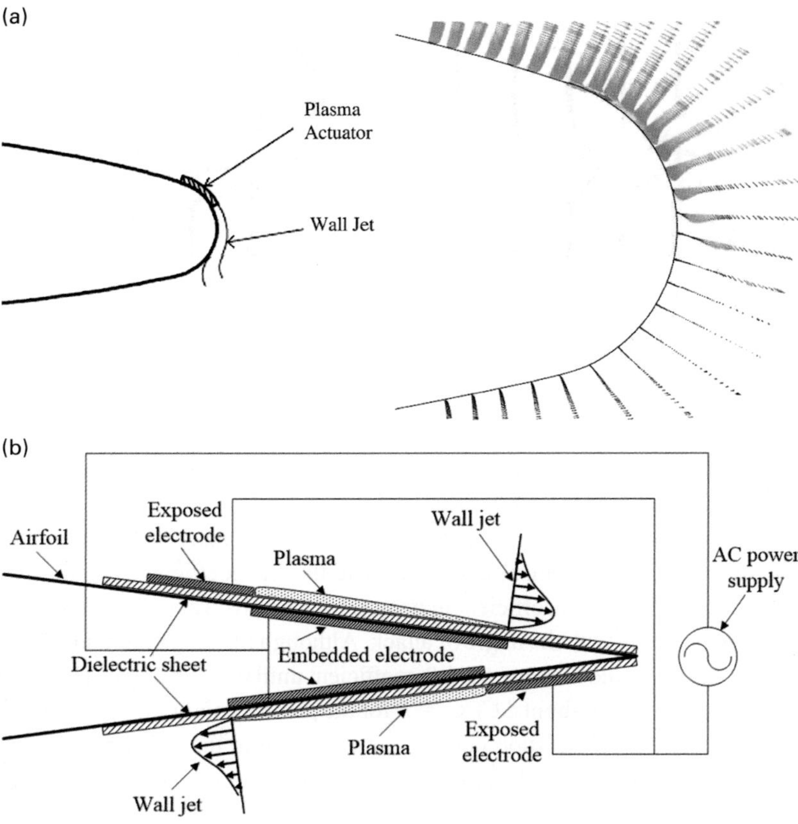

Figure 9.37 Schematic of different conceptions of plasma circulation control. (a) Plasma circulation control for the airfoil with rounded trailing edge (left figure) and the induced flow around the trailing edge (right figure) (Zhang et al. 2010b), reproduced with permission, copyright © The American Institute of Aeronautics and Astronautics. (b) Plasma circulation control for the airfoil with sharp trailing edge (Feng et al. 2013). Reproduced with permission from Chinese Journal of Theoretical & Applied Mechanics.

The circulation control technique is constructed from two DBD plasma actuators, as shown in Figure 9.37(b). One is attached to the suction surface near the trailing edge, with its induced wall jet the same direction as the free stream. The other is attached to the pressure surface near the trailing edge, with its induced wall jet the opposite direction to the free stream. The experimental model was similar with that used in Feng et al. (2015). The plasma actuator was powered at a peak-to-peak voltage of 8 kV with a frequency of 21 kHz. The experiment was conducted at $U_\infty = 3.0$ m/s, corresponding to Reynolds numbers of $Re = 2 \times 10^4$ and plasma forcing momentum coefficient of $C_\mu = 1.7\%$.

The mean streamwise velocity profiles without and with plasma circulation control are compared in Figure 9.38. At $\alpha = 0°$, it is obvious that plasma control increases the velocity over the suction while reducing the velocity over the pressure surface, which is evidence for the increase in circulation over the airfoil. A similar variation can also be

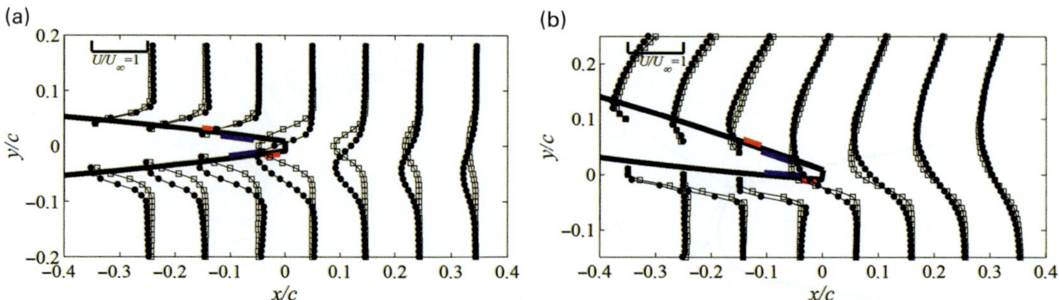

Figure 9.38 Time-averaged streamwise velocity profiles of the airfoil without (-□-) and with (-●-) plasma circulation control at $\alpha = 0°$ (a) and $\alpha = 12°$ (b) (Feng et al. 2013). Reproduced with permission from Chinese Journal of Theoretical & Applied Mechanics.

observed at $\alpha = 12°$ for the near wall region, though the velocity far away from the airfoil surface changes less in comparison with the natural case.

The aerodynamic forces on the airfoil without and with control are shown in Figure 9.39. Plasma control could increase the lift coefficient for all angles of attack with the maximum lift coefficient increased by about 16% at $\alpha = 10°$, but the improvement seems to decrease with the angle of attack. Although the drag coefficient is also increased, plasma control could increase the lift coefficient until the stall angle. The lift augmentation efficiency at $\alpha = 0°$ is about $\Delta C_L/C_\mu = 9$ for the present configuration. The lower efficiency than conventional circulation control is mainly due to the difference in the airfoil used. However, it is still greater than that achieved by continuous blowing at the trailing edge by Traub et al. (2004).

9.4.4   Plasma Vortex Generator

The conventional definition of a vortex generator is a device that could induce a streamwise vortex in the free stream. It has also been found that the interaction of a free jet with the crossflow could also induce a streamwise vortex. Thus, it is possible to use the DBD plasma actuator to induce similar vortical structures. A schematic of the plasma vortex generator proposed by Jukes and Choi (2013) is shown in Figure 9.40. The electrodes of the plasma actuator are placed along the free-stream direction, thus inducing a wall jet normal to the free stream. The interaction leads to the roll up of the vortex stretching in the streamwise direction. In addition, the direction of the electrode and the induced wall jet can also be varied, as long as it can generate a streamwise vortex.

Similar to mechanical vortex generators, plasma vortex generators could also be arranged in co-rotating (Figure 9.41(a)) and counter-rotating (Figure 9.41(b)) manners. Plasma vortex generators have been applied to control the flow around a flat plate (Barckmann et al. 2015), a circular cylinder (Kozlov and Thomas 2011a), a ramp (Jukes and Choi 2012), an airfoil (Jukes et al. 2013), and a wind turbine (Jukes 2015). Similar control effects to those induced by the mechanical vortex generators have been found.

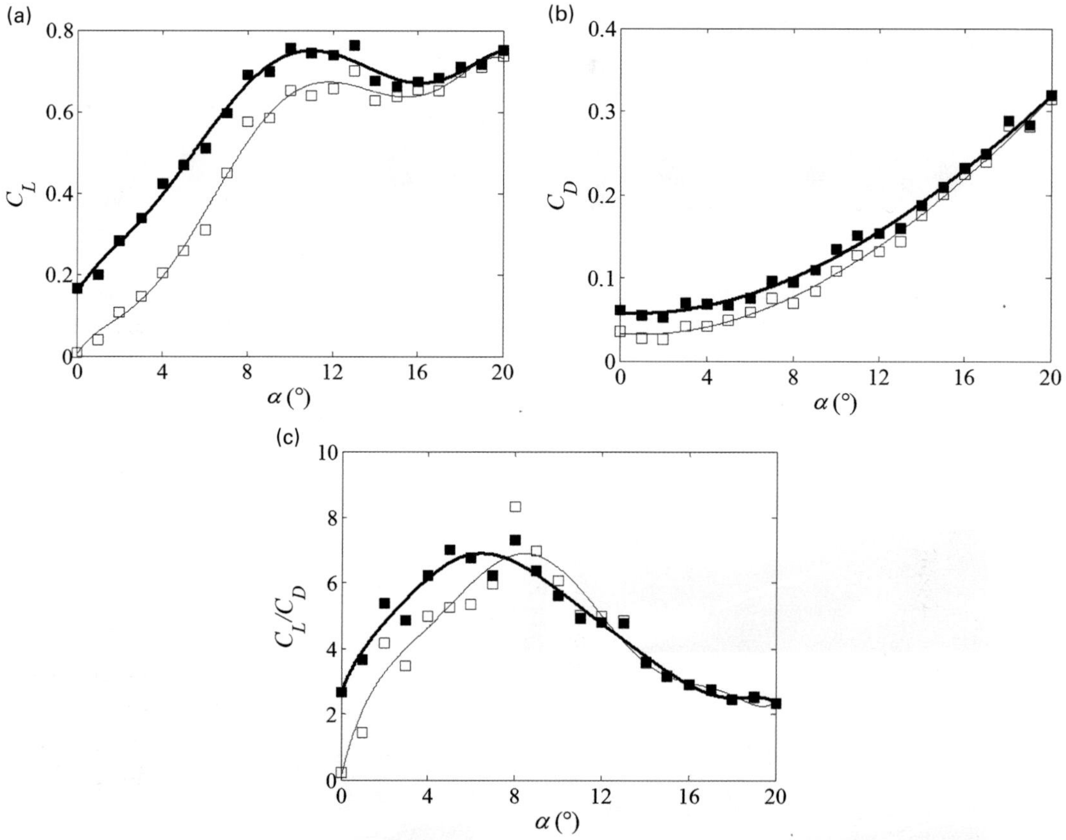

Figure 9.39 Aerodynamic forces of the airfoil without (-□-) and with (-■-) plasma circulation control. (a) Lift coefficient; (b) drag coefficient; and (c) lift-to-drag ratio (Feng et al. 2013). Reproduced with permission from Chinese Journal of Theoretical & Applied Mechanics.

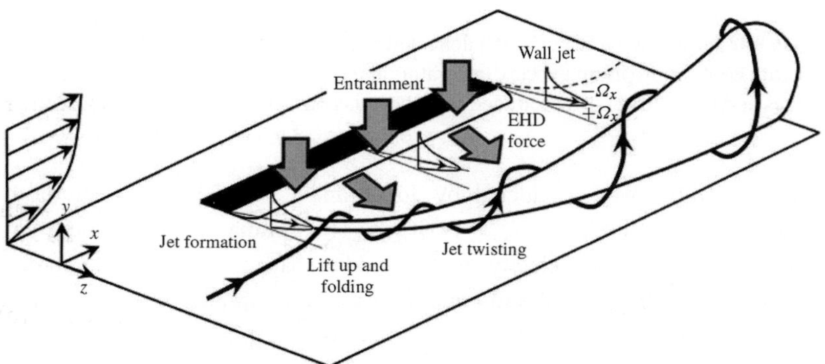

Figure 9.40 Schematic of the plasma vortex generator (Jukes and Choi 2013). Reproduced with permission, copyright © Cambridge University Press 2013.

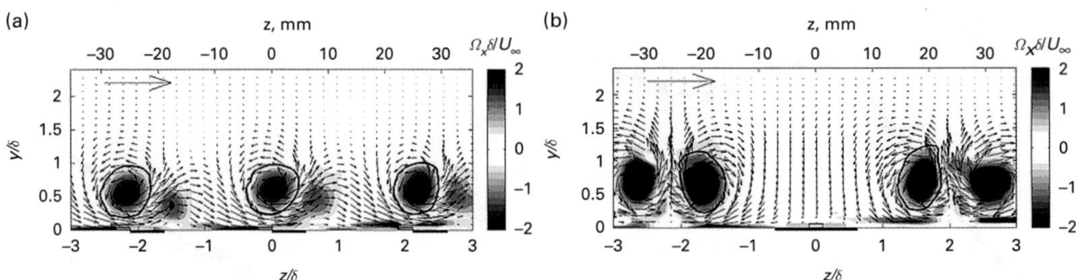

Figure 9.41 Different arrangements of the plasma vortex generator. (a) Co-rotating arrangement; (b) counter-rotating arrangement (Jukes and Choi 2012). Reproduced with permission, copyright © Springer-Verlag 2011.

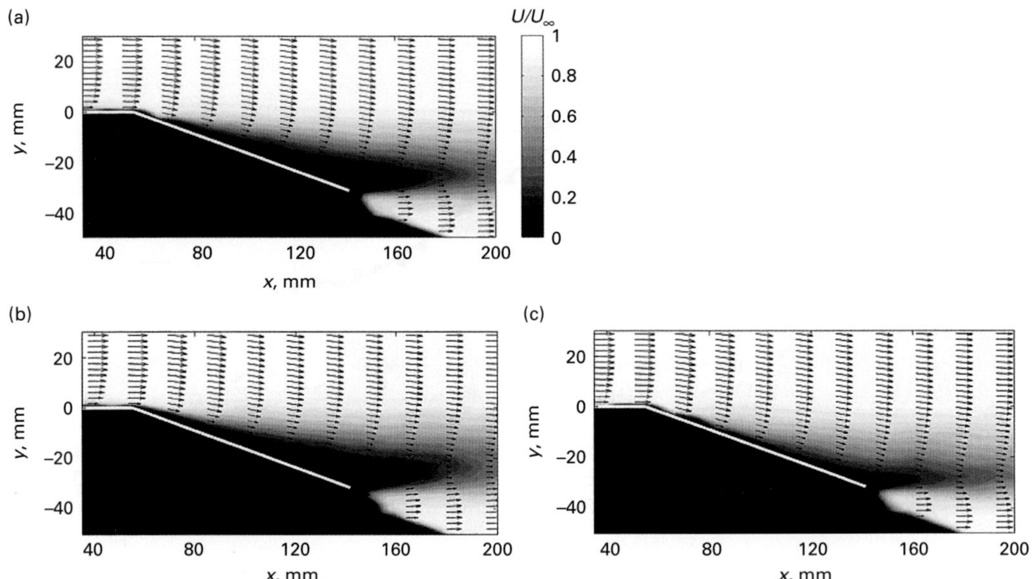

Figure 9.42 Streamwise velocity of a deflected ramp at $z = 0$ mm, $U_\infty = 14.9$ m/s. (a) without control; (b) with co-rotating plasma vortex generator; and (c) with counter-rotating plasma vortex generator (Jukes and Choi 2012). Reproduced with permission, copyright © Springer-Verlag 2011.

One example was conducted by Jukes and Choi (2012) to control the flow over a deflected ramp model at $U_\infty = 14.9$ m/s. The plasma-induced velocity ratio was 0.07. The mean streamwise velocity at the middle span of the deflected ramp is shown in Figure 9.42 for the natural and control cases. A separation region occurs over the ramp for the natural case (Figure 9.42(a)), the co-rotating plasma vortex generator could only slightly reduce the separation region (Figure 9.42(b)), while the counter-rotating one could partially avoid flow separation (Figure 9.42(c)). Jukes and Choi (2012) and Jukes et al. (2013) concluded that the counter-rotating arrays behaved better but were less homogenous than co-rotating arrays, because the vortices induced by the latter case naturally swept across the span by the induced velocity.

9.5   Concluding Remarks

The plasma actuator is an efficient and simple device that could be used for active flow control. Based on their configurations, there are different kinds of plasma actuators, such as the DBD plasma actuator, surface corona discharge actuator, and plasma spark-jet actuator. In this chapter, we have introduced the fundamental characteristics of the different plasma actuators and their applications in various fields, with particular attention being paid to the DBD plasma actuator. Two main features of the DBD plasma actuator are that it can induce a wall jet and a starting vortex, which can be used for efficient flow control.

First, the conventional applications of the plasma actuators are introduced. The plasma actuator can delay laminar to turbulence transition of a boundary layer, and thus reduce the friction drag. It could also delay or eliminate flow separation, and thus improve aerodynamic performance of airfoils, straight wings, delta wings, aircraft, and bluff bodies. In particular, plasma actuators have significant great ability for momentum control, proving a potential application to replace conventional movable control surfaces on future aircraft.

Second, some novel flow control conceptions have also been proposed based on the plasma actuators simulating some of the traditional control functions, such as plasma synthetic jet, plasma Gurney flap, plasma circulation control, plasma vortex generator. It has been indicated that those techniques could achieve a similar effect to the traditional ones. Meanwhile, plasma-based techniques facilitate real-time and unsteady active control, and are more convenient to implement than conventional techniques.

One disadvantage of the DBD plasma actuators is the limited velocity induced by the device, which may restrict their scope of application. There are three ways to solve this problem. First, it is suggested that the plasma actuators could be placed at the sensitive position of the controlled body, thus a small excitation may trigger the global instability of the flow to enlarge the applied velocity. Second, one can further improve the capability of electrodes and dielectric material and increase the applied voltage, thus to increase the induced velocity. Third, novel configurations of the plasma actuator should be developed, such as the plasma spark-jet actuators, in order to enhance the control ability.

References

Barckmann, K., Tropea, C., and Grundmann, S. Attenuation of Tollmien–Schlichting waves using plasma actuator vortex generators. *AIAA Journal*, 2015, 53(5): 1384–1388

Benard, N., Jolibois, J., and Moreau, E. Lift and drag performances of an axisymmetric airfoil controlled by plasma actuator. *Journal of Electrostatics*, 2009, 67(2–3): 133–139

Bhattacharya, S. and Gregory, J. W. Effect of three-dimensional plasma actuation on the wake of a circular cylinder. *AIAA Journal*, 2015a, 53(4): 958–967

Bhattacharya, S. and Gregory, J. W. Investigation of the cylinder wake under spanwise periodic forcing with a segmented plasma actuator. *Physics of Fluids*, 2015b, 27(1): 014102

Boesch, G., Vo, H. D., Savard, B., Wanko-Tchatchouang, C., and Mureithi, N. W. Flight control using wing-tip plasma actuation. *Journal of Aircraft*, 2010, 47(6): 1836–1846

Corke, T. C., Enloe, C. L., and Wilkinson, S. P. Dielectric Barrier Discharge Plasma Actuators for Flow Control. *Annual Review of Fluid Mechanics*, 2010, 42(42): 505–529

Cybyk, B. Z., Wilkerson, J. T., Grossman, K. R., and Van Wie, D. M. Computational assessment of the sparkjet flow control actuator. AIAA Paper 2003–3711

Duchmann, A., Simon, B., Tropea, C., and Grundmann, S. Dielectric barrier discharge plasma actuators for in-flight transition delay. *AIAA Journal*, 2014, 52(2): 358–367

Feng, L. H., Choi, K. S., and Wang, J. J. Control of flow around a circular cylinder by plasma synthetic jet. 11th International Conference on Fluid Control, Measurements and Visualization, December 5–9, 2011, Taiwan, China

Feng, L. H., Choi, K. S., and Wang, J. J. Flow control over an airfoil using virtual Gurney flaps. *Journal of Fluid Mechanics*, 2015, 767: 595–626

Feng, L. H., Jukes, T. N., Choi, K. S., and Wang, J. J. Flow control over a NACA 0012 airfoil using dielectric-barrier-discharge plasma actuator with a Gurney flap. *Experiments in Fluids*, 2012, 52(6): 1533–1546

Feng, L. H., Wang, J. J., and Choi, K. S. A novel concept on the plasma Gurney flap. 29th Congress of the International Council of the Aeronautical Sciences, September 7–12, 2014, St. Petersburg, Russia

Feng, L. H., Wang, J. J., and Choi, K. S. Experimental investigation on lift increment of a plasma circulation control airfoil. *Chinese Journal of Theoretical & Applied Mechanics*, 2013, 45(6): 815–821 (in Chinese)

Feng, L. H. and Wang, J. J. Circular cylinder vortex-synchronization control with a synthetic jet positioned at the rear stagnation point. *Journal of Fluid Mechanics*, 2010, 662(7): 232–259

Greenblatt, D., Kastantin, Y., Nayeri, C. N., and Paschereit, C. O. Delta-wing flow control using dielectric barrier discharge actuators. *AIAA Journal*, 2008, 46(6): 1554–1560

Gregory, J. W., Porter, C. O., and McLaughlin, T. E. Circular cylinder wake control using spatially distributed plasma forcing. AIAA Paper 2008–4198

Grundmann, S. and Tropea, C. Active cancellation of artificially introduced Tollmien–Schlichting waves using plasma actuators. *Experiments in Fluids*, 2008, 44(5): 795–806

Grundmann, S. and Tropea, C. Experimental transition delay using glow-discharge plasma actuators. *Experiments in Fluids*, 2007, 42(4): 653–657

Huang, X., Zhang, X., and Li, Y. Broadband flow-induced sound control using plasma actuators. *Journal of Sound & Vibration*, 2010, 329(13): 2477–2489

Im, S., Do, H., and Cappelli, M. A. Dielectric barrier discharge control of a turbulent boundary layer in a supersonic flow. *Applied Physics Letters*, 2010, 97(4): 041503

Jukes, T. N. Smart control of a horizontal axis wind turbine using dielectric barrier discharge plasma actuators. *Renewable Energy*, 2015, 80: 644–654

Jukes, T. N. and Choi, K. S. Dielectric-barrier-discharge vortex generators: characterisation and optimisation for flow separation control. *Experiments in Fluids*, 2012, 52(2): 329–345

Jukes, T. N. and Choi, K. S. Control of unsteady flow separation over a circular cylinder using dielectric-barrier-discharge surface plasma. *Physics of Fluids*, 2009a, 21(9): 094106

Jukes, T. N. and Choi, K. S. Flow control around a circular cylinder using pulsed dielectric barrier discharge surface plasma. *Physics of Fluids*, 2009b, 21(8): 084103

Jukes, T. N. and Choi, K. S. Long lasting modifications to vortex shedding using a short plasma excitation. *Physical Review Letters*, 2009c, 102(25): 254501

Jukes, T. N. and Choi, K. S. On the formation of streamwise vortices by plasma vortex generators. *Journal of Fluid Mechanics*, 2013, 733(6): 370–393

Jukes, T. N., Segawa, T., and Furutani, H. Flow control on a NACA 4418 using dielectric barrier discharge vortex generators. *AIAA Journal*, 2013, 51(2): 452–464

Kozlov, A. V. and Thomas, F. O. Bluff-body flow control via two types of dielectric barrier discharge plasma actuation. *AIAA Journal*, 2011a, 49(9): 1919–1931

Kozlov, A. V. and Thomas, F. O. Plasma flow control of cylinders in a tandem configuration. *AIAA Journal*, 2011b, 49(10): 2183–2193

Léger, L., Moreau, E., Artana, G., and Touchard, G. Influence of a DC corona discharge on the airflow along an inclined flat plate. *Journal of Electrostatics*, 2001, 51(1): 300–306

Li, Y., Zhang, X., and Huang, X. The use of plasma actuators for bluff body broadband noise control. *Experiments in Fluids*, 2010, 49(2): 367–377

Liu, A. B., Zhang, P. F., Yan, B., Dai, C. F., and Wang, J. J. Flow characteristics of synthetic jet induced by plasma actuator. *AIAA Journal*, 2011, 49(3): 544–553

Mabe, J. H., Calkins, F. T., Wesley, B., Woszidlo, R., Taubert, L., and Wygnaanski, L. Single dielectric barrier discharge plasma actuators for improved airfoil performance. *Journal of Aircraft*, 2009, 46(3): 847–855

McLaughlin, T. E., Munska, M. D., Vaeth, J. P., Dauwalter, T. E., Goode, J. R., and Siegel, S. Plasma-based actuators for cylinder wake vortex control. AIAA Paper 2004–2129

McLaughlin, T., Felker, B., Avery, J., and Enloe, C. Further experiments in cylinder wake modification with dielectric barrier discharge forcing. AIAA Paper 2006–1409

Moreau, E., Léger, L., and Touchard, G. Effect of a DC surface-corona discharge on a flat plate boundary layer for air flow velocity up to 25 m/s. *Journal of Electrostatics*, 2006, 64(3–4): 215–225

Moreau, E. Airflow control by non-thermal plasma actuators. *Journal of Physics D: Applied Physics*, 2007, 40(3): 605

Munska, M. D. and McLaughlin, T. E. Circular cylinder flow control using plasma actuators. AIAA Paper 2005–141

Rizzetta, D. P. and Visbal, M. R. Large eddy simulation of plasma-based control strategies for bluff body flow. *AIAA Journal*, 2009, 47(3): 717–729

Roth, J. R., Tsai, P. P., Liu, C., Laroussi, M., and Spence, P. D. One atmosphere, uniform glow discharge plasma. United States Patent 5414324, 1995

Sidorenko, A. A., Budovskiy, A. D., Maslov, A. A., Postnikov, B. V., Zanin, B. Y., Zverkov, I. D., and Kozlov, V. V. Plasma control of vortex flow on a delta wing at high angles of attack. *Experiments in Fluids*, 2013, 54(8): 1585

Sosa, R. and Artana, G. Steady control of laminar separation over airfoils with plasma sheet actuators. *Journal of Electrostatics*, 2006, 64(7–9): 604–610

Sosa, R., Artana, G., Benard, N., and Moreau, E. Mean lift generation on cylinders induced with plasma actuators. *Experiments in Fluids*, 2011, 51(3): 853–860

Sosa, R., Artana, G., Moreau, E., and Touchard, G. Stall control at high angle of attack with plasma sheet actuators. *Experiments in Fluids*, 2007, 42(1): 143–167

Roupassov, D. V., Nikipelov, A. A., and Nudnova, M., and Starikovskii, A. Flow separation control by plasma actuator with nanosecond pulse discharge. *AIAA Journal*, 2009, 47(1): 168–185

Sung, Y., Kim, W., Mungal, M. G., and Cappelli, M. A. Aerodynamic modification of flow over bluff objects by plasma actuation. *Experiments in Fluids*, 2006, 41(3): 479–486

Thomas, F. O., Kozlov, A., and Corke, T. C. Plasma actuators for cylinder flow control and noise reduction. *AIAA Journal*, 2008, 46(8): 1921–1931

Traub, L. W., Miller, A. C., and Rediniotis, O. Comparisons of a Gurney and jet-flap for hinge-less control. *Journal of Aircraft*, 2004, 41(2): 420–423

Vinogradov, I. andHuang, X. Bluff body flow-induced noise control with sliding plasma actuators. *Chinese Science Bulletin*, 2011, 56(28–29): 3079–3081

Vorobiev, A. N., Rennie, R. M., Jumper, E. J., and McLaughlin, T. E. Experimental investigation of lift enhancement and roll control using plasma actuators. *Journal of Aircraft*, 2008, 45(4): 1315–1321

Wang, J. J., Choi, K. S., Feng, L. H., Jukes T. N., and Whalley, R. D. Recent developments in DBD plasma flow control. *Progress in Aerospace Sciences*, 2013, 62(4): 52–78

Wang, L., Xia, Z., Luo, Z., and Chen, J. Three-electrode plasma synthetic jet actuator for high-speed flow control. *AIAA Journal*, 2014, 52(4): 879–882

Whalley, R. D. and Choi, K. S. The starting vortex in quiescent air induced by dielectric-barrier-discharge plasma. *Journal of Fluid Mechanics*, 2012, 703(1): 192–203

Wu, Y. and Li, Y. H. Progress and outlook of plasma flow control. *Acta Aeronautica et Astronautica Sinica*, 2015, 36(2): 381–405 (in Chinese)

Wu, Y., Li, Y. H., Jia, M., Liang, H., and Song, H. M. Experimental investigation of nanosecond discharge plasma aerodynamic actuation. *Chinese Physics B*, 2012, 21(4): 045202

Wu, Y., Li, Y. H., Jia, M., Song, H. M., and Liang, H. Optical emission characteristics of surface nanosecond pulsed dielectric barrier discharge plasma. *Journal of Applied Physics*, 2013, 113 (3): 033303

Zhang, P. F., Wang, J. J., Feng, L. H., and Wang, G. Z. Experimental study of plasma flow control on highly swept delta wing. *AIAA Journal*, 2010a, 48(1): 249–252

Zhang, P. F., Yan, B., Liu, A. B., and Wang, J. J. Numerical simulation on plasma circulation control airfoil. *AIAA Journal*, 2010b, 48(10): 2213–2226

Zhang, P. F., Liu, A. B., and Wang, J. J. Aerodynamic modification of NACA 0012 airfoil by trailing-edge plasma Gurney flap. *AIAA Journal*, 2009, 47(10): 2467–2474

Zhang, P. F., Liu, A. B., and Wang, J. J. Numerical simulation of flow induced by plasma actuator based on phenomenological model. *Journal of Beijing University of Aeronautics and Astronautics*, 2010c, 36(1): 52–56 (in Chinese)

Zhao, G. Y., Li, Y. H., Hua, W. Z., Liang, H., Han, M. N., and Niu, Z. G. Experimental study of flow control on delta wings with different sweep angles using pulsed nanosecond DBD plasma actuators. *Journal of Aerospace Engineering*, 2015a, 229(11): 1966–1974

Zhao, G., Li, Y., Liang, H., Han, M. H., and Hua, W. Z. Control of vortex on a non-slender delta wing by a nanosecond pulse surface dielectric barrier discharge. *Experiments in Fluids*, 2015b, 56(1): 1864

Zhao, G. Y., Li, Y. H., Liang, H., Han, M. H., and Wu, Y. Flow separation control on swept wing with nanosecond pulse driven DBD plasma actuators. *Chinese Journal of Aeronautics*, 2015c, 28(2): 368–376

Zhao, G. Y., Li, Y. H., Liang, H., Hua, W. Z., and Han, M. H. Phenomenological modeling of nanosecond pulsed surface dielectric barrier discharge plasma actuation for flow control. *Acta Physica Sinica*, 2015d, 64(1): 015101

Zhu, Y. F., Wu, Y., Cui, W., Li, Y. H., and Jia, M. Modelling of plasma aerodynamic actuation driven by nanosecond SDBD discharge. *Journal of Physics D: Applied Physics*, 2013, 46(35): 355205

Zong, H. H., Cui, W., Wu, Y., Zhang, Z. B., Liang, H., Jia, M., and Li, Y. H. Influence of capacitor energy on performance of a three-electrode plasma synthetic jet actuator. *Sensors and Actuators A: Physical*, 2015, 222: 114–121

10 Lorentz Force

10.1 Background

The subject of flow control with Lorentz force (or electromagnetic body force) is originated from electromagnetism. It is known that a charged particle moving in the presence of an electric field and a magnetic field will experience a Lorentz force. A similar phenomenon also happens when a wire carrying an electric current is placed in a magnetic field. The direction of the Lorentz force can be determined by the left-hand rule. If the fingers of the left hand are extended to point in the electric current direction, and the magnetic field is passing towards the palm, then the extended thumb will point in the direction of the Lorentz force.

The Lorentz force can be generated in fluids with specifically designed devices, and one typical design is shown in Figure 10.1. The required Lorentz force can be created by placing electrodes and magnets side by side, parallel to one another. The direction of the Lorentz force can be determined by both the electric and magnetic fields. Thus, one can change the direction of the Lorentz force by changing the direction of either the electric field or magnetic field, and change its magnitude by adjusting the strength of either the electric or magnetic field. In particular, one can produce a Lorentz force varying in the sinusoidal waveform. Note that the Lorentz force can be produced in either spanwise direction or in the streamwise direction.

It was reported by Berger et al. (2000) that probably the first application of the Lorentz force for flow control was conducted by Gailitis and Lielausis in 1961, who applied streamwise Lorentz force to a laminar boundary layer to increase the thrust force and delay transition to turbulence of the flow. However, more detailed works were conducted from the 1990s, and the streamwise or spanwise Lorentz force was used for flow control of various fields.

In this chapter, we will introduce the recent applications of the Lorentz force in the boundary layer over a flat plate and the flow around airfoils and bluff bodies. In addition, the flow control mechanism and parameter influence will be discussed.

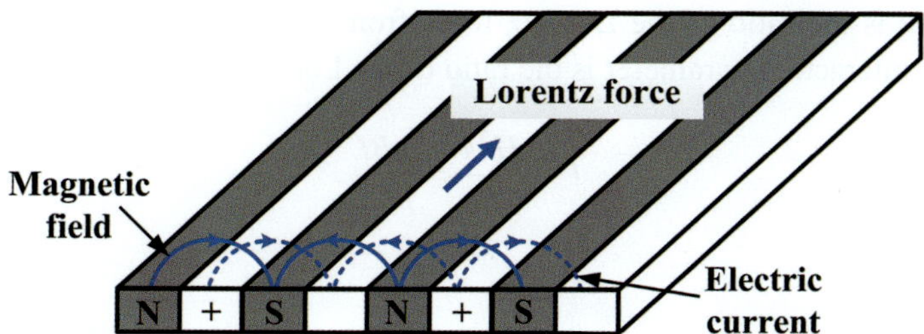

Figure 10.1 Schematic of an arrangement for the creation of a Lorentz force (adapted from Berger et al. 2000). Reproduced with permission from AIP Publishing.

10.2 Boundary Layer

10.2.1 Important Parameters

When the voltage potential is applied across the electrodes in a conducting fluid, the electric current density lines are shown in Figure 10.1. The current density vector is proportional to the electrical conductivity of the fluids. Then, the cross product of the electric field vectors ($\mathbf{j}$) with the magnetic field line vectors ($\mathbf{B}$) produce the Lorentz force normal to both field lines, namely

$$\mathbf{F} = \mathbf{j} \times \mathbf{B}. \tag{10.1}$$

For a specific flow field with electromagnetic (EMHD) forcing, there are some important dimensionless parameters, namely the Reynolds number Re, the Hartmann number H, and the ratio of the applied current to induced current κ, which denote the parameters for the flow field, the magnetic field and the electric field, respectively.

Reynolds number:

$$Re = \frac{U_\infty L}{v}, \tag{10.2}$$

where U_∞ is the free-stream velocity, L is the typical length scale and v is the kinematic viscosity.

Hartmann number:

$$H = BL\sqrt{\frac{\sigma}{v\rho}}, \tag{10.3}$$

where B is the magnetic field, σ is the electric conductivity and ρ is the fluid density.

The ratio of the applied current to induced current:

$$\kappa = \frac{E}{U_\infty B}, \tag{10.4}$$

where E is the electric field, and there is a relationship $j = \sigma E$.

Here, $H^2\kappa$ is the ratio of the Lorentz force from the applied current to the viscous forces. The interaction parameter is the ratio of the Lorentz force to the inertial force

$$I = \frac{H^2\kappa}{Re} = \frac{\sigma BL}{\rho U_\infty^2}.$$

(10.5)

10.2.2   Streamwise Forcing

Henoch and Stace (1995), Crawford and Karniadakis (1997) used the streamwise Lorentz force to control a turbulent boundary by experimental and numerical approaches, respectively. The flow control device was similar to that shown in Figure 10.1, but the force was in the free-stream direction. The experiment was conducted in a seawater flat plate boundary layer, while the numerical simulation followed similar parameters.

In Crawford and Karniadakis's (1997) study, the simulation was conducted for seawater in a channel flow with a dimensionless length scale in the x, y, and z directions of 5.61, 2.0, and 2.0, respectively. Other typical parameters were selected as $L = \delta = 5$ cm, $U_\infty = 0.1$ m/s, $\sigma = 4$ mhos/m, and $\rho = 1000$ kg/m, corresponding to the Reynolds number of $Re = 55\,000$ ($Re_\tau \approx 200$) based on the channel half-height δ. Two control cases were tested at the interaction parameters $I = 0.4$ (case 1) and 0.1 (case 2). For case 1, a time-dependent forcing was imposed by switching on and off the electrodes every 1 convective time unit. For case 2, two pulsing frequencies with 1 and 0.01 convective time units were used.

The mean streamwise velocity contours for the $I = 0.4$ control case and the dimensionless streamwise Lorentz force are shown in Figure 10.2. It is indicated that the Lorentz force shows a periodic variation in the spanwise direction. It creates a series of near-wall jets, which is also periodic varying in the spanwise direction. This has also been detected by Henoch and Stace (1995). In addition, the flow over the upper wall is also influenced.

The time history of the drag at both lower and upper walls is shown in Figure 10.3 for the natural and control cases. Though the Lorentz force is produced on the lower wall surface, it is also effective in influencing the drag development over the upper surfaces. It is indicated that the Lorentz force could significantly increase the drag on both lower and upper walls in comparison with the no control case. In general, the $I = 0.4$ case results in a greater drag than the $I = 0.1$ case.

Turbulence statistics were analyzed for three typical locations for the control case. $z = 0$ and 0.125 stand for the locations where the Lorentz force is the largest and the lowest, respectively, while $z = 0.0625$ is the middle position between them. The profiles of mean streamwise velocity and RMS streamwise velocity fluctuation are shown in Figure 10.4. The mean velocity profiles for the control case show significant differences to the natural case (Figure 10.4(a)). It is indicated that the fluid jet at the wall extends to $y^+ = 20$ for $I = 0.4$ and the jet is strongest at the center of the Lorentz force ($z^+ = 0$). The velocity profiles at different spanwise locations start to collapse from $y^+ \geq 20$. The Lorentz force could reduce the RMS streamwise velocity fluctuation greatly in comparison with the natural case (Figure 10.4(b)). The minimum value occurs at $z = 0.125$. In addition, the

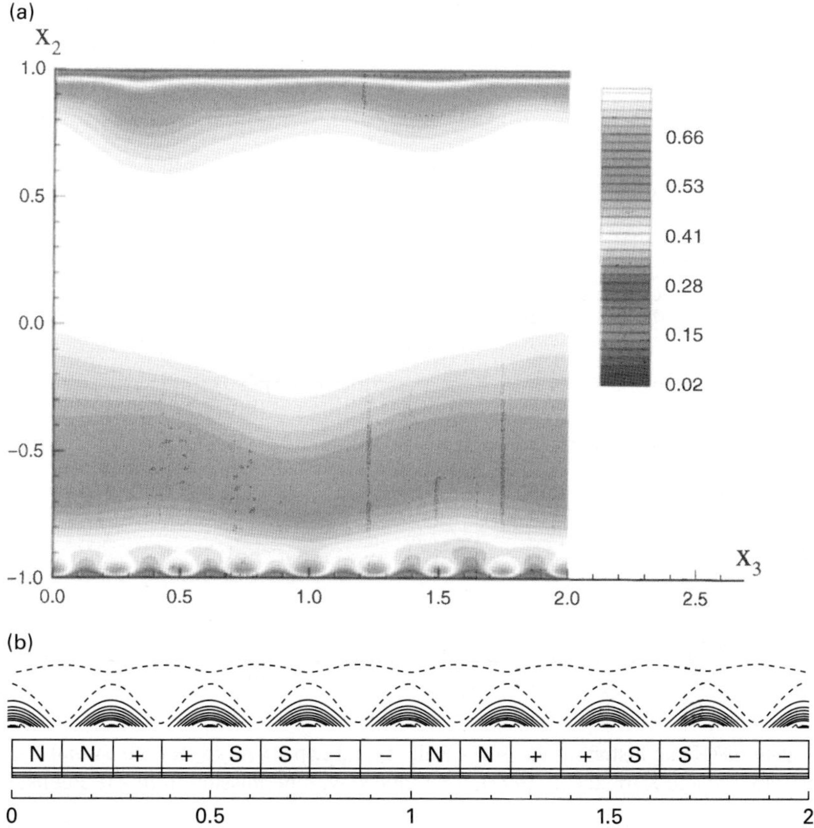

Figure 10.2 (a) Mean streamwise velocity contours for $I = 0.4$ control case, and (b) the dimensionless streamwise Lorentz force at the lower wall $y = -1$ (Crawford and Karniadakis 1997). Reproduced with permission from AIP Publishing.

near wall RMS of the vertical and spanwise velocity fluctuations is also reduced. A similar variation has also occurred for the $I = 0.1$ control case. These results also suggest that suppression of velocity fluctuations does not imply a drag reduction.

The variation in the velocity fluctuations suggests that the fundamental structure of the near-wall turbulence has been changed by the Lorentz force, as shown in Figure 10.5. Low-speed streaks denoted by darker shade can be observed for both cases. However, with Lorentz force, there are 8 streaks in the field of view with an average lateral spacing of about 90 wall units, which is smaller than the natural case whose average lateral spacing exceeds 100 wall units. Also, they are straighter and much longer for the control case than the natural case.

10.2.3   Spanwise Forcing

Besides the applications of the streamwise Lorentz force for boundary layer control, several groups started to use the spanwise Lorentz force for flow control (Berger et al. 2000; Pang and Choi 2004; Breuer et al. 2004). The device configuration is

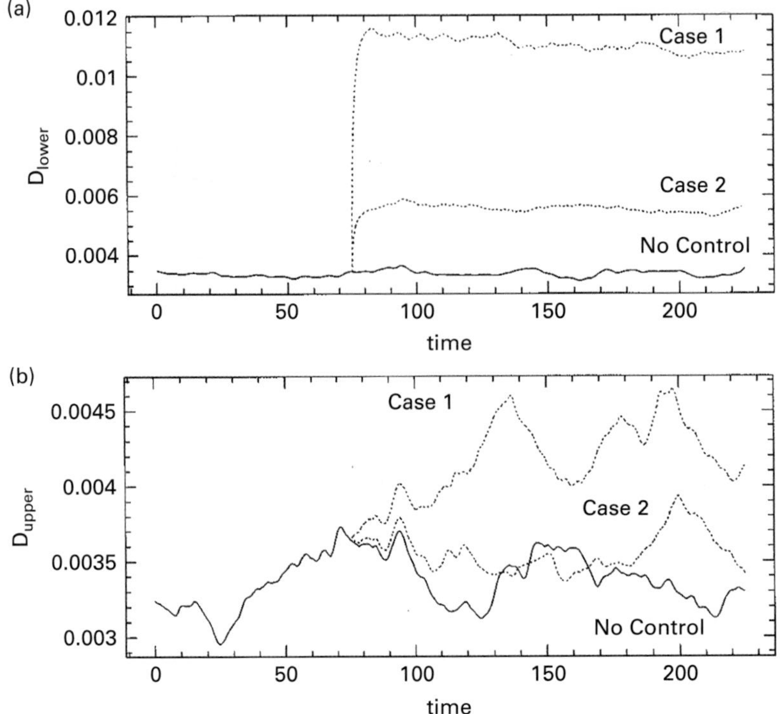

Figure 10.3 Drag history at the lower wall (a) and upper wall (b) for the natural and control cases (Crawford and Karniadakis 1997). Reproduced with permission from AIP Publishing.

similar to that shown in Figure 10.1. In Breuer et al.'s (2004) experiment, a sinusoidal driving voltage was used to produce the sinusoidal Lorentz force. We can see the effects of this force in Figure 10.6 for the spanwise velocity profiles at different phases, indicating that the velocity profiles vary periodically in the spanwise direction. It is indicated that the velocity reaches a maximum at about $y = 1$ mm from the wall surface, which then decreases to zero at about $y = 4$ mm. In particular, the maximum velocity generated by the Lorentz force is a function of the forcing voltage and frequency. It increases with an increase in the forcing voltage as well as a decrease in the forcing frequency.

Two dimensionless parameters may influence the effects of the sinusoidal Lorentz force, namely the Stuart number F^+ and the forcing period T^+. The Stuart number F^+ stands for the strength of the magnetic force relative to the inertial force of the fluid,

$$F^+ = EB\delta/\rho u_e^2, \tag{10.6}$$

where δ and u_e are the thickness and velocity for the undisturbed boundary layer, respectively. The dimensionless forcing period is defined as

$$T^+ = T u_e^2/\nu, \tag{10.7}$$

where T is the oscillation period.

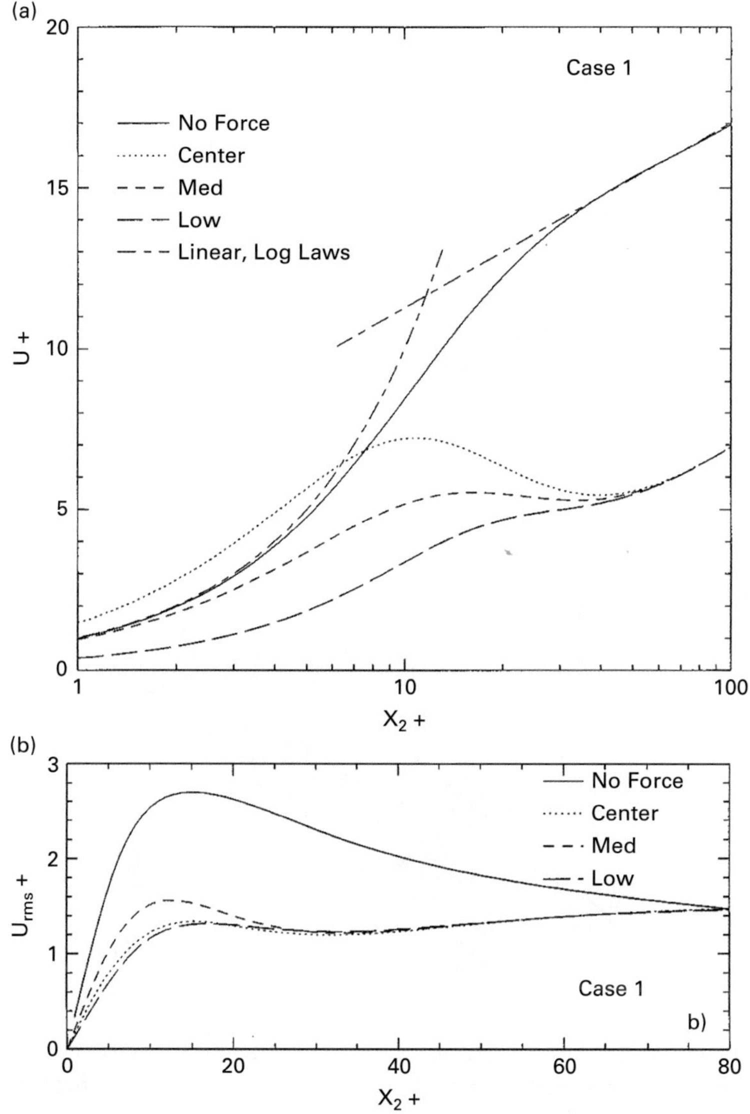

Figure 10.4 Profiles of mean streamwise velocity (a) and RMS streamwise velocity fluctuation (b) for the natural and $I = 0.4$ control cases (Crawford and Karniadakis 1997). Reproduced with permission from AIP Publishing.

The velocity profile of the turbulent boundary layer for the natural and one typical control case is shown in Figure 10.7. Here, both u^+ and y^+ for the natural and control cases are nondimensionalized using their own friction velocities. Thus, the near-wall velocity profiles for natural and control cases in the viscous sublayer coincide with each other. However, the logarithmic velocity profile for the control case is shifted upwards, suggesting that the viscous sublayer is thickened as a result of the friction drag reduction.

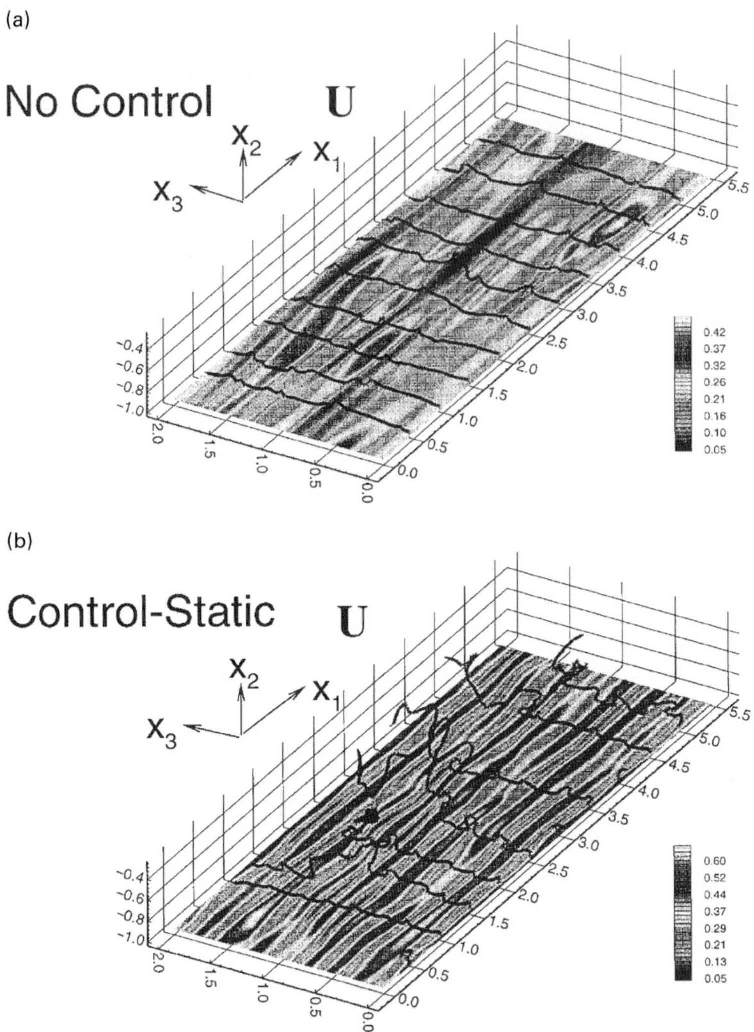

Figure 10.5 Instantaneous vortex lines and contours of streamwise velocity at $y = -0.97$ for the natural (a) and $I = 0.4$ control (b) cases (Crawford and Karniadakis 1997). Reproduced with permission from AIP Publishing.

The spanwise Lorentz force can lead to a considerable drag reduction, which is summarized in Figure 10.8(a) as a function of the Stuart number F^+ and the forcing period T^+. The maximum drag reduction of about 40% can be obtained near $F^+ = 210$ and $T^+ = 100$. Note that the variation in the Stuart number and the forcing period can both result in a variation of the spanwise wall velocity. Thus, Pang and Choi (2004) proposed an equivalent spanwise wall velocity as a new scaling parameter, namely

$$W^+ = F^+ T^+ / 2\pi Re. \tag{10.8}$$

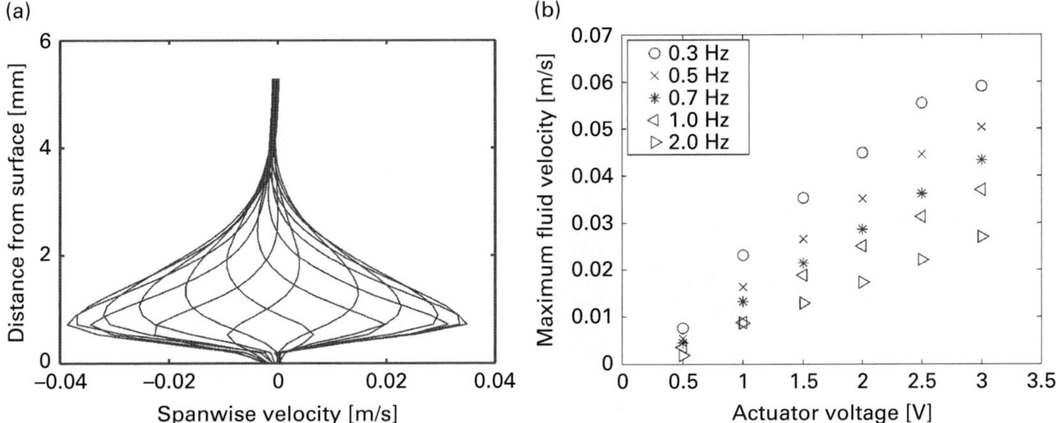

Figure 10.6 (a) Velocity profiles due to oscillatory spanwise forcing, where 16 profiles are plotted at equally spaced phases during the forcing cycle. (b) Maximum velocity generated by the Lorentz force by varying the forcing voltage (Breuer et al. 2004). Reproduced with permission from AIP Publishing.

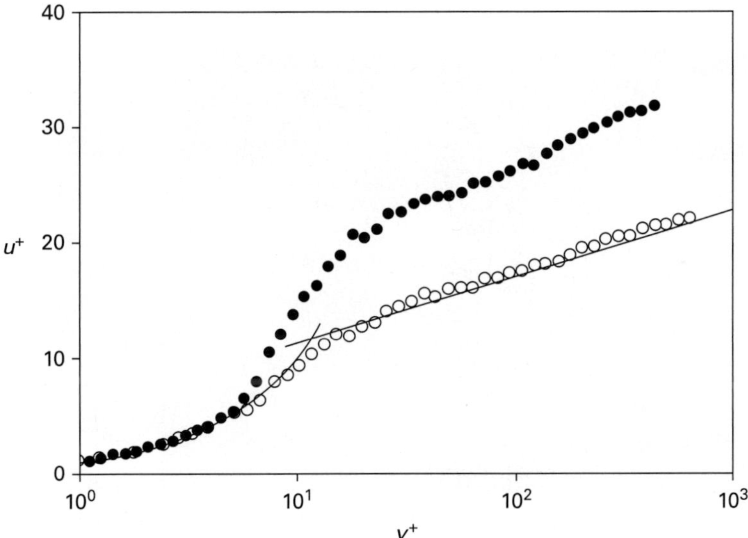

Figure 10.7 Velocity profiles of the turbulence boundary layer for the natural case (hollow circle) and $F^+ = 209$, $T^+ = 115$ control case (filled circles). Velocities were measured at 5 mm ($x^+ = 30$) downstream of the end of actuator array (Pang and Choi 2004). Reproduced with permission from AIP Publishing.

When the drag reduction is plotted against the spanwise wall velocity, the data can be fitted around a curve (Figure 10.8(b)). The maximum drag reduction occurs at around $W^+ = 15$. Note that the variations of the drag reduction data for the Lorentz force agree well with those for the spanwise wall oscillation, indicating the similarity of the inherent flow physics induced by the two control approaches.

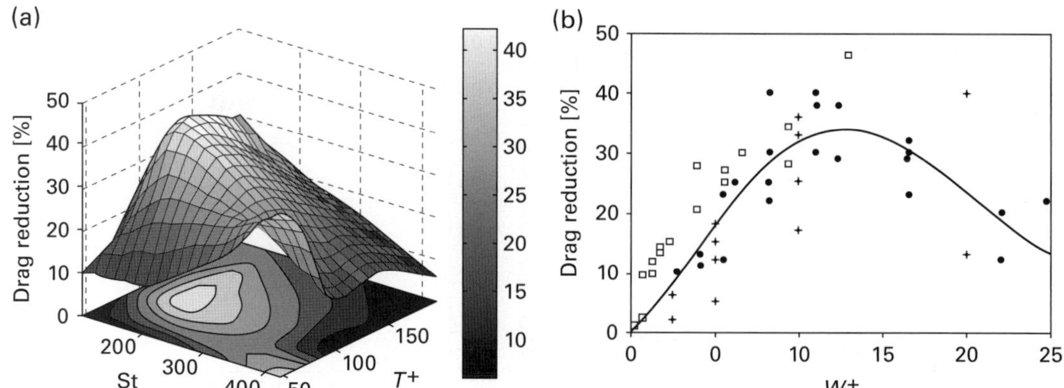

Figure 10.8 (a) Drag reduction as a function of the Stuart number and the forcing period with a contour map. (b) Drag reduction as a function of the spanwise velocity W^+: ● Pang and Choi (2004) and + Berger et al. (2000) for the Lorentz force; □ Choi et al. (1998) for spanwise wall oscillation. The plots are from Pang and Choi (2004). Reproduced with permission from AIP Publishing.

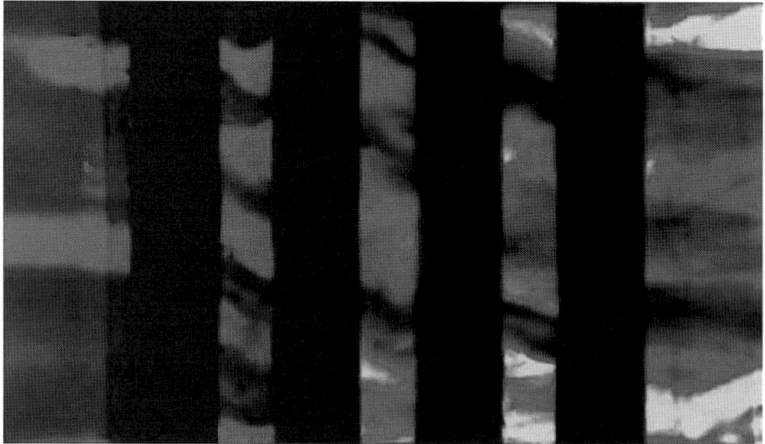

Figure 10.9 Flow visualization of near-wall turbulence structures. The flow is from left to right of the picture, 74 mm wide ($x^+ = 437$) and 43 mm high ($z^+ = 254$) (Pang and Choi 2004). Reproduced with permission from AIP Publishing.

Figure 10.9 shows the flow visualization of the turbulent boundary layer. It is indicated that the low-speed streaks are periodically twisted into the spanwise direction by the oscillating Lorentz force. The spanwise vortices may weaken the turbulence activities, which is beneficial for drag reduction.

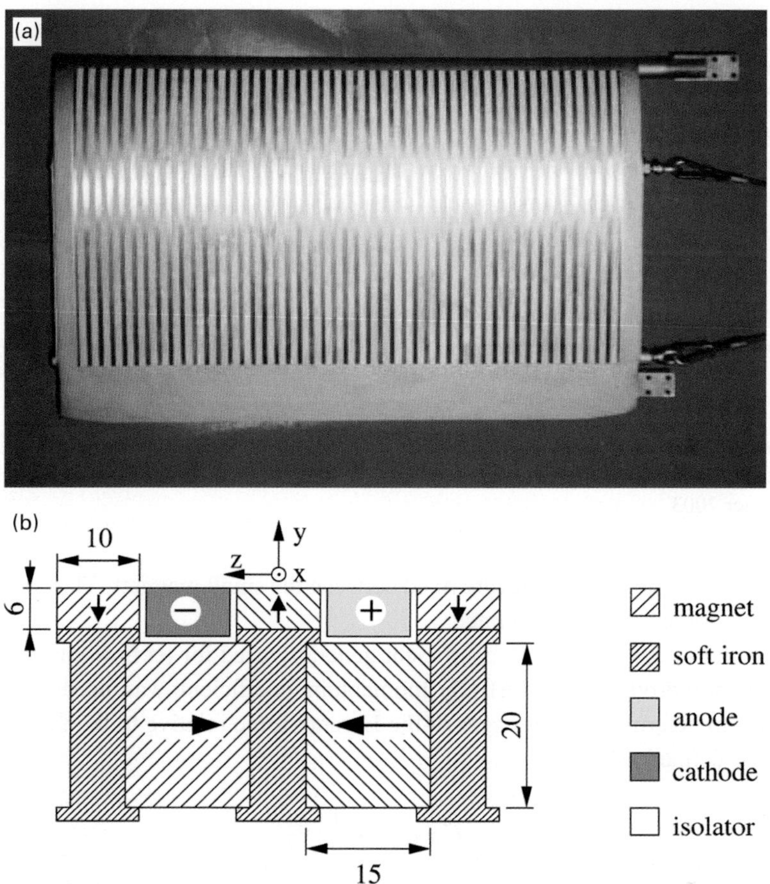

Figure 10.10 (a) NACA 0015 airfoil model used for Lorentz force control; and (b) sketch of the magnets and electrodes system used at the NACA 0015 airfoil (Weier et al. 2003). Reproduced with permission, copyright © Springer 2003.

10.3 Airfoil

10.3.1 Forcing over the Entire Suction Surface

The Lorentz force could be used to control the flow over an airfoil by attaching the device over the suction surfaces. One approach is to place the electrode-magnet system over the entire suction surface, which has been studied by Weier et al. (2003) and Chen et al. (2009). Weier et al. (2003) used a NACA 0015 airfoil with chord length of $c = 667$ mm and span of 1088 mm, as shown in Figure 10.10(a). Figure 10.10(b) presents a sketch of the magnets and electrodes system, where a soft H-iron is mounted between two Nd-Fe-B permanent magnets to concentrate the magnetic flux. Additional smaller magnets are glued on top of the H-iron to further increase the magnetic induction.

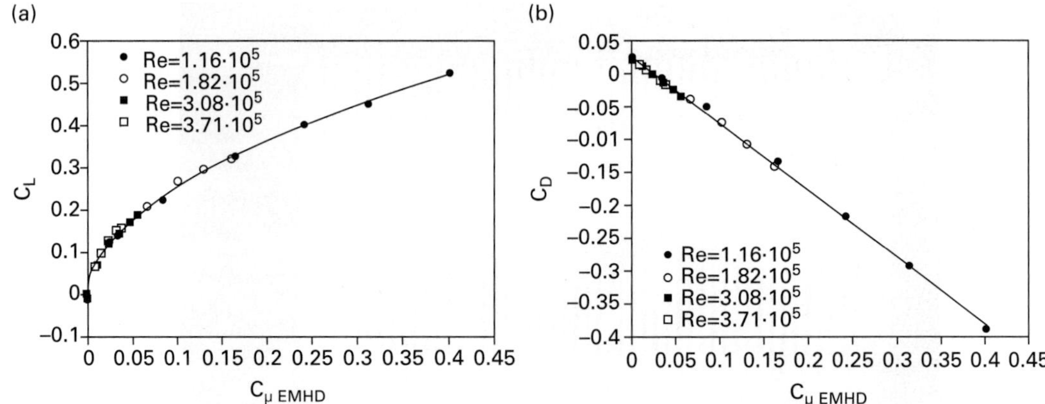

Figure 10.11 Lift (a) and drag (b) coefficient for the NACA 0015 versus $C_{\mu\text{EMHD}}$ at $\alpha = 0°$ and different Reynolds numbers (Weier et al. 2003). Reproduced with permission, copyright © Springer 2003.

The stainless steel electrodes are fixed between the small magnets. Thus, the streamwise Lorentz force could be generated.

Similar to the momentum coefficient generally used for jet flow control, the electro-magneto-hydrodynamic (EMHD) momentum coefficient was used by Weier et al. (2003) to determine the strength of the Lorentz force relative to the free stream. It can also be defined as

$$C_{\mu\text{EMHD}} = \frac{1}{2} \cdot \frac{wjB}{\rho U_\infty^2} \cdot \frac{l}{c}, \tag{10.9}$$

where w is the electrode width, j is the applied current density σE, and l is the fraction of the chord covered with electrodes and magnets.

Figure 10.11 shows the variations of the lift and drag coefficients with the momentum coefficient $C_{\mu\text{EMHD}}$ at $\alpha = 0°$. It is indicated that the lift coefficient increases with the momentum coefficient. A scaling of approximately $C_L \propto \sqrt{C_{\mu\text{EMHD}}}$ has been found, namely

$$C_L = 0.843 C_{\mu\text{EMHD}}^{0.521}. \tag{10.10}$$

Such a variation is very similar to that induced by circulation control. On the other hand, the Lorentz force could decrease the drag coefficient in a linear relationship to the momentum coefficient, namely

$$C_D = 0.0239 - 1.01 C_{\mu\text{EMHD}}. \tag{10.11}$$

In particular, the Lorentz force could generate thrust to the airfoil, exhibiting the appearance of negative drag.

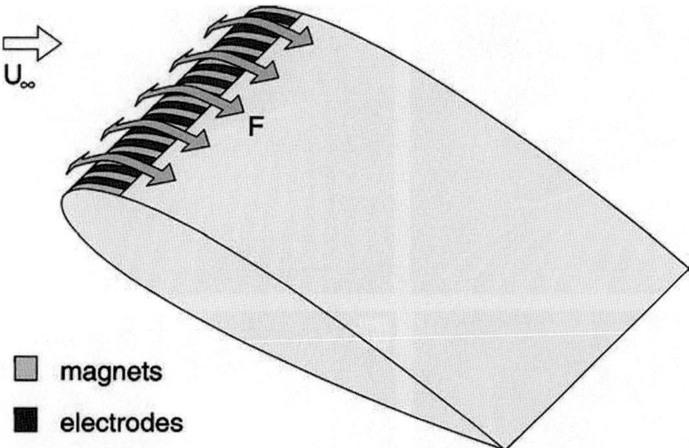

Figure 10.12 NACA 0015 airfoil model used for Lorentz force control near the leading edge (Weier and Gerbeth 2004). Reproduced with permission, copyright © 2004 Elsevier Masson. All rights reserved.

10.3.2 Forcing near the Leading Edge

Though the Lorentz force can be used over the entire suction surface of the airfoil, more recent works have indicated that the streamwise Lorentz force just positioned at a portion of the airfoil near the leading edge could also lead to considerable effects for the flow over inclined flat plates (Weier and Gerbeth 2004; Weier et al. 2008; Cierpka et al. 2008) and airfoils (Weier and Gerbeth 2004; Mutschke et al. 2006; Cierpka et al. 2010) even at post stall condition.

One work was conducted by Weier and Gerbeth (2004), with the experimental model shown in Figure 10.12, consisting of a NACA 0015 airfoil model with chord length $c = 160$ mm and span of 240 mm. The actuator extended from the leading edge to approximately 9% of the chord length. The experiments were carried out in an electrolyte tunnel at low Reynolds numbers of $10^4 \leq Re \leq 10^5$. They used a time-periodic Lorentz force for flow control, and thus two dimensionless parameters were used. One is the forcing Strouhal number based on the excitation frequency f_e, which is defined as

$$St_e = \frac{f_e c}{U_\infty}.$$ (10.12)

The other is the momentum coefficient similar to that used by Weier et al. (2003),

$$C_{\mu\mathrm{EMHD}} = \frac{1}{2} \cdot \frac{wB}{\rho U_\infty^2} \cdot \frac{l}{c} \cdot \sqrt{\frac{1}{T} \int_0^T j(t)^2 dt},$$ (10.13)

where $j(t)$ is the time-dependent applied current density and $T = 1/f_e$ is the forcing period.

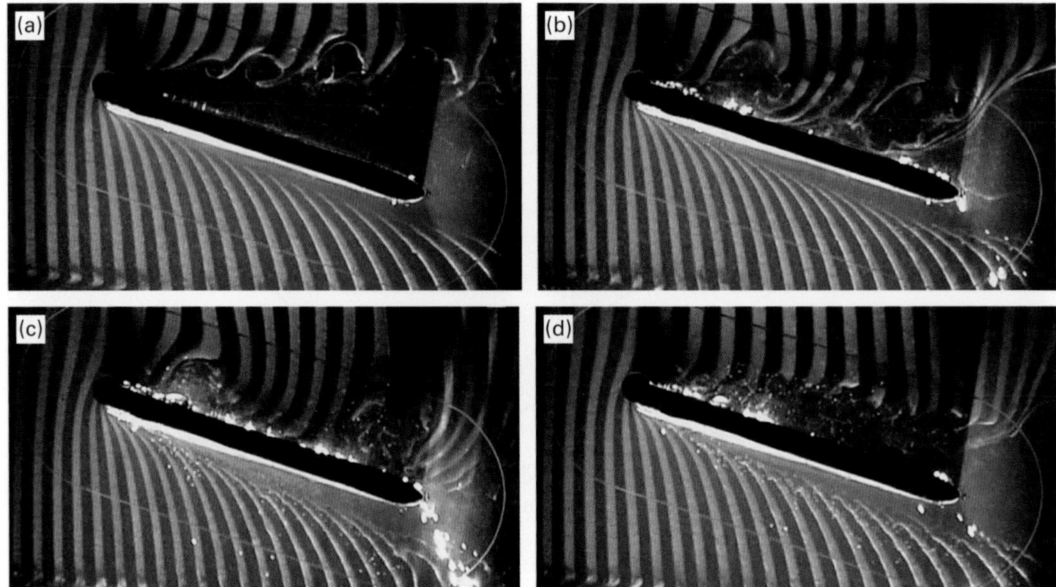

Figure 10.13 Flow visualization of a 15° inclined flat plate without and with Lorentz force at the first 8% chord at $Re = 1.4 \times 10^4$. (a) Natural case; (b) $C_{\mu\text{EMHD}} = 4.4\%$, $St_e = 1.4$; (c) $C_{\mu\text{EMHD}} = 8.9\%$, $St_e = 1.4$ and (d) $C_{\mu\text{EMHD}} = 8.9\%$, $St_e = 5.9$ (Weier and Gerbeth 2004). Reproduced with permission, copyright © 2004 Elsevier Masson. All rights reserved.

Weier and Gerbeth (2004) first tested the effects of the leading-edge Lorentz force on an inclined flat plate, as shown in Figure 10.13 by flow visualization. Flow has separated for the baseline case at $\alpha = 15°$, and one can see a series of Kelvin–Helmholtz vortices generating from the shear layer (Figure 10.13(a)). The Lorentz force at $C_{\mu\text{EMHD}} = 4.4\%$, $St_e = 1.4$ reattaches the flow to the plate in an averaged sense (Figure 10.13(b)), while further increasing the momentum coefficient to $C_{\mu\text{EMHD}} = 8.9\%$ could lead to a larger attached flow region (Figure 10.13(c)). However, when the momentum coefficient maintains $C_{\mu\text{EMHD}} = 8.9\%$ and the Strouhal number increases to $St_e = 5.9$ (Figure 10.13(d)), the flow over the plate is completely separated again, though the separation region is still smaller than that for the natural case. Besides, the discrete vortices are no longer detectable, indicating the effects of the unsteady forcing at high frequency.

Figure 10.14 shows the lift coefficient over the airfoil at different momentum coefficients. Before stall angle, the Lorentz force nearly has no influence on the lift coefficient, since the flow is already fully attached. An increase in the lift coefficient during the linear stage could occur when the circulation over the airfoil is increased, which can be achieved by arranging the Lorentz force over the entire suction surface. However, at post stall angles of attack, the Lorentz force could considerably increase the lift coefficient, while the stall angle is also greatly delayed. The control effects increase with the momentum coefficient.

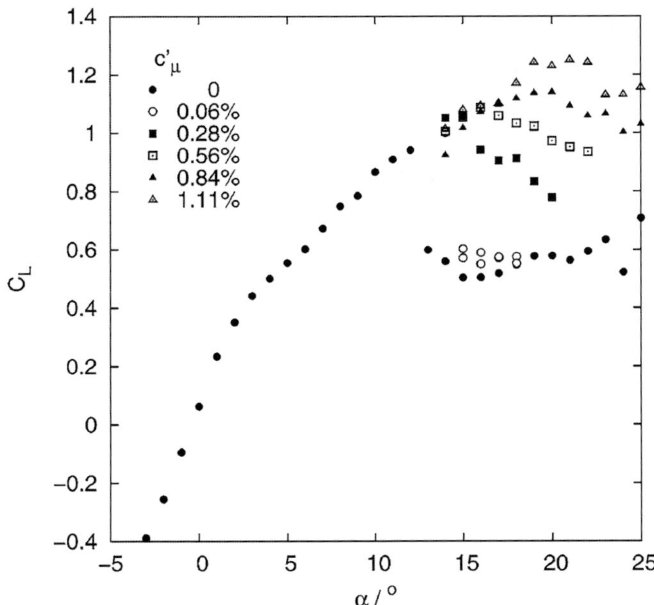

Figure 10.14 Lift coefficient of the NACA 0015 airfoil at $Re = 5.2 \times 10^4$, $St_e = 0.5$ for different momentum coefficients (c, d) (Weier and Gerbeth 2004). Reproduced with permission, copyright © 2004 Elsevier Masson. All rights reserved.

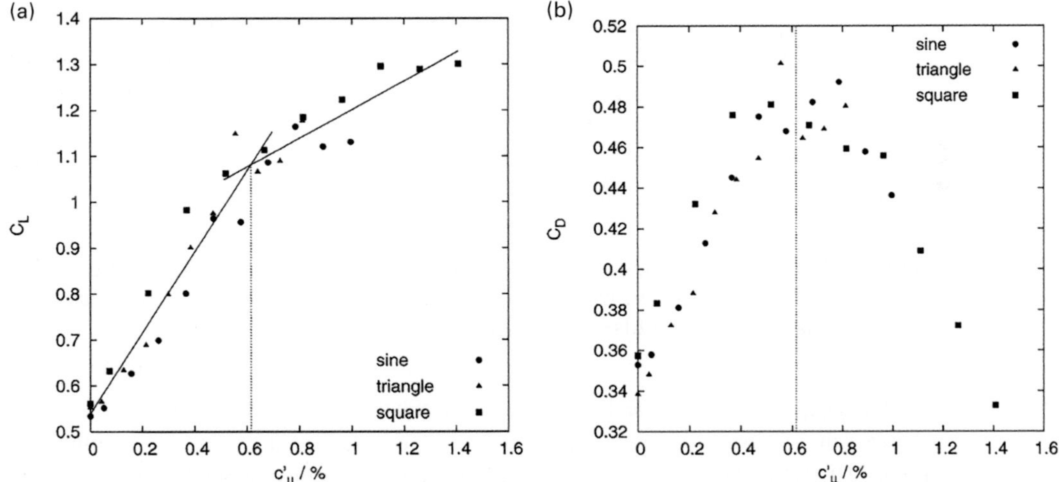

Figure 10.15 Lift coefficient (a) and drag coefficient (b) versus momentum coefficient at $Re = 5.2 \times 10^4$, $\alpha = 20°$, $St_e = 0.5$ (Weier and Gerbeth 2004). Reproduced with permission, copyright © 2004 Elsevier Masson. All rights reserved.

The influence of the momentum coefficient on the lift and drag coefficients is shown in Figure 10.15. The lift coefficient (Figure 10.15(a)) increases with the momentum coefficient. Two straight lines with different slopes can fit the data, which intersect at

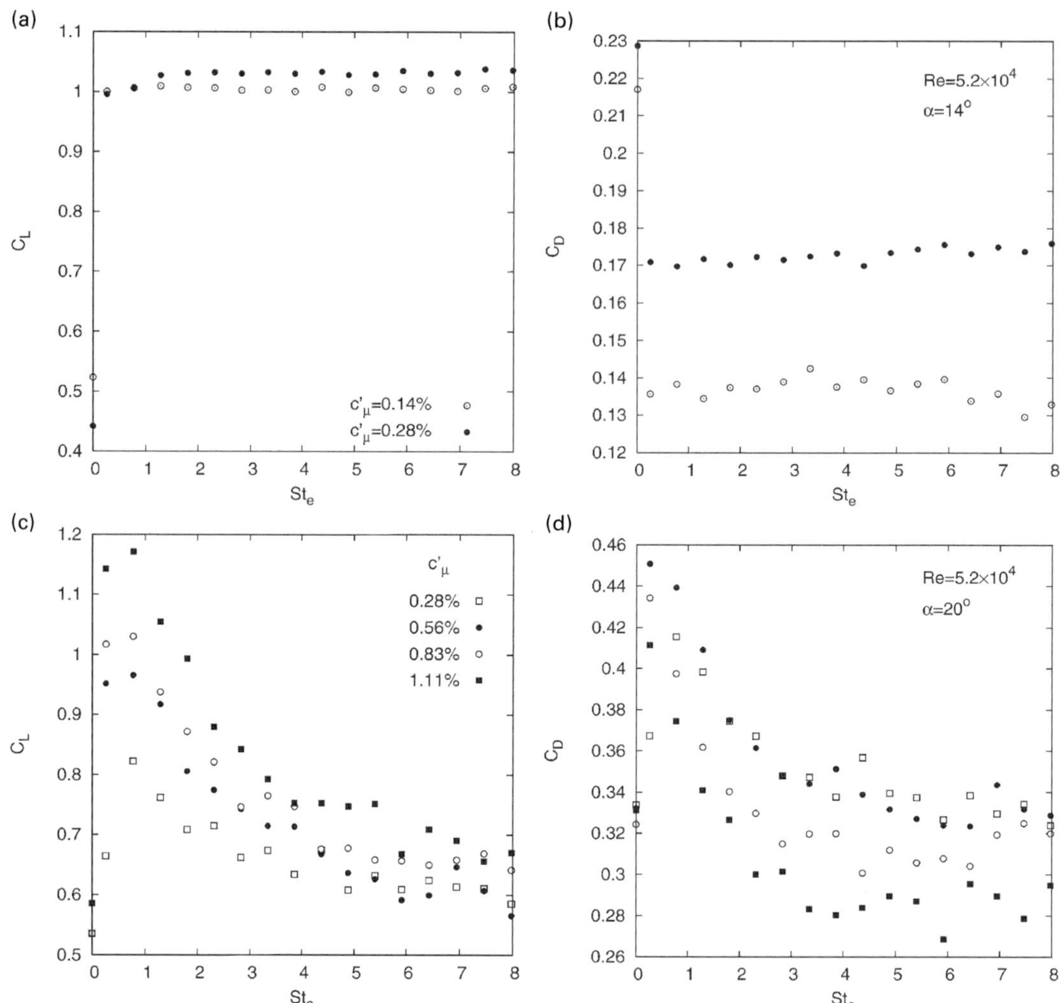

Figure 10.16 Lift coefficient (a, c) and drag coefficient (b, d) versus excitation frequency at $Re = 5.2 \times 10^4$, $\alpha = 14°$ (a, b) and $\alpha = 20°$ (Weier and Gerbeth 2004). Reproduced with permission, copyright © 2004 Elsevier Masson. All rights reserved.

about $C_{\mu\text{EMHD}} = 0.61\%$. The initial faster lift enhancement is coupled to a drag rise, whereas the following slower enlargement is accompanied by a decreasing drag (Figure 10.15(b)). The maximum drag coefficient also occurs at about $C_{\mu\text{EMHD}} = 0.61\%$. Here, the results presented are mainly by using sinusoidal forcing. However, the Lorentz force can also be driven by some other waveforms, such as triangular and square waveforms (such as Weier et al. 2008; Cierpka et al. 2010). Similar effects can also be achieved by these waveforms, which can also be seen in Figure 10.15.

The influence of the excitation frequency on the aerodynamic force is strongly dependent on the flow state, as shown in Figure 10.16. At $\alpha = 14°$ with slight stall, namely 1° after stall, the lift and drag coefficient are nearly constant with the variation of

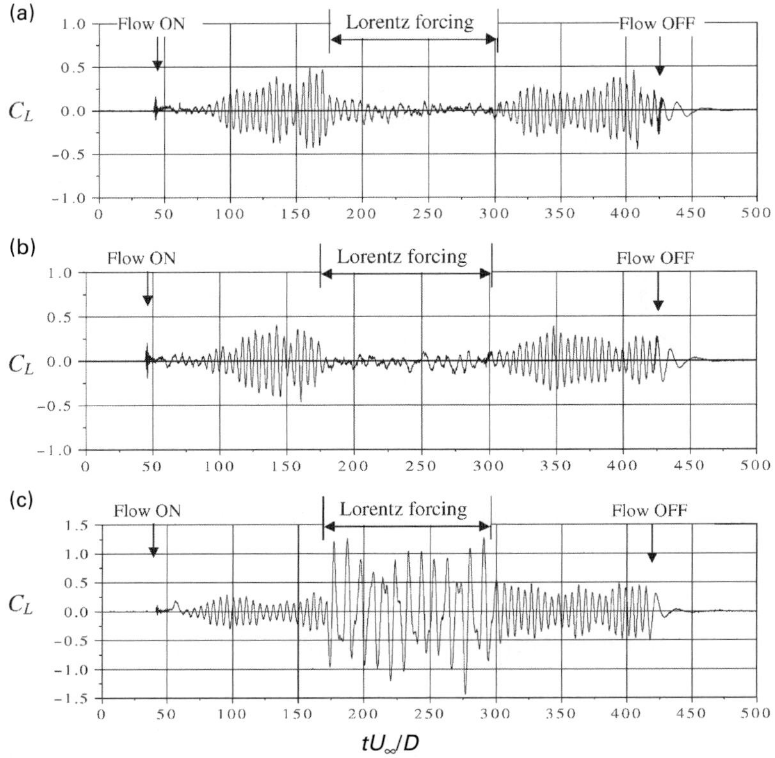

tU_∞/D

Figure 10.17 Time history of lift force under Lorentz force at $N = 2.4$, $Re = 5 \times 10^3$. (a) Continuous positive Lorentz force, (b) continuous negative Lorentz force, and (c) rotary oscillating Lorentz force where the excitation frequency is half of the natural frequency (Kim and Lee 2001). Reproduced with permission, copyright © 2001 The Japan Society of Fluid Mechanics and IOP Publishing Ltd.

excitation frequency. However, at deep post stall angles of attack, such as $\alpha = 20°$, the lift coefficient increases to a maximum at around $St_e = 0.75$, while the drag coefficient increases to a maximum at about $0.25 \le St_e \le 0.75$, then both of them decrease with an increase in the excitation frequency.

10.4 Bluff Body

10.4.1 Stationary Cylinder

The Lorentz force has also been used by many researchers to control the flow around a circular cylinder. One representative work was conducted by Kim and Lee (2001). The electrodes and the permanent magnets were mounted on the cylinder surface side by side in the order of positive electrode, N-pole magnet, negative electrode, S-pole magnet. This sequence was repeated in the spanwise direction. The forcing covered an angular region between 70° and 130°from the front stagnation point for both the upper

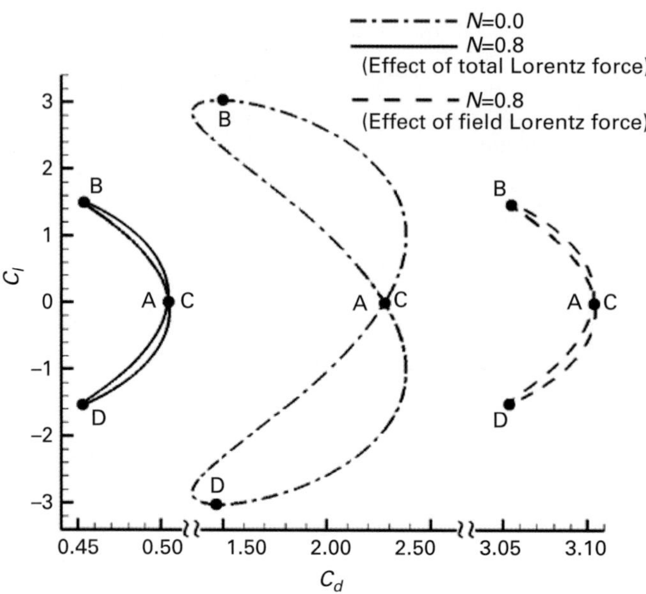

Figure 10.18 The C_L–C_D phase diagram for flexible supported cylinder with Lorentz force control (Zhang et al. 2011). Reproduced with permission, copyright © Springer 2011.

and lower sides of the cylinder. Three types of Lorentz forces were applied. The first two were called continuous positive and negative Lorentz forces, where both upper and lower Lorentz forces were parallel to the flow or opposite to the flow, respectively. The third mode was an oscillatory rotating Lorentz force in an ordered frequency, namely a pulsed positive Lorentz force. For all cases presented here, the Lorentz force was kept at $jBD/\rho U^2_\infty = 2.4$.

The time history of the lift coefficient is shown in Figure 10.17. The flow is first turned on for a certain time, then the Lorentz force is implemented, and finally it is switched off and the flow over the circular cylinder without control continues for a certain time. It is indicated that the continuous positive (Figure 10.17(a)) and negative (Figure 10.17(b)) Lorentz forces could substantially reduce the fluctuations of the lift coefficient, while the rotary oscillating mode will greatly increase the lift fluctuations (Figure 10.17(c)). The flow visualization suggested that the positive Lorentz force could suppress the periodic vortex shedding, while the negative Lorentz force could increase the width of the wake. Both of these modifications could lead to a weakened mutual interaction between upper and lower wake vortices, which is beneficial for the suppression of lift fluctuations.

10.4.2   Flexible Supported Cylinder

Since the Lorentz force can reduce the lift fluctuations, it could also be an effective approach to suppress the vortex-induced vibration, such as in the work conducted by Zhang et al. (2011). The numerical simulation was performed at $Re = 150$. They used the wall Lorentz force and the field Lorentz force for flow control. The wall

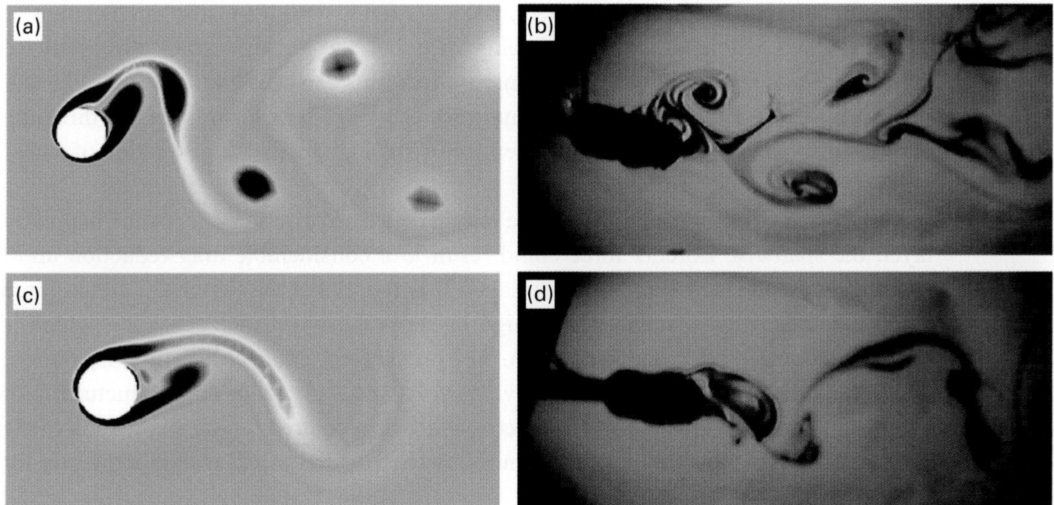

Figure 10.19 Snapshot of spanwise vorticity (a, c) and flow visualization (b, d) for the natural (a, b) and Lorentz force control (c, d) cases at phase A shown in Figure 10.18 (Zhang et al. 2011). Reproduced with permission, copyright © Springer 2011.

Lorentz force is the same as that introduced above, which acts on the cylinder surface. The field Lorentz force influences the flow field in the boundary layer and thus changes the aerodynamic forces on the cylinder. The Lorentz force was fixed at $0.5jBD/\rho U^2_\infty = 0.8$.

The C_L–C_D phase diagram to show the vortex-induced vibration of a circular cylinder is presented in Figure 10.18. In comparison with the vortex-induced vibration diagram for the natural case, the field Lorentz force makes the phase diagram shrink and move to the right, indicating that the lift fluctuation is reduced while the drag is increased; while the wall Lorentz force causes the diagram move to the left, indicating a decrease in both the lift fluctuation and the drag coefficient. For all cases, the decrease in the lift fluctuation suggests a reduction in the vortex-induced vibration.

Snapshots of vortex dynamic for both the natural and control cases are shown in Figure 10.19. For the natural case, large-scale wake vortices can be observed, which are shedding downstream alternately. With Lorentz force, the scale of the wake vortices is decreased, and their transverse dynamic range is greatly reduced. The variations in the vortex dynamics can interpret the reduction in the vortex-induced vibration with Lorentz force.

10.5 Concluding Remarks

In this chapter, we introduce the main characteristics of the Lorentz force and its applications in the boundary layer and the flow over airfoils and cylinders. The Lorentz force is a novel active flow control device originating from electromagetism. It can be generated in electric fluids with a specific arrangement of electrodes and magnets, with the forcing in either the streamwise or the spanwise direction. Sinusoidal, triangular, and square waveforms can be used to drive the electrodes and magnets device, thus the Lorentz force can change its magnitude and direction periodically. The boundary layer can therefore be changed with the generation of a near wall jet.

Though the streamwise Lorentz force may increase the friction drag of the boundary layer, the spanwise Lorentz force could result in a considerable drag reduction up to about 40% for the turbulent boundary layer. It is found that the periodic Lorentz force could induce spanwise vortices that might weaken the turbulence activities, which is beneficial for drag reduction. In addition, the Lorentz force could reduce flow separation, thus it can increase the lift coefficient over airfoils and also reduce the lift fluctuation of a circular cylinder. Considering that electric fluids are required for the application of the Lorentz force, it is suitable for flow control of sea vehicles, such as drag reduction for submarines.

References

Berger, T. W., Kim, J., Lee, C, Lim, J. Turbulent boundary layer control utilizing the Lorentz force. *Physics of Fluids*, 2000, 12(3): 631–649

Breuer, K. S., Park, J., and Henoch, C. Actuation and control of a turbulent channel flow using Lorentz forces. *Physics of Fluids*, 2004, 16(4): 897–907

Chen, Y. H., Fan, B. C., Chen, Z. H., and Li, H. Z. Flow pattern and lift evolution of hydrofoil with control of electro-magnetic forces. *Science in China Series G: Physics, Mechanics and Astronomy*, 2009, 52(9): 1364–1374

Choi, K. S., DeBisschop, J. R., and Clayton, B. R. Turbulent boundarylayer control by means of spanwise-wall oscillation. *AIAA Journal*, 1998, 36(7): 1157–1163

Cierpka, C., Weier, T., and Gerbeth, G. Evolution of vortex structures in an electromagnetically excited separated flow. *Experiments in Fluids*, 2008, 45(5): 943–953

Cierpka, C., Weier, T., and Gerbeth, G. Synchronized force and particle image velocimetry measurements on a NACA 0015 in poststall under control of time periodic electromagnetic forcing. *Physics of Fluids*, 2010, 22(7): 075109

Crawford, C. H. and Karniadakis, G. E. Reynolds stress analysis of EMHD-controlled wall turbulence. Part I. *Physics of Fluids*, 1997, 9(3): 788–806

Henoch, C. and Stace, J. Experimental investigation of a salt water turbulent boundary layer modified by an applied streamwise magnethydrodynamic body force. *Physics of Fluids*, 1995, 7 (6): 1371–1383

Kim, S. J. and Lee, C. M. Control of flows around a circular cylinder: suppression of oscillatory lift force. *Fluid Dynamics Research*, 2001, 29(1): 47–63

Mutschke, G., Gerbetha, G., Albrecht, T., and Grundmann, R. Separation control at hydrofoils using Lorentz forces. *European Journal of Mechanics-B/Fluids*, 2006, 25(2): 137–152

Pang, J. and Choi, K. S. Turbulent drag reduction by Lorentz force oscillation. *Physics of Fluids*, 2004, 16(5): L35-L38

Weier, T., Cierpka, C., and Gerbeth, G. Coherent structure eduction from PIV data of an electromagnetically forced separated flow. *Journal of Fluids and Structures*, 2008, 24(8): 1339–1348

Weier, T., Gerbeth, G., Mutschke, G., Lielausis, O., and Lammers, G. Control of flow separation using electromagnetic forces. *Flow, Turbulence and Combustion*, 2003, 71(1–4): 5–17

Weier, T. and Gerbeth, G. Control of separated flows by time periodic Lorentz forces. *European Journal of Mechanics-B/Fluids*, 2004, 23(6): 835–849

Zhang, H., Fan, B. C., and Li, H. Z. Suppression of vortex-induced vibration of a circular cylinder by Lorentz force. *Science China Physics, Mechanics and Astronomy*, 2011, 54(12): 2248–2259

11 Closed-Loop Control

11.1 Background

As was explained in the Introduction, flow control techniques can be classified into passive and active based on energy input. Passive flow control techniques need no energy and are easy to implement, but cannot be changed during the control process. Active techniques need extra energy, can be adjusted during the control process, and thus can implement real-time and unsteady flow control. So the control effectiveness and efficiency for active techniques are usually more significant than for passive ones.

Another unique advantage of active control techniques is that they can be used with closed-loop control. In this case, the control parameters depend on information from the control system which in turn depends on the control. A schematic of the closed-loop control system is shown in Figure 11.1. A control system includes fluid system, actuator, sensor, and control algorithm. The fluid system is the flow field to be controlled. The actuator refers to the active flow control techniques, which could introduce excitation to the original flow. The sensor measures variables from the controlled flow, which are used as a reference to adjust the actuator parameters. The control algorithm is used to guide the operation of the actuator. Such a control approach is also referred to as feedback control. It is different from feedforward control, where the sensor signal is based on the oncoming flow not the controlled flow.

The control algorithm is the essential factor for flow control. The control is usually based on the signal from the sensor. There are two approaches, namely the direct method and the indirect method. The direct method is directly based on the measured signal by the sensor, such as velocity, pressure, and force information. The threshold for the closed-loop control is set at some value based on an analysis of the signal, such as the amplitude and the power spectrum. However, the indirect method also deals with the signal with a reduced-order model. Then the threshold for the controller is based on reduced-order information. One of the most frequently used reduced-order model methods is proper orthogonal decomposition (POD), which can decompose the original flow into low-dimensional modes based on the energy contributions. Joslin and Miller (2009) have presented a detailed introduction to closed-loop control.

In general, the closed-loop is directly based on the measured signal which is taken from one point, such that the signal should be well represented for the global flow field. This approach is suitable for experimental studies. On the other hand, closed-loop based on

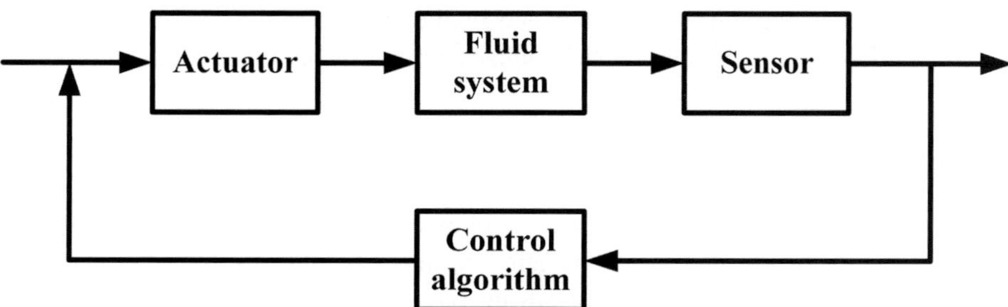

Figure 11.1 Schematic of the closed-loop control algorithm.

a reduced-order model could consider global variations in the flow field, but the major problem of this approach is the accuracy of the reduced-order model in representing the true controlled flow. It is suitable for numerical study of flow control since the control algorithm can be coupled into the numerical code, but it is difficult for experimental study.

In this chapter, the main ideas of closed-loop control will be first introduced, and then some examples of closed-loop control based on the reduced-order model and the measured variables will be presented.

11.2   Closed-Loop Based on Reduced-Order Model

One example presented here is by Siegel et al. (2006), who controlled a cylinder with cross-stream oscillation at $Re = 100$ by direct numerical simulation. A limited number of sensors placed in the wake were used to estimate the state of the flow that was characterized using a reduced-order model based on POD. The fluctuating velocity component $u'(x, y, t)$ of the streamwise velocity $u(x, y, t)$ measured by the sensors was used for POD decomposition. The decomposition of the velocity field is,

$$u(x, y, t) = U(x, y) + u'(x, y, t) \tag{11.1}$$

where U is the mean flow velocity. Then, the fluctuating component could be expanded by POD as

$$u'(x,y,t) = \sum_{k=1}^{n} a_k(t)\phi_i^k(x,y) \tag{11.2}$$

where $a_k(t)$ is the time-dependent POD coefficients and $\phi_i^{(k)}(x, y)$ are the nondimensional spatial eigenfunctions.

Usually, the first few modes represent the large-scale vortical structures that dominate the main features of the global flow. The POD coefficients also vary periodically to reflect the periodic shedding process of the Kármán vortex. The variation of the first few POD coefficients closely depends on the strength of the vortex shedding. For example, the amplitude of the POD coefficient is related to that of the lift coefficient. On the other hand, it was found by Siegel et al. (2006) that a smaller amplitude for the first few POD modes corresponds to the case with smaller drag coefficient.

Thus, the control algorithm acts on the estimate of the amplitude of mode 1. A closed-loop control strategy for the displacement of the cylinder could be proposed as

$$Y_{cy} = K_p \cdot a_1 + K_d \cdot \frac{da_1}{dt},$$ (11.3)

where y_{cy} denotes the vertical amplitude imposed on the cylinder, a_1 is the first POD coefficient, K_p and K_d are the gains. K_p and K_d could be specified or expressed in terms of an overall gain K and a phase advance φ

$$K_p = K \cdot \cos(\varphi), \ K_d = K \cdot \sin(\varphi)/2\pi f_0,$$ (11.4)

where f_0 is the natural frequency for Kármán vortex shedding.

Siegel et al. (2006) further proposed a variable-gain strategy, where the flow field was modified from its original state. Since mode 5 was found to be a good measure of the change in the mean flow, the feedback gain K and phase advance φ_0 could be adjusted from the initial values K_0 and φ_0 in proportion to the change in mode 5 by applying a phase-advance factor K_φ and a gain-change factor K_A, namely

$$K_d = K_0 + K_A(a_5 - a_5|_{t=0}), \varphi = \varphi_0 + K_\varphi(a_5 - a_5|_{t=0}),$$ (11.5)

where a_5 is the fifth POD coefficient.

Siegel et al. (2006) turned on the controller at 3.03 s and off at 8 s. It was indicated that the maximum amplitude for modes 1 to 4 decreased gradually while that for mode 5 increased gradually until a time of about 6 s, and then the variations showed a stabilized periodic behavior. Similar variations could also be found for the lift and drag coefficients and the cylinder displacement. In the final stabilized stage, the actuation amplitude was maintained at about 5%D, while the cylinder motion developed a limit cycle with mode 1 amplitude. Meanwhile, the vortex shedding frequency was reduced by about 23%, the drag coefficient was reduced by about 15%, and the unsteady lift amplitude was reduced by about 90%, in comparison with the no control case.

In addition, there have been numerous works about closed-loop control based on the reduced-order model. The fields include the flow around a rotary circular cylinder (Bergmann et al. 2005), control of flow around a square cylinder by jets (Weller et al. 2009), control of flow around an airfoil by synthetic jets (Pinier et al. 2007), and flow around a pitching and plunging flat plate (Brunton et al. 2014), etc. It has been found that closed-loop control has produced favorable control results.

11.3   Closed-Loop Based on Measured Variables

11.3.1   Based on Pressure Variation

There are different variables, such as pressure and force, measured from the flow field that could be used as the feedback signal. The pressure is measured from one single point or different positions, while the force is an integral result. Thus, the pressure signal could be more sensitive to the flow separation if the measured position is well chosen. For the unsteady flow separation, the pressure coefficient shows dynamic variation with time, providing us with a way to forecast flow separation, such as in the studies by Pastoor et al. (2008), Tewes et al. (2011), and Benard et al. (2011). Here, we present Benard et al.'s (2011) work as an example to introduce the closed-loop control process.

Benard et al. (2011) conducted closed-loop control of an NACA 0015 airfoil by using a plasma actuator. The airfoil model had a chord length of $c = 200$ mm and a span of 400 mm. The plasma actuator was placed at the leading edge. The width of the exposed electrode and the embedded electrode was 10 mm and 40 mm, respectively, and there was a gap of 5 mm. The applied voltage amplitude was in the range of a few kV to 18 kV and the driven frequency was 1000 Hz. The pressure was measured at a sampling frequency of 1000 Hz by six pressure sensors located at $x/c = 0.22$, 0.3, 0.4, 0.5, 0.6, and 0.7 on the suction side of the model and the force was measured by a force balance. The free-stream velocity was fixed at $U_\infty = 15$ m/s, corresponding to $Re = 1.9 \times 10^5$.

Figure 11.2 shows variation of lift and pressure coefficients with the applied voltage of the plasma actuator at $\alpha = 13°$. The voltage amplitude to the exposed electrode first continuously increases from 6 to 18 kV, in steps of 1 kV for 20 s duration, and then subsequently decreases. The lift coefficient increases with the voltage until 11 kV, when the flow is reattached. Then, a further increase in the applied voltage does not improve the lift coefficient. The flow remains attached even as the voltage is subsequently reduced until 7 kV. It was found by Benard et al. (2011) that the plasma discharge was even not clearly formed between 7 and 8 kV. The hysteresis effects of the plasma forcing are clearly shown, which can also be concluded from the variations of the pressure coefficient. This phenomenon allows the optimization of the plasma control such that control is maintained while the actuator voltage could be reduced.

Thus, closed-loop control could be achieved based on the sensor of dynamic variation of the pressure coefficient. Typical variations of the pressure coefficient with plasma on and off are shown in Figure 11.3(a) and (b), respectively. For both actuations, the flow

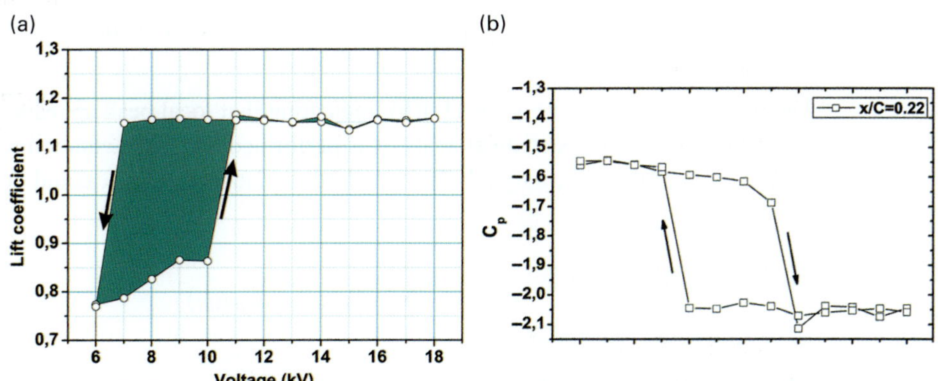

Figure 11.2 Lift coefficient (a) and pressure coefficient (b) for increasing and then decreasing voltage amplitude at $\alpha = 13°$ (Benard et al. 2011). Reproduced with permission from AIP Publishing.

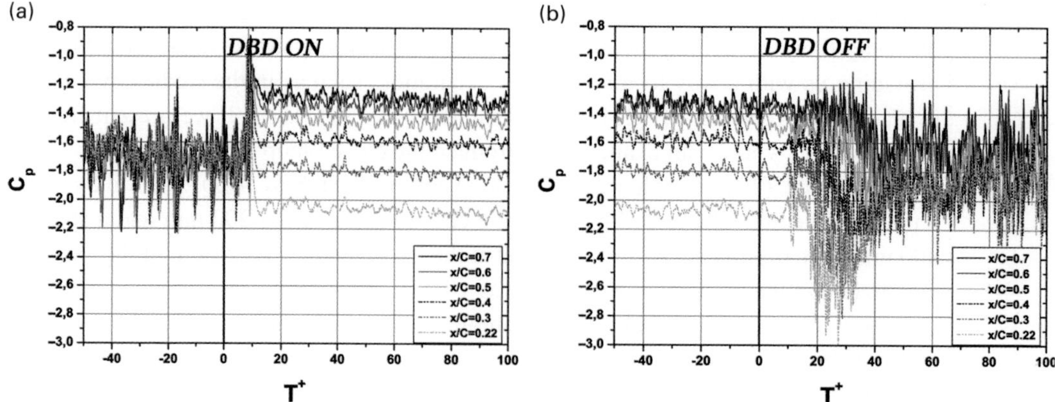

Figure 11.3 Time evolution of the pressure coefficient for (a) forced reattachment (12 kV) and (b) natural separation at $\alpha = 13°$ (Benard et al. 2011). Reproduced with permission from AIP Publishing.

takes some time to become attached or separated, which can be detected from the value of the pressure coefficient. Here, the pressure coefficient at $x/c = 0.22$ was used as the feedback information and the value of $C_{pref} = -1.6$ was selected as the threshold to identify the states of flow separation or attachment. The flow is attached when $C_p < -1.6$, otherwise it is separated. The duration for flow attachment is shorter than that for separation. Thus, it is possible to implement closed-loop control to attach the flow before the occurrence of flow separation.

A second approach was based on the further data processing of the pressure signal. A typical pressure signal is shown in Figure 11.4(a), where the plasma is turned off at time 0, and the flow separates at about $t^+ = tU_\infty/c = 215$. First, the sliding RMS of the pressure coefficient could be calculated based on a sliding average $\overline{C_p(t^+)}$, computed over a normalized time interval $\Delta t^+ = 0$, namely

$$C_{pRMS}(t^+) = \sqrt{\frac{1}{\Delta t^+} \int_{t^+ - \Delta t^+}^{t^+} [C_p(t^+) - \overline{C_p(t^+)}]^2 dt^+} \qquad (11.6)$$

The results are shown in Figure 11.4(b), indicating that local peak appears before flow separation. Secondary, statistical analysis is performed on a segmentation with interval of $\Delta t^+ = 7.5$ of the C_{pRMS} signal. The standard deviation of C_{pRMS} could be defined as $\sqrt{\text{var}}$ with

$$Var = \frac{1}{N} \sum_{i=1}^{N} (C_{pRMS_i} - \overline{C}_{pRMS})^2 \qquad (11.7)$$

where $N = 100$ is the number of C_{pRMS} values in the $\Delta t^+ = 7.5$ interval, $\overline{C_{pRMS}}$ is the mean value of C_{pRMS} over the considered time segmentation. Then, the coefficient of variation (CV), which is a normalized quantity defined as the ratio of the standard deviation to the mean value, could be calculated by

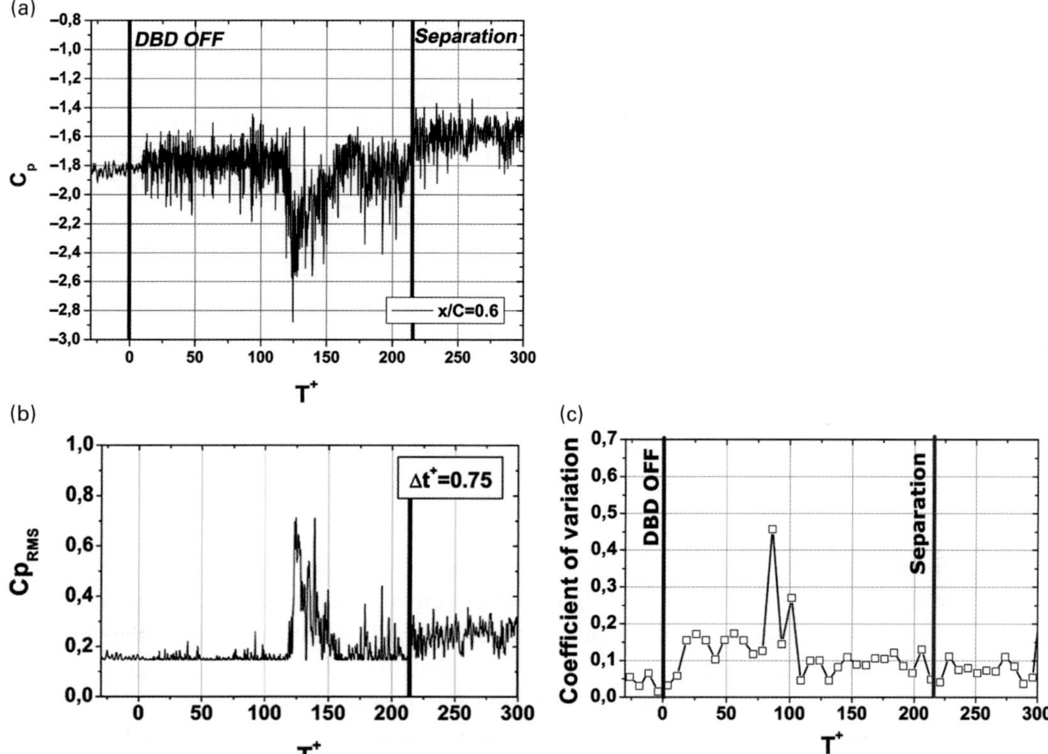

Figure 11.4 (a) Time evolution of the pressure coefficient for natural separation. (b) Sliding RMS C_{pRMS} of the pressure coefficient. (c) Coefficient of variation (CV) of the pressure coefficient (Benard et al. 2011). Reproduced with permission from AIP Publishing.

$$CV = \frac{\sqrt{\text{var}}}{C_{pRMS}}. \tag{11.8}$$

The results are presented in Figure 11.4(c). A local peak also appears before flow separation, but the value has been normalized in the range of 0 to 1. The predefined threshold for the coefficient of variation is fixed at 0.2 in the subsequent design. It is a more robust criterion which can withstand limited influence from the flow condition and the measured pressure location.

Accordingly, the control algorithm can be designed, as shown in Figure 11.5. Two closed-loop controllers are tested, one with Cp threshold and the other with CV threshold. When no flow separation is detected, the applied voltage amplitude is reduced incrementally after a prescribed plateau duration $T_{plateau}$. When the flow separation is detected, the voltage amplitude is autonomously set to the maximum voltage of 18 kV to reattach the flow.

The results of closed-loop control with Cp threshold are shown in Figure 11.6. The control process follows that designed in Figure 11.5(a). However, the lift coefficient may also decrease suddenly when the flow separation is detected,

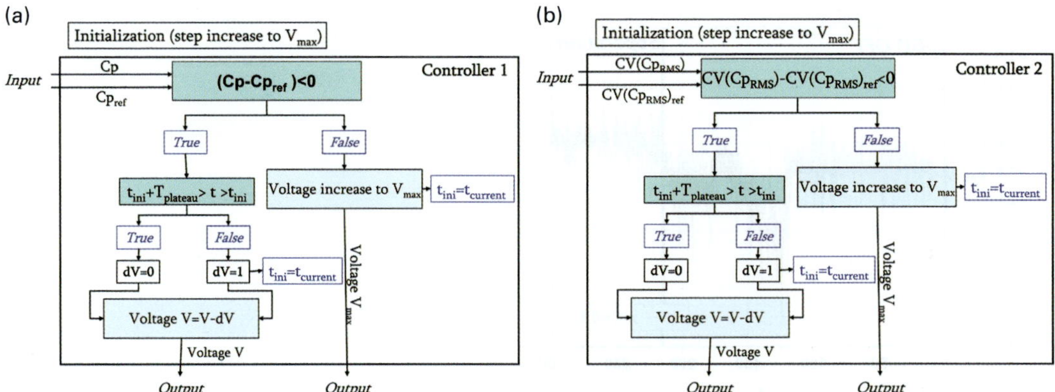

Figure 11.5 Schematics of two closed-loop controllers. (a) Closed-loop control with Cp threshold; and (b) closed-loop control with coefficient of variation (CV) threshold (Benard et al. 2011). Reproduced with permission from AIP Publishing.

indicating that the algorithm fails to maintain an attached flow in a robust manner. When flow stall occurs, the algorithm attempts to eliminate flow separation by immediately increasing the input voltage to 18 kV, which is sufficient to recover a high lift coefficient.

The results of closed-loop control with CV threshold are shown in Figure 11.7. The algorithm detects that the flow is about to separate when the CV is above 0.2, and thus increases the applied voltage to the maximum value of 18 kV. In comparison, there might be no flow separation detected from the pressure signal. Thus, the lift coefficient mostly stays at high values, while no actual stall occurs. It is indicated that closed-loop control with CV threshold is more sensitive than that with Cp threshold.

It is also indicated that when the applied voltage is decreased to the minimum value, such as 6 kV, the flow is easy to separate. Thus, the minimum voltage amplitude can be adjusted manually or autonomously. For example, when flow separation is detected, the controller increases the voltage to the maximum voltage, but now the minimum voltage is automatically increased by 1 kV. Such a control strategy makes the closed-loop algorithm more robust.

11.3.2   Based on Dominant Frequency

Corke and his co-workers conducted studies on the closed-loop control of a NACA 0015 airfoil base on the spectral analysis of the pressure signal (such as Patel et al. 2007, Corke et al. 2011). The plasma actuator was also placed at the leading edge of the airfoil. A static pressure port with response frequency of 500 Hz was located at $x/c = 0.05$ on the suction side of the airfoil. Force balance was used to measure the lift and drag coefficients.

Typical flow visualization over the airfoil without and with plasma control at $\alpha = 16°$ is shown in Figure 11.8. For the no control case, the boundary layer separates from the leading

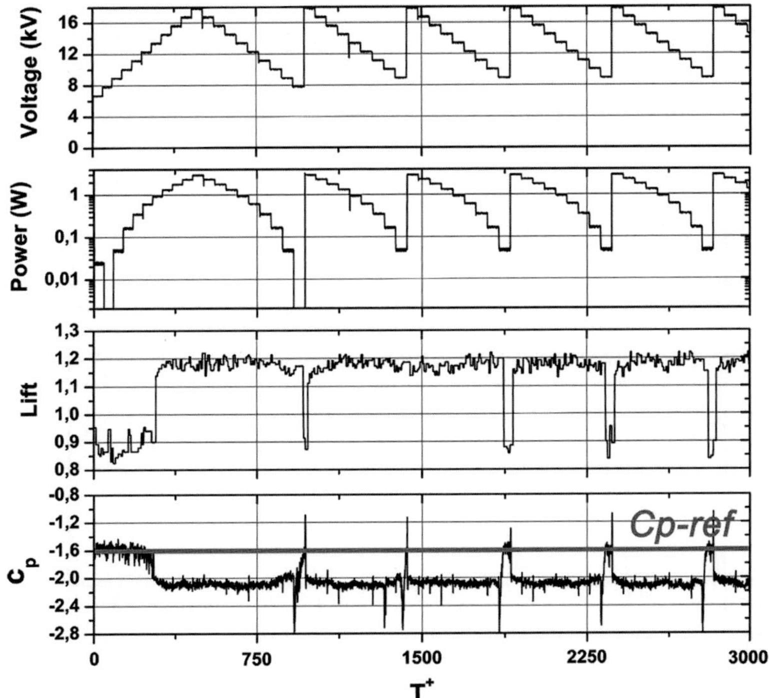

Figure 11.6 Results of the closed-loop control with *Cp* threshold for voltage amplitude steps of 1 kV and plateau duration of $T_{plateau}$ = 45 (Benard et al. 2011). Reproduced with permission from AIP Publishing.

edge and does not reattach to the airfoil. With plasma control, the boundary layer is forced to attach.

Figure 11.9 shows the power spectral density of pressure measured at x/c =0.05 for two angles of attack at 6° and 14°. The plasma actuator is turned on and off at 80 Hz. At α = 6°, the pressure spectra without and with plasma control are indistinguishable from each other, and no evident peak is detected. It is indicated that disturbances introduced by the plasma actuator at the leading edge are not sensed from the pressure fluctuations at x/c = 0.05. However, when the angle of attack increases to 14°, one degree before stall, spectral peaks at the plasma excitation frequency and its harmonics are both clearly observed.

The difference in spectral amplitude between the two angles, which are respectively related to the flow with full attachment and near separation, provides us with an approach to build the closed-loop control algorithm, as shown in Figure 11.10. The leading-edge plasma actuator works at a normalized excitation frequency of $St = f_e c/U_\infty$ = 1. Then, a spectral analysis is performed on the time series of the pressure to determine if there is a spectral peak at St =1. If it is detected, the flow is separated or about to separate, and therefore the plasma actuator voltage is increased to a higher level to reattach the flow or prevent it from separating. If it is not detected, the flow is already attached, and therefore the plasma voltage is maintained at the current level.

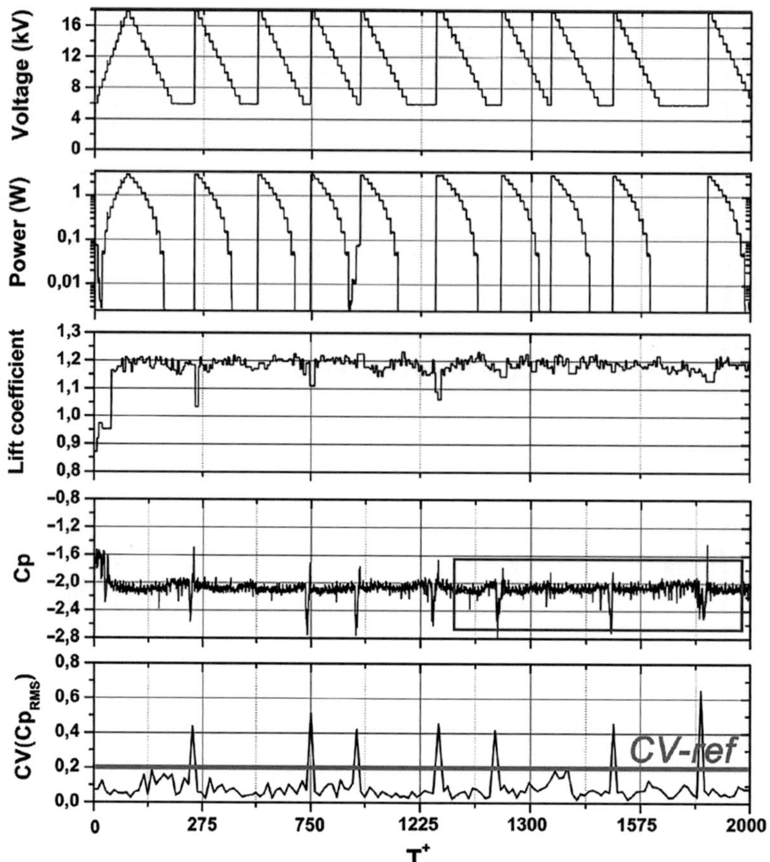

Figure 11.7 Results of closed-loop control with CV threshold for voltage amplitude steps of 1 kV and plateau duration of $T_{plateau} = 15$ (Benard et al. 2011). Reproduced with permission from AIP Publishing.

The mean lift coefficient as a result of the closed-loop control is shown in Figure 11.11. The airfoil stalls at $\alpha = 15°$, and the lift coefficient drops suddenly. With closed-loop control, no flow separation is detected before $\alpha = 14°$, thus the lift coefficient is very similar to the natural case. After that, imminent leading-edge separation is detected and the plasma actuator voltage is increased to the control level. Thus, the lift coefficient is increased by plasma control, with the maximum lift coefficient increased by about 9% and the stall angles postponed to $\alpha = 17°$. Thus, the control effectiveness of the proposed closed-loop control algorithm is validated.

11.3.3   Based on Spectrum Amplitude

In addition to using the dominant frequency from the power spectrum, the pressure amplitude was also used by Patel et al. (2003, 2007) for closed-loop control. This method firstly calculated the power amplitude of the pressure signal distributed with frequency.

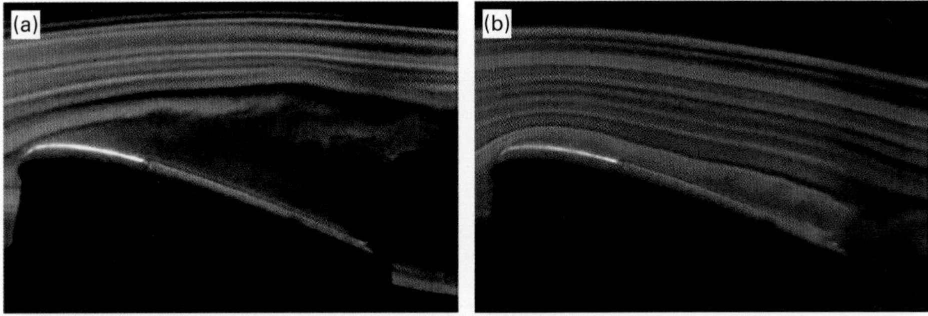

Figure 11.8 Flow visualization over the NACA 0015 airfoil without and with plasma control at a post-stall angle of attack $\alpha = 16°$ (Corke et al. 2011). Reproduced by permission of the Royal Society.

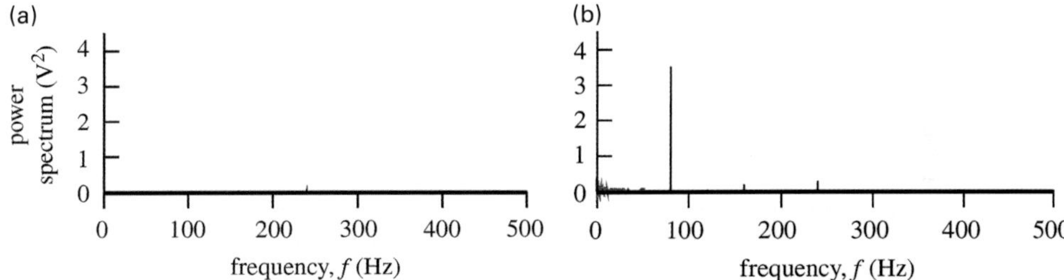

Figure 11.9 Power spectrum of the NACA 0015 airfoil without (red line) and with (black line) plasma control at $\alpha = 6°$ (a) and $14°$ (b) (Corke et al. 2011). Reproduced by permission of the Royal Society.

A characteristic rise in the amplitude levels of pressure frequencies was usually detected before stall, which could be used as a precursor to the onset of stall phenomena.

A closed-loop control algorithm was therefore proposed by Patel et al. (2007). The spectral analysis of the pressure signal was first performed by using FFT. The standard deviation of the FFT signal within a narrow frequency range was calculated to scale the spectrum amplitude of this frequency bin. This was performed for each frequency bin. Each standard deviation was compared to a predetermined threshold. If more than a predetermined fraction of the frequency bins exceeded their respective thresholds, then the actuator state was turned on, or else the actuator was set to off. Their results indicated that the controller accurately detected the stall of an airfoil at $\alpha = 15°$ and thus turned on the plasma actuator immediately. Thus, the flow was still attached to the airfoil until $\alpha = 22°$ with plasma control, while the lift coefficient was increased significantly in comparison with the no control case.

11.4   Concluding Remarks

The main features and implementation of closed-loop control have been introduced in this chapter. The closed-loop control parameters depend on information from the control system which in turn depend on the control. Thus, it can adapt to the flow condition and

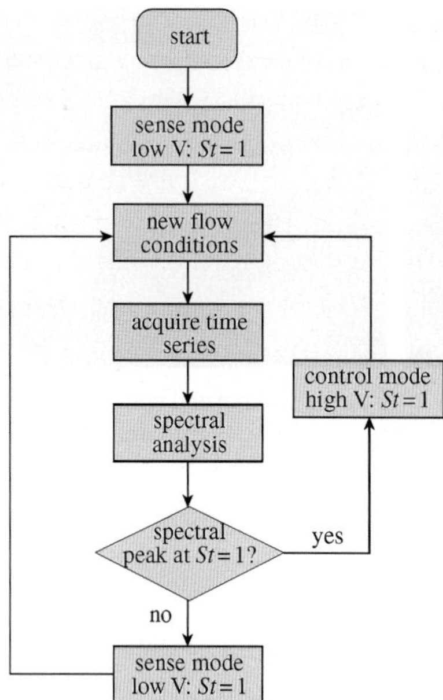

Figure 11.10 Schematics of the closed-loop control algorithm based on spectral analysis (Corke et al. 2011). Reproduced by permission of the Royal Society.

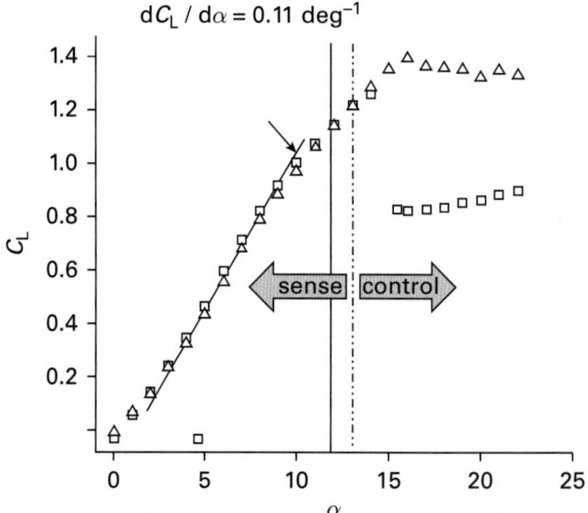

Figure 11.11 Mean lift coefficient of the NACA 0015 airfoil without ($\square$) and with ($\triangle$) closed-loop control (Corke et al. 2011). Reproduced by permission of the Royal Society.

thus contribute an intelligent and robust system, which meets the real application requirements of the active flow control techniques.

A closed-loop control system includes fluid system, actuator, sensor, and control algorithm. The control algorithm is the essential factor for flow control, which could be based on the reduced-order model or the measured signal from the flow field. The reduced-order model is usually achieved by advanced data processing methods, such as POD. The flow is decomposed into a series of modes. Then closed-loop control is used based on the dominant modes which could approximate the original flow. This is usually applied for numerical study. On the other hand, the measured variables, such as pressure and force coefficients, could also be used for the design of the control algorithm. The indicator to adjust the control parameters could be based on a statistic analysis of the signal, such as the mean value, the fluctuation and the power spectrum. These methods have been introduced to inform readers of the main process of closed-loop control. It is suggested that realization of closed-loop control is important for engineering applications of active flow control techniques.

References

Benard, N., Cattafesta III, L. N., Moreau, E., Griffin, J., and Bonnet, J. P. On the benefits of hysteresis effects for closed-loop separation control using plasma actuation. *Physics of Fluids*, 2011, 23(8): 894–894

Bergmann, M., Cordier, L., and Brancher, J. P. Optimal rotary control of the cylinder wake using proper orthogonal decomposition reduced-order model. *Physics of Fluids*, 2005, 17(9): 336–346

Brunton, S. L., Dawson, S. T. M., and Rowley, C. W. State-space model identification and feedback control of unsteady aerodynamic forces. *Journal of Fluids and Structures*, 2014, 50: 253–270

Corke, T. C. Bowles, P. O., He, C., and Matlis, E. H. Sensing and control of flow separation using plasma actuators. *Philosophical Transactions*, 2011, 369(1940): 1459–1475

Joslin, R. D. and Miller, D. N. Fundamentals and applications of modern flow control. American Institute of Aeronautics and Astronautics, 2009

Pastoor, M., Henning, L., Noack, B. R., King, R., and Tadmor, G. Feedback shear layer control for bluff body drag reduction. *Journal of Fluid Mechanics*, 2008, 608: 161–196

Patel, M. P., Sowle, Z. H., Corke, T. C., and He, C. Autonomous sensing and control of wing stall using a smart plasma slat. *Journal of Aircraft*, 2007, 44(2): 516–527

Patel, M. P., Tilmann, C. P., and Ng, T. T. Active transparent stall control system for air vehicles. *Journal of Aircraft*, 2003, 40(5): 993–997

Pinier, J. T., Ausseur, J., Glauser, M. N., and Higuchi, H. Proportional closed-loop feedback control of flow separation. *AIAA Journal*, 2007, 45(1): 181–190

Siegel, S. G., Cohen, K., and McLaughlin, T. Numerical simulations of a feedback-controlled circular cylinder wake. *AIAA Journal*, 2006, 44(6): 1266–1276

Tewes, P., Wygnanski, I., and Washburn, A. E. Feedback controlled forcefully attached flow on a stalled airfoil. *Journal of Aircraft*, 2011, 48(3): 940–951

Weller, J., Camarri, S., and Iollo, A. Feedback control by low-order modelling of the laminar flow past a bluff body. *Journal of Fluid Mechanics*, 2009, 634: 405–418

INDEX